Lecture Notes in Computer Science 7307

Commenced Publication in 1973
Founding and Former Series Editors:
Gerhard Goos, Juris Hartmanis, and Jan van Leeuwen

W0193056

Zhenjiang Hu   Juan de Lara (Eds.)

Theory and Practice of Model Transformations

5th International Conference, ICMT 2012
Prague, Czech Republic, May 28-29, 2012
Proceedings

 Springer

Volume Editors

Zhenjiang Hu
National Institute of Informatics
2-1-2 Hitotsubashi, Chiyoda-ku
Tokyo 101-8430, Japan
E-mail: hu@nii.ac.jp

Juan de Lara
Universidad Autónoma de Madrid
Department of Computer Science
Campus de Cantoblanco
28049 Madrid, Spain
E-mail: juan.delara@uam.es

ISSN 0302-9743   e-ISSN 1611-3349
ISBN 978-3-642-30475-0   e-ISBN 978-3-642-30476-7
DOI 10.1007/978-3-642-30476-7
Springer Heidelberg Dordrecht London New York

Library of Congress Control Number: 2012937671

CR Subject Classification (1998): D.2, F.3, D.3, C.2, H.3-4, K.6, D.2.4

LNCS Sublibrary: SL 2 – Programming and Software Engineering

Typesetting: Camera-ready by author, data conversion by Scientific Publishing Services, Chennai, India

Printed on acid-free paper

Springer is part of Springer Science+Business Media (www.springer.com)

Preface

This volume contains the papers presented at the International Conference on Model Transformation (ICMT 2012) held during May 28–29, 2012, in Prague, Czech Republic.

Modeling is essential in reducing the complexity of software systems during their development and maintenance. Model transformations are essential for elevating models from documentation elements to first-class artifacts of the development process. Model transformation includes model-to-text transformation to generate code from models, text-to-model transformations to parse textual artifacts to model representations, model extraction to derive higher-level models from legacy code, and model-to-model transformations to normalize, weave, optimize, simulate and refactor models, as well as to translate between modeling languages.

ICMT is the premier forum for contributions advancing the state of the art in the field of model transformation and aims to bring together researchers from all areas of model transformation. Model transformation encompasses a variety of technical spaces, including modelware, grammarware, and XML-ware, a variety of transformation representations including graphs, trees, and DAGs, and a variety of transformation paradigms including rule-based graph transformation, term rewriting, and implementations in general-purpose programming languages. The study of model transformation includes foundations, semantics, structuring mechanisms, and properties (such as modularity, composability, and parameterization) of transformations, transformation languages, techniques, and tools. To achieve an impact on software engineering practice, tools and methodologies to integrate model transformation into existing development environments and processes are required.

This was the fifth edition of the conference, and this year we accepted two kinds of contributions: research and tool papers. We received 69 abstracts, of which 56 materialized as full papers, and 16 research papers and 2 tool papers were selected, with an acceptance rate of 32%. In addition to 18 contributed papers, the program included an invited talk by James Terwilliger (Microsoft) on a unifying theory for incremental bidirectional model transformation.

ICMT 2012 was made possible by the collaboration of many people. We were supported by a great team. As Web Chair, Davide Di Ruscio kept the information up to date on the conference website, and Jesús Sánchez-Cuadrado actively publicized the conference, in his role of Publicity Chair. The Steering Committee was very helpful and provided advice when we needed it. We thank the TOOLS teams headed by Bertrand Meyer for taking care of the organization of the conference. Finally, we would like to thank all the members of the ICMT 2012 Program Committee for the tremendous effort they put into their reviews and deliberations, and all the external reviewers for their invaluable contributions. The submission and review process was managed using the EasyChair system.

March 2012   Zhenjiang Hu
                                                                Juan de Lara

Organization

Program Committee

Vasco Amaral	Universidade Nova de Lisboa, Portugal
Don Batory	University of Texas at Austin, USA
Xavier Blanc	Bordeaux 1 University, France
Jordi Cabot	INRIA-École des Mines de Nantes, France
Antonio Cicchetti	Mälardalen University, Sweden
Benoît Combemale	IRISA, Université de Rennes 1, France
Alcino Cunha	Universidade de Minho, Portugal
Davide Di Ruscio	Università degli Studi dell'Aquila, Italy
Alexander Egyed	Johannes Kepler University, Austria
Gregor Engels	University of Paderborn, Germany
Claudia Ermel	Technische Universität Berlin, Germany
Jesus Garcia-Molina	Universidad de Murcia, Spain
Sebastien Gerard	CEA, LIST, France
Esther Guerra	Universidad Autónoma de Madrid, Spain
Reiko Heckel	University of Leicester, UK
Soichiro Hidaka	National Institute of Informatics, Japan
Gerti Kappel	Vienna University of Technology, Austria
Gabor Karsai	Vanderbilt University, USA
Dimitrios Kolovos	The University of York, UK
Thomas Kuehne	Victoria University of Wellington, New Zealand
Ivan Kurtev	University of Twente, The Netherlands
Ralf Laemmel	Universität Koblenz-Landau, Germany
Leen Lambers	Hasso-Plattner-Institut, Universität Potsdam, Germany
Ivan Porres	Åbo Akademi University, Finland
Bernhard Rumpe	RWTH Aachen University, Germany
Andy Schürr	Technische Universität Darmstadt, Germany
Jim Steel	University of Queensland, Australia
Perdita Stevens	University of Edinburgh, UK
Gabriele Taentzer	Philipps-Universität Marburg, Germany
Pieter Van Gorp	Eindhoven University of Technology, The Netherlands
Daniel Varro	Budapest University of Technology and Economics, Hungary

Janis Voigtländer   University of Bonn, Germany
Hironori Washizaki   Waseda University, Japan
Jules White   Virginia Tech, USA
Yijun Yu   The Open University, UK
Haiyan Zhao   Peking University, China
Jianjun Zhao   Shanghai Jiao Tong University, China

Additional Reviewers

Abbors, Fredrik   Horst, Andreas
Alexandru, Cristina-Adriana   Jackel, Sebastian
Anjorin, Anthony   Jay, Barry
Arendt, Thorsten   Khan, Ali Hanzala
Arifulina, Svetlana   Kusel, Angelika
Barais, Olivier   Langer, Philip
Barroca, Bruno   Lauder, Marius
Becker, Basil   Machado, Rodrigo
Boskovic, Marko   Mantz, Florian
Büttner, Fabian   Mir Seyed Nazari, Pedram
Cheng, Xiao   Naeem, Muhammad
Demuth, Andreas   Patzina, Sven
Domingues, Rui   Rauf, Irum
Donyina, Adwoa   Reder, Alexander
Douib, Saadia   Schrijvers, Tom
Dubois, Hubert   Schönböck, Johannes
Duddy, Keith   Soltenborn, Christian
Espinazo, Javier   Spijkerman, Michael
Espinazo-Pagán, Javier   Sánchez Cuadrado, Jesús
Garfinkel, Simson   Süß, Jörn Guy
Gerth, Christian   Tisi, Massimo
Golas, Ulrike   Varanovich, Andrei
Guy, Clément   Varro, Gergely
Gómez, Cristina   Widl, Magdalena
Güldali, Baris   Willink, Ed
Hebig, Regina   Wortmann, Andreas
Hermann, Frank   Wozniak, Ernest
Hermerschmidt, Lars   Zhang, Sai
Hildebrandt, Stephan

Table of Contents

Transformation Languages, Virtual Machines

Pattern Matching

Transformations in Modelling, Reutilization

How Clean Is Your Sandbox?

Towards a Unified Theoretical Framework
for Incremental Bidirectional Transformations

James F. Terwilliger[1], Anthony Cleve[2], and Carlo A. Curino[3]

[1] Microsoft Corporation
[2] PReCISE Research Center, University of Namur
[3] Yahoo! Research

Abstract. Bidirectional transformations (*bx*) constitute an emerging mechanism for maintaining the consistency of interdependent sources of information in software systems. Researchers from many different communities have recently investigated the use of *bx* to solve a large variety of problems, including relational view update, schema evolution, data exchange, database migration, and model co-evolution, just to name a few. Each community leveraged and extended different theoretical frameworks and tailored their use for specific sub-problems. Unfortunately, the question of how these approaches actually relate to and differ from each other remains unanswered. This question should be addressed to reduce replicated efforts among and even within communities, enabling more effective collaboration and fostering cross-fertilization. To effectively move forward, a systematization of these many theories and systems is now required. This paper constitutes a first, humble yet concrete step towards a unified theoretical framework for a tractable and relevant subset of *bx* approaches and tools. It identifies, characterizes, and compares tools that allow the *incremental* definition of bidirectional mappings between software artifacts. Identifying similarities between such tools yields the possibility of developing practical tools with wide-ranging applicability; identifying differences allows for potential new research directions, applying the strengths of one tool to another whose strengths lie elsewhere.

1 Introduction

Data management has long been a key pillar of any information-technology infrastructure, but the advent of the web, social networks, business intelligence, digitalization of health care, the explosion of sensor generated data, cloud computing, and the upcoming explosion of data markets is significantly increasing the pressure to tackle the many issues related to transforming data across data models, and schemas—few examples of such data transformations include information extraction, exchange, querying, evolution, migration, provenance, etc.

Furthermore, software systems are increasingly large and complex, with various interdependent artifacts employing different models and levels of abstraction. The complexity of the problem scales from a simple two-body problem of an application and its database to a full constellation of artifacts requiring synchronization. And, as those

Z. Hu and J. de Lara (Eds.): ICMT 2012, LNCS 7307, pp. 1–23, 2012.

artifacts evolve, one must also contend with different versions of each as well. Managing the consistency of those artifacts and versions is at the heart of software system development and evolution.

One solution to this consistency problem that has been widely studied is the *bidirectional transformation*. The basic idea of a bidirectional transformation is simple: there exist two models or schemas S and T (sometimes referred to as a *source* and *target* model), and a mapping between them M. The mapping serves as a bridge to allow operations and data to flow between the two models, and must conform to bidirectional properties that govern the quality of the synchronization between S and T. How the mapping M operates, how it is specified, and what services it offers consumers of the target schema T differ greatly depending on what technology is used to build M. However, regardless of how M is built, it must ensure that any consumers of either S or T are able to access their data with no surprises.

Bidirectional transformations have been studied for several decades now. Perhaps the most widely known example is the updatable view [12], where a mapping is expressed using SQL or some other relational language. In that time, many different bidirectional tools have been created and evaluated, each developed according to its own requirements and supported scenarios. As a result, each tool may support different operations and satisfy different formal properties. However, these tools have a tantalizing amount of overlapping capabilities as well, which invites the question of whether a common framework or formalism for incrementally-specified mappings could exist.

The problem of examining and unifying Bidirectional Mappings (abbreviated *bx*) in general has begun to be explored. Through a collection of seminars and workshops, researchers from four computer science disciplines (Programming Languages, Software Engineering, Graph Transformation, and Databases) have been working on establishing collaboration, common terminology and research directions in the bidirectional space. The work began at a meeting in Tokyo in 2008 [11], and was continued at a seminar at Dagstuhl in 2011 [23,24]. These two meetings primarily allowed the participants to present the state of the art in *bx* in each discipline, and allowed for some initial discussion and collaboration. Also, after the Dagstuhl meeting, workshops dedicated to bidirectional transformations were set up at various venues. One such workshop, itself entitled *BX*, was intended to start a series of workshops at conferences[1]. Another workshop, entitled *CSXW*, was set more generally on coupled transformations and was held in concert with the GTTSE summer school[2].

These seminars and workshops have been fruitful in developing insights and collaboration amongst specific research projects across disciplines. However, the area of *bx* is vast, and more workshops and meetings are planned as the space of possible research collaborations is just beginning to be populated. Each *bx* tool has been developed in its own sandboxed set of supported scenarios in which that particular tool can excel. In this paper, we propose a concrete opportunity to align several *specific* research efforts across disciplines — to explore outside the sandbox, if you will.

Over the past decade, a number of tools have been developed that one can use to specify a bidirectional mapping *incrementally*. Using such a tool, one constructs a

---

[1] http://www.program-transformation.org/BX12/

[2] http://www.di.univaq.it/CSXW2011/

mapping between two models as a combination of elements from some set of mapping primitives with known properties. In practice, one will begin with one of the two models and apply mapping primitives one at a time until one arrives at the other model. Contrast this approach with a *declaratively* specified mapping, where one uses a language like SQL or predicate calculus to specify the relationship between the models. This incremental approach has been used frequently in software development tools like Extract-Transform-Load (ETL) tools, which implies that one could use a similar approach to build a general-purpose tool for bidirectional mappings.

Our investigation into incremental bidirectional mappings proceeds as follows. First, in Section 2, we introduce some criteria by which we can compare different incremental tools. Next, in Section 3, we investigate and analyze a set of five incremental *bx* tools; this set of tools is not intended to be comprehensive, but sufficiently representative to illustrate typical scenarios currently addressed by such tools and to show that some potential for unification already exists. Section 4 charts some new research directions that we see as potentially useful consequences of our analysis. We conclude the paper with some final thoughts about the state of the overarching discussion about *bx*.

2   An Initial Taxonomy

To better characterize the various incremental tools considered in this paper, we begin with a brief taxonomy of several features and capabilities that such tools may exhibit. This taxonomy is not intended to be complete, but rather a starting point for analysis and discussion.

Information Capacity

When considering a bidirectional mapping tool, it is useful to consider whether the tool can construct a mapping that alters the information capacity of the models on which it operates. In particular, one can ask two basic questions:

Can a mapping decrease information capacity? If a mapping can decrease information capacity, even if each source model state maps uniquely to a state of the target model, the reverse may not be true. For such mappings, each target state can still be translated uniquely into a source model state when combined with the "old" state of the source model to be used as context. For instance, consider a source model state s translated to target model state t, then updated to target model state t'. The new source model state s' can be computed as a function of t' and s.

Can a mapping increase information capacity? A mapping may, for instance, add new columns or tables or otherwise add new data capacity to a model's structure. A source model state applied as input to a capacity-increasing mapping may not be able to map uniquely to a target model state. Because the target model in some tools is virtual, there may not be a target model instance to provide additional context. Thus the tool may need to provide alternative means for filling the empty capacity in the target.

Note that support for increasing capacity need not be correlated with support for decreasing capacity. Note also that a mapping tool that can neither increase nor decrease the information capacity of models can only support bijective mappings. In other words, the mapping between two models may follow a strictly isomorphic relationship. Each state of the target model maps uniquely to a state of the source model and vice versa. This type of relationship has very nice functional properties, but is limited in its expressive capabilities.

One orthogonal question to ask about a tool, when it supports alteration of information capacity, is this:

Is a mapping's operation strictly a functional dependency of its inputs? Consider a case when a mapping between model S and target T has a relationship where no data must ever be deleted from S but rather be "deprecated" to maintain audit trails. The modeling of deprecated data includes the time and active user at the time of deprecation. This mapping must turn all attempts to delete or update data into new records, and those records must include environment data about the system time that cannot be inferred from the data itself. Still, such a mapping must still must maintain correctness properties with respect to how data appears in the target model, displaying the correct data and omitting stale, deprecated items.

Concrete Versus Virtual State

Another major consideration in the construction of a mapping tool is the expected mode of operation of the target model. In particular, the tool designer must decide whether the target model states will be materialized or not. The answer to that question largely determines the kinds of scenarios the tool will support.

Materialized Target State. In this scenario, the entire state of the source model is transformed into its target-model counterpart according to the mapping. Effectively, the mapping constructs an instance of the target model by applying a function to an instance of the source model. Data retrieval from the target in this mode of operation is simple, as one has a full concrete instance to work with.

With a concrete target state, updates against the target modify the state in place. Propagating the update back to a source state consists of applying the reverse process of the mapping to the modified target state.

Virtual Target State. In this scenario, one constructs a declarative query against the target model, and that query is translated into a corresponding query or set of instructions against the source model, this is also referred to as query/update rewriting. Using this paradigm, one typically does not ever have a complete instance of the target model, but may have access to portions of it in case the complete instance is too large to reasonably work with.

For updates against a virtual target state, one constructs an update expressed in terms of a declarative statement against the target model. The typical INSERT, UPDATE, and DELETE statements in SQL are examples of such statements.

Model Evolution

An orthogonal question to answer is how a tool expects to respond to the evolution of either of its models. Such a scenario is commonplace over the many versions of an application over its lifespan. Given an evolution to either the target model T or the source model S, all other artifacts in the system must reconcile themselves somehow. For instance, consider a situation where a data-driven application interacts with a target schema T, with database schema S and mapping M connecting them. The application designer now makes changes to the application where new fields are required, old ones are no longer needed, and some existing fields now employ a different data type. In this situation, these changes induce a new target model T'.

The central question in this circumstance is how M and S evolve to compensate, if at all, to continue to have a working database backing the application. There are two choices, centered around one question: Does the source model need to remain invariant?

Source-Invariant. In this case, one constructs a new mapping that connects the old model T to the new model T', and then constructs a mapping from S to T' by composition, $M' \circ M$. This scenario is common when dealing with legacy data sources that need to remain unchanged to support older (and possibly numerous) versions of software.

Co-Evolution. Alternatively, one can construct a new mapping M' and a new source model S' such that M' maps S' to T', and the relationship between S and S' is correct with respect to some sense of consistency against the original model S and the changes made to T. The tool must be clear what its notion of "consistency" is. Also, when evolving model S into S', one must also consider the impact on any services that rely on that model, in particular any data states that conform to S.

3 Incremental Transformation Languages

The five tools that we investigate in this section come from two disciplines of computer science: programming languages and databases. They form a representative sample of the design space in that they cover most of the options in Section 2.

3.1 Lenses

A *lens* is a mathematical abstraction centered around a pair of functions called *get* and *put*. If a lens ℓ is a transformation from model S to model T, then function *get* describes how to get data from an instance of S to populate a state in T, while function *get* describes how to put data back into the state of S [15].

Lenses were originally introduced as a programming language construct for building bidirectional transformation programs over strings and tree structures. As an example, one can use a lens to build a program that extracts log data from XML files into flat structures, and allow updates to those structures to re-update the XML. Lenses were also explored in the relational context, where one can use lenses to build an updatable view over a database. *Relational Lenses* roughly correspond to relational algebra operators like selection, projection, and join, where each operator is adorned with an update policy [3].

Background: Over time, different categories of lenses have been introduced to cover different programming scenarios. The original formulation, now as a retronym called either *classical lenses* or *asymmetric lenses* to disambiguate from newer forms, are an encapsulation of the idea of an updatable view. Functions *get* and *put* in a classical lens have the following formulations:

$$get : S \rightarrow T \qquad\qquad put : T \times S \rightarrow S$$

In short, function *get* translates a store state into a target state without any other context required, while function *put* uses the original store state as context when translating an updated target state back to a new store state. An alternative but equivalent formulation is to consider the part of the store model S that is not relevant to function *get*, and call this part of the model its *complement* C. Then:

$$get : S \rightarrow T \times C \qquad\qquad put : T \times C \rightarrow S$$

This formulation has some appealing symmetry to it, a symmetry that one can use to extend the formalism to a scenario that is itself symmetric. A classical lens has a directionality to it, with a target state and a store state. However, lenses can also be used to synchronize two different models T and T', with the expectation that neither model subsumes the other in information capacity. In a *symmetric lens*, the two functions are allowed to each have a complement [22]:

$$putl : T \times C \rightarrow T' \times C \qquad\qquad putr : T' \times C \rightarrow T \times C$$

where *putl* and *putr* are the two constituent functions, renamed as such in the literature to reflect that neither model is "dominant" over the other. Also, the complement C spans the unmapped portions of both models T and T', which provides some formal convenience and an implementation simplification where a single storage point is required to maintain the complement state.

The literature on lenses has described three different methods for constructing a lens. First, one can construct a lens by using a domain-specific language whose primitives are themselves lenses [15]. Such a combinator approach is similar to the other tools listed in this paper. Second, one can use a standard language, restricted to a subset with known properties, to define either the *get* or *put* function and automatically infer the other one from the syntax of the defining expression for *get* [25]. For example, one can use SQL to define views (the *get* function) that are updatable (the *put* function) if one is limited to a syntactic subset of the language where the FROM clause only has one table reference, there are no aggregates or set operations, and so forth. Finally, one can use a standard language to define a *get* function and, without the syntactic restriction, infer the *put* function based on evaluating the *get* function over a cleverly-selected set of inputs [35]. A hybrid approach between the syntactic approach and the input analysis approach has also been considered [36].

Formal Properties: Much of the formalism around lenses centers around the subset that are both total and well-behaved. A *total* lens is one whose pair of functions are both total on their inputs. A *well-behaved* lens must, in addition, satisfy two roundtripping policies that match the same intuition as classical updatable views. First, there is the intuition that, given a target state, if one pushes that state to the store and retrieves it again, one always gets the original state as a result. Using the original formulation of a classical lens, this property — also called "PutGet" — amounts to:

$$\forall_{(t \in T)} \forall_{(s \in S)} get(put(t, s)) = t$$

Second, there is the intuition that, given a store state, if one applies a lens to it and immediately pushes the result back, one gets back the original store state. This property — also called "GetPut" — amounts to:

$$\forall_{(s \in S)} put(get(s), s) = s$$

Analogs to these formulae exist for the symmetric case as well, corresponding to the intuition that unaltered data can flow back and forth through the lens functions without altering state.

Several proposals have been made for further formal properties for lenses. In particular, consider the "PutPut" rule:

$$\forall_{(t,t' \in T)} \forall_{(s \in S)} put(t', put(t, s)) = put(t', s)$$

This rule states that the *put* function, when applied to a sequence of target states, will always result in the store state that would have occurred if only the final target state was applied. The intuition here is that for a lens that satisfied PutPut, the target state can be updated incrementally or holistically and will yield the same result. The literature gives several examples of lenses that satisfy PutPut, as well as many useful ones that do not; as a result, well-behaved lenses that also satisfy PutPut are given their own category, *very well-behaved*.

One notable additional flavor of lens, the *quotient lens*, relaxes the PutGet and GetPut properties slightly:

$$\forall_{(t \in T)} \forall_{(s \in S)} get(put(t, s)) \equiv_T t$$

$$\forall_{(s \in S)} put(get(s), s) \equiv_S s$$

A quotient lens treats bidirectionality as being relative to equivalence classes over the underlying models [16]. These equivalence classes allow lenses to treat certain kinds of differences in model states as being insignificant, such as the whitespace in XML documents.

Bidirectional Characteristics: Different flavors of lenses have somewhat different characteristics relative to their bidirectionality. In addition to the basic classical and symmetric lenses, and whether a lens is merely well-behaved or *very* well-behaved, there have been other categories of lenses introduced that play with adjusting the formal properties to suit new scenarios. What follows is a characterization of many of those categories.

- Both classical and symmetric lenses are capable of **decreasing information capacity**.
- A classical lens is not capable of increasing information capacity, as it is only a function of its source model input. A symmetric lens, however, is capable of **increasing information capacity** through the means of a complement.
- Lenses are by their very construction **functional** with respect to the input states of the source and target models.
- A lens retrieves data through its *get* function applied to the source state. It is therefore, by definition, based on **materialized target states** according to the taxonomy of Section 2. A lens translates an update when the user updates a target state in place, and the *put* function replaces the updated state back to a new source state.

Some efforts have been made to adjust this basic mode of operation to be more like a virtual state. For instance, an *edit lens* is a variant of a symmetric lens where instead of translating states, the lens translates descriptions of updates tailored to the structure of the underlying model [21]. A list structure would then support operations that insert entries at the end, delete an entry at a given position, update an entry at a given position, or reorder entries.

– A lens ℓ from model S to model T does not have any built-in capabilities for handling a case where S or T evolves. However, if one can construct a lens ℓ' from S (or T) to an evolved model S' (or T'), and compose ℓ with ℓ' to form a new lens from S' to T (or S to T', respectively). In this way, lenses can support **source-invariant model evolution**.

Note that the composed lens may need to be symmetric where the original lens ℓ did not. For instance, consider the case where one evolves the target model T to add a new column. This new column would participate in the model complement as it is unmapped.

3.2 Schema Modification Operators

Schema Modification Operators (SMO) and Integrity Constraint Modification Operators (ICMO), have been designed to provide a simple operational language for database administrators to describe the evolution of a relational schema S_1 with integrity constraints IC_1 into a new schema S_2 with integrity constraints IC_2. The SMO/ICMO syntax is heavily influenced by SQL, but each operator is atomic and unambiguous for evolution purposes.

By design, SMOs provide structural transformations of the schema, that thanks to careful handling of integrity constraints are guaranteed to be invertible (and thus information preserving). ICMOs are responsible for integrity constraint manipulation, i.e., do not affect the schema structure, but modify its information capacity (thus they are not trivially invertible, and not always information preserving). This division of responsibilities manages complexity by handling structural changes and information-capacity changes separately. The above property also provides a significant practical advantage: SMOs can be inverted automatically, so the user is only required to specify an inverse for ICMOs (more precisely for a subset of ICMOs).

This operational language is leveraged in [9,10,28,29] as a paradigm for interacting with the user and capturing the intended evolution semantics. Each operator can then be automatically translated into:

– An equivalent set of SQL statement capable of migrating the data from schema S_1 to schema S_2—thus accomplishing data migration functionalities.
– A logical mapping M based on Disjunctive Embedded Dependencies (DED) between schema S_1 and S_2—necessary for query/update rewriting.

Thanks to the invertibility properties of SMOs, the above benefits are bidirectional.

The choice of DEDs to capture M (and its inverse M') is based on the expressiveness of this language capable of covering all cases representable by SMOs and ICMOs. Moreover, powerful theoretical results have been proven about soundness and completeness of the chase and back-chase algorithm [13] for DEDs.

In PRISM++ [10] the chase-and-backchase algorithm is used to accomplish query rewriting across SMO-based evolution steps, while new algorithms have been devised to handle rewriting through ICMOs steps. The theory around chase-and-backchase is extended to handle in a sound way rewritings of queries with negation. The problem tackled is akin to view-update, and the combination of the SMO/ICMO language and the algorithms provides a sound solution, for three practically very useful subclasses of the general view-update problem. Decreasing information capacity is challenging for data migration (there is information loss), but trivial for query and update rewriting (the target schema has more stringent constraints, on the contrary increasing information capacity is simple for data migration but makes query and update rewriting more challenging. The three sub-classes considered for cases in which the information capacity of new schema S_2 is strictly larger than the one in S_1, the user can choose between three semantics of the rewriting:

- If the data instance I_2 under S_2 also (still) satisfies all constraints of S_1 queries and updates can be executed on S_2, provided updates do not introduce new violations (this can be checked via queries as pre and post conditions), the query/update rewriting fails otherwise.
- I_2 violates the constraints of S_1, the scope of queries and updates is limited to the portion of I_2 that satisfy S_1 constraints, and we updates should not introduce or remove constraints, the rewriting fails otherwise.
- I_2 violates the constraints of S_1, but the user is ok with queries and updates operating with side effects on a less stringent schema S_2, all rewritings succeed but side effects are not always revertible.

While SMO/ICMOs can be used to capture general mappings between relational schemas, their design is mostly conducive to schema evolution scenarios, where the user naturally thinks in terms of changes of the input schema. The underlying DED framework on the contrary is rather general.

Bidirectional Characteristics: Relative to the questions and categories laid out in Section 2, SMOs have the following properties:

- Sequences of SMOs/ICMOs **can increase/decrease information capacity**, in particular, ICMOs are responsible for modifying information capacity. Only the three special cases defined above can be handled, this however cover the vast majority of practical scenarios. The use of User Defined Functions in the SMOs can be leveraged to embed a limited form of **non-functional generation of new values**, however this is far from fully general.
- SMOs/ICMOs support **virtual target state** by chase-and-back-chase rewritings (combined with dedicated algorithms for ICMO, updates and negation). SMOs / ICMOs support a form of **materialized target state** for data migration, i.e., operators are translated into corresponding SQL scripts and data are physically moved across schemas. This can be leveraged in other contexts if deemed necessary, however the virtual version is typically more efficient.
- In terms of model evolution, SMOs/ICMOs are focused on single schema and concerned with its evolution. Thus PRISM++ allows a **source-invariant** form of evolution by concatenating mappings.

3.3 Channels

A *channel* [32] is a bidirectional transformation from S to T derived from the following design requirements:

- The target model is virtual. Therefore, data retrieval is done through queries, and updates are done through declarative statements.
- In addition to being able to update the data in the target model, one must be able to evolve the target model T itself. For instance, one can add a new column to a table in the target model, and the channel must propagate that change to a corresponding change against the store model.
- The mapping from S to T must be surjective, both in terms of model states and in terms of schema elements. Every state in the target model must map to at least one store model state, and all data elements in the target model must be somehow "backed" by elements in the store model.

Background: Channels are part of a larger framework called *Guava* designed to facilitate simplified development of applications with persistent database backing [31]. Using Guava, a developer implements a graphical application, from which Guava derives a data model using the application's data entry widgets and windows as a guide. The channel is the tool that connects this implied data model — itself a target model, since it represents the virtual state of the application — to a database.

Even though a channel can be considered to be an abstraction from a store model S to a target model T, the specification is done in the reverse order, from T to S. The specification order is due to the application-centric nature of the Guava framework; one starts with the target model inferred from the graphical application, then applies one transformation at a time until one arrives at the desired database schema (as shown in Figure 1).

The application-driven nature of Guava also explains why schema translation is a requirement for channels. Since in Guava, the target model is derived from an application, when the application is modified (say, because a new textbox was added), the target model will also be modified, which will induce changes through the channel to the store model in turn.

Formal Properties: A channel is itself made up of discreet transformations. A *channel transformation* (CT) is a 4-tuple of functions (S, I, Q, U) that operate on *S*chema, data *I*nstances, *Q*ueries, and *U*pdates respectively. The S and I functions are never employed in practice; they provide the semantic backbone for a CT, demonstrating how the CT would operate if it were to function on fully materialized instances.

In practice, a channel operates much the same way as an updatable view in ordinary relational database systems. The target model represents a virtual (non-physical) data state, but the user can construct and execute declarative query and update statements against it in exactly the same way as if it were a physical, stored state.

There are several key differences between channels and updatable views, though. First, since all individual channel transformations are updatable, the target schema of any channel is also updatable by composition. Second, an updatable view is typically defined by a query, which returns only a single table; the target model of a channel

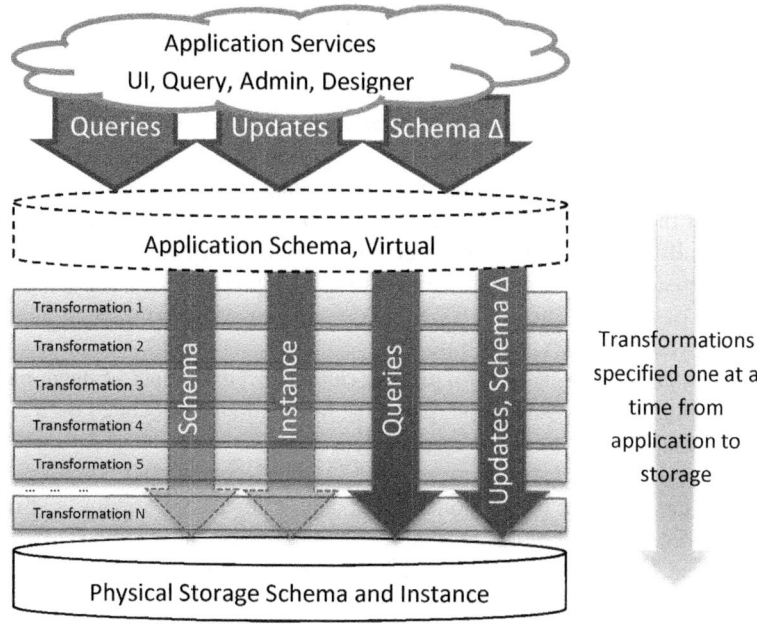

Fig. 1. The basic operation of a channel with respect to queries and updates

may be a fully capable schema, including multiple tables, as well as primary key and referential constraints. Finally, the only way to update the schema of a relational view is to edit the underlying relational query. When using a channel, the user may evolve a target schema directly. Such evolutions propagate through a channel and to the store schema to evolve it in turn.

One evaluates the correctness of a CT based on three properties, loosely stated:

- If one runs a query q against what would be a fully materialized instance i of a target model T, one will get the same result as if one runs $Q(q)$ (the translated query) against $I(i)$ (the translated instance). This property establishes a correlation between functions I and Q.
- Running a translated update against a translated schema is equivalent to running the update first, then translating the result. This property establishes a correlation between functions S and U.
- Given a query q and and update u, running $Q(q)$ against the store model after running $U(u)$ will give the same result as running q against the target model after running u. This property establishes a correlation between functions S, Q, and U.

Put together, these three conditions formalize smooth function of an application. The intuition is that if someone is using a database application and issues the same query twice, they will get the same result, and if the application updates data or schema between queries, the new query result will have the correctly updated data.

Bidirectional Characteristics: Relative to the questions and categories laid out in Section 2, channels have the following properties when considered as a mapping from source model to target model:

- A channel **can decrease information capacity** in some limited cases. The set of CTs allowed by a channel fall into two categories. One category qualifies as fully bidirectional and includes transformations like horizontal and vertical partitioning, as well as pivoting. The second category **supports transformations that are not functionally dependent on inputs**, and includes transformations that can add store columns that track environment information, or "deprecates" rows rather than deleting them or updating them. A CT in this second category can reduce information capacity from source to target model.
- A channel **cannot increase information capacity**, a byproduct of the application-based scenarios that channels support. The target model associated with an application must be entirely backed by persistent storage.
- A channel supports a **virtual target state**. As mentioned earlier, the user interacts with the target model using a query language, at the target model is strictly not materialized. In the literature on channels, translation of updates is strictly speaking done based on insert, update, and delete statements that key off of primary key values. However, channels support a "looping" construct as well, which can be used to support arbitrary update statements as well.
- A channel supports **model co-evolution**, where an incremental change to the target model induces and incremental change to the store model.

3.4   DB-MAIN

Background: DB-MAIN is a programmable CASE environment that provides a large transformational toolkit supporting such processes as database design, reverse engineering, evolution and migration. It is based on the *Generic Entity/Relationship* (GER) model [17] to represent data structures at all abstraction levels and according to all popular modeling paradigms. The GER model includes, among others, the concepts of schema, entity type, domain, attribute, relationship type, keys, as well as various constraints. A GER schema is made up of specification constructs which can be, for convenience, classified into the usual three abstraction levels, namely *conceptual*, *logical* and *physical*. The use of the GER as *pivot* model for schema representation allows *inter-model* transformations to reduce to *intra-model* ones.

Formal Properties: DB-MAIN relies on a *transformational* approach, considering most database engineering processes as chains of *schema transformations*. A schema transformation consists in deriving a target schema S' from a source schema S by replacing construct C (possibly empty) in S with a new construct C' (possibly empty) [18]. C (resp. C') is empty when the transformation consists in adding (resp. removing) a construct. More formally, a schema transformation Σ is a couple of mappings $\langle T, t \rangle$ such that : $C' = T(C)$ and $c' = t(c)$, where c is any instance of C and c' the corresponding instance of C'. *Structural mapping* T explains how to modify the schema while *instance mapping* t states how to compute the instance set of C' from instances of C.

Any transformation Σ can be given an *inverse* transformation $\Sigma^{-1} = \langle T^{-1}, t^{-1} \rangle$ such that $T^{-1}(T(C)) = C$. If, in addition, there also exists an instance mapping t^{-1} such that: $t^{-1}(t(c)) = c$, then Σ (and Σ^{-1}) are said *semantics-preserving* or *reversible*. If $\langle T^{-1}, t^{-1} \rangle$ is also reversible, Σ and Σ^{-1} are called *symmetrically reversible*. Intuitively, if a schema transformation is reversible, then the source schema can be replaced with the target one without *loss of information*. Table 1 provides a semantic classification of the main GER schema transformations, that can be *semantics-increasing* (S^+), *semantics-decreasing* (S^-) or *semantics-preserving* ($S^=$)[3].

Table 1. Semantic classification of GER schema transformations

GER schema construct	Semantic impact of construct transformation		
	S^+	S^-	$S^=$
Entity type	add	remove	rename convert to attribute convert to rel. type split/merge
Relationship type	add	remove	rename convert to ent. type convert to attribute
Role	create increase max. card. decrease min. card. add entity type	delete decrease max. card. increase min. card. remove entity type	rename
Is-a relationship	add change type	remove change type	
Attribute	add increase max. card. decrease min. card. extend domain change type	remove decrease max. card. increase min. card. restrict domain change type	rename convert to ent. type aggregate disaggregate (if compound) instantiate (if multi-valued) concatenate (if multi-valued)
Identifier	add add component	remove remove component	rename change type
Constraints	add add component change type	remove remove component change type	rename
Access key	add add attribute	remove remove attribute	rename
Collection	add add entity type	remove remove entity type	rename

The chain of transformations that are successively applied to a source schema S to produce a target schema S' is used as a formal basis to specify/derive forward and backward mappings between S and S'. Such mappings can, in turn, be exploited to develop and automate incremental, bi-directional approaches to:

1. Co-evolving conceptual schema, logical schema, physical schema and data [20];
2. Co-evolving database schemas and application programs [7,6];

---

[3] A *semantics*-preserving (resp. -decreasing, -increasing) transformation is also called *information-capacity*-preserving (resp. -decreasing, -increasing).

Co-evolving conceptual schema, logical schema and data. The problem addressed in this scenario can be summarized as follows: how must a change in a given schema be propagated to the other level schemas and to the data instances? The DB-MAIN approach assumes that the conceptual CS, logical LS, and physical schemas PS exist, and that the transformational histories of the conceptual-to-logical (CL) and logical-to-physical (LP) processes are available. Three evolution scenarios are then considered, depending on the schema *initially* modified:

- *Conceptual modifications* typically translate evolutions in the functional requirements into conceptual schema change;
- *Logical modifications* do not modify the requirements but adapt their platform-dependent implementation in the logical schema.
- *Physical modifications* aim at adapting the physical schema to new or evolving technical requirements, like data access performance.

The problem addressed in the first scenario is the propagation of conceptual schema modifications towards the logical, physical, and data layers, and to the revised histories. In the second scenario, the conceptual schema is kept unchanged, but the logical design history CL must be updated and the modifications must be propagated in the physical and data layers. In the third scenario, only the physical design history LP must be updated.

Co-evolving database schemas and application programs. When modeling schema evolutions as schema transformations, the schema-program co-evolution problem translates as follows. Given a semantics-preserving schema transformation Σ applicable to data construct C, how can it be propagated to the database queries that select, create, delete and update instances of construct C? The DB-MAIN approach consists in associating with structural mapping T of Σ, in addition to instance mapping t, a query rewriting mapping stating how to adapt the related queries accordingly. In other words, the concept of *schema transformation* is extended to the more general term of *database co-transformation*, formally defined below.

Let S denote the set of all possible database schemas, $\mathcal{D}(S)$ the set of all possible database states complying with schema S, $Q(S)$ the set of all possible queries that can be expressed on schema S, $\mathcal{R}(S)$ the set of all possible results of reading queries on S. A database query q expressed on a schema S is then defined as a function that takes a database state $d \in \mathcal{D}(S)$ as input, and returns a (possibly updated) database state $d' \in \mathcal{D}(S)$ together with a (possibly empty) result $r \in \mathcal{R}(S)$[4].

A database co-transformation $\Phi = \langle T, t_d, t_q \rangle$ is a 3-tuple of transformations where:

- T transforms a schema $S \in S$;
- t_d transforms related data instances such that $\forall d \in \mathcal{D}(S) : t_d(d) \in \mathcal{D}(T(S))$;
- t_q transforms related queries, such that: $\forall q \in Q(S) : t_q(q) \in Q(T(S)) : \forall d \in \mathcal{D}(S) : t_q(q)(t_d(d)) = t_d(q(d))$.

---

[4] Reading primitives typically leave the database state unchanged ($d' = d$), in contrast with modification primitives (create, delete, update) that do affect the database contents ($d' \neq d$).

We refer to [5] for a set of generic co-transformation rules corresponding to eleven semantics-preserving schema transformations. Those generic rules have been used as formal basis to develop automated support to (1) schema refactoring via program transformation [7], (2) database platform migration via wrapper generation [19,8] and (3) database design and evolution via conceptual data manipulation API (re)generation [6].

Bidirectional Characteristics: Following the considerations of Section 2, DB-MAIN can be characterized as follows:

- In the context of inter-schema co-evolution, a GER schema transformation **can decrease information capacity** and, under particular conditions, still be correctly propagated to the other schema levels and to the data instances. As far as the schema-program co-evolution scenario is concerned, only **information-capacity preserving** schema transformations are automatically supported by the DB-MAIN co-transformational approach.
- The same holds for transformations that **increase information capacity**, which can also be propagated to other schemas and to data, depending on the nature of the relationships between the additional and pre-existing schema constructs.
- DB-MAIN mainly follows a **virtual target state** scenario, for both schema-schema and schema-program co-evolution scenarios. When co-evolving a conceptual, a logical and a physical schema, the only schema that is *materialized* is the physical schema, while the two other schemas can be considered higher-level *virtual views* defined over it. Similarly, when propagating logical schema evolutions to programs, the translated queries (or the generated APIs/wrappers) can be seen as a means to *virtually simulate* the source logical schema on top of the target one.
- DB-MAIN supports **model co-evolution** processes, according to which incremental changes to a schema at a given abstraction level propagates, when possible, as incremental changes to the other-level schemas, to the database contents and to the program queries.

3.5   Both-as-View

Background: *Both-as-View* (BAV) is an approach to data integration that is based on the use of reversible schema transformation sequences [26]. Data integration is the process through which several databases, with their associated *local schemas*, are integrated in order to form a single, *virtual* database, with a corresponding *global schema*. In this context, BAV constitutes a unifying framework for *Global-as-View* (GAV) and *Local-as-View* (LAV) approaches, while combining their respective benefits. Using BAV, it is indeed possible to express the local schemas as *views* over the global schema, and to process queries over the global schema by rewriting queries using those views (as in LAV). Conversely, BAV allows the definition of the global schema as a *view* over the local schemas, and to rewrite queries over the global schema into distributed queries over the local databases (as in GAV). Furthermore, a major advantage of the BAV approach over GAV and LAV is that it readily supports the *evolution* of both local and global schemas.

Formal Properties: BAV relies on a general formal framework supporting schema transformation and integration [30], which consists of (1) a low-level hypergraph-based data model called HDM [4], and (2) a set of *primitive* schema transformations defined for this model. HDM serves as a common model in terms of which *higher-level* data models (e.g., relational, ER and UML) and their basic schema transformations can be defined[5].

The BAV approach, when applied to simple relational data models[6], considers the following set of primitive schema transformations:

- $addRel(\langle\langle R, k_1, ..., k_n\rangle\rangle, q)$, which adds a new relation R to the schema with primary key attributes $k_1, ..., k_n, n \geq 1$. Query q expresses the set of primary key values (i.e., a set of n-tuples) belonging to the extent of the R in terms of the already existing schema constructs.
- $addAtt(\langle\langle R, a\rangle\rangle, c, q)$, that adds to an non-primary key attribute a to an existing relation R in the schema. Parameter c is either *null* or *notnull*. Query q specifies the extent of the relationship between the new attribute a and the primary key attributes of R in terms of *pre-existing* schema constructs. This extent consists of a set of pairs.
- $deleteRel(\langle\langle R, k_1, ..., k_n\rangle\rangle, q)$, that deletes from the schema relation R with primary key attributes $k_1, ..., k_n$, under the condition that all its non-primary key attributes have already been deleted. Query q indicates how the set of primary key values in the extent of R can be restored from the *remaining* schema constructs.
- $deleteAtt(\langle\langle R, a\rangle\rangle, c, q)$ allows to delete non-primary key attribute a of relation R from the schema. Query q specifies the way the relationship between the primary key attribute(s) of R can be restored from the *remaining* schema constructs.

Each of primitive schema transformation t described above has an automatically derivable *reverse* transformation t^{-1}, as shown in Table 2.

Table 2. Primitive schema transformations in BAV, and their respective reverse transformation

$t : S \to S'$	$t^{-1} : S' \to S$
$addRel(\langle\langle R, k_1, ..., k_n\rangle\rangle, q)$	$deleteRel(\langle\langle R, k_1, ..., k_n\rangle\rangle, q)$
$addAtt(\langle\langle R, a\rangle\rangle, c, q)$	$deleteAtt(\langle\langle R, a\rangle\rangle, c, q)$
$deleteRel(\langle\langle R, k_1, ..., k_n\rangle\rangle, q)$	$addRel(\langle\langle R, k_1, ..., k_n\rangle\rangle, q)$
$deleteAtt(\langle\langle R, a\rangle\rangle, c, q)$	$addAtt(\langle\langle R, a\rangle\rangle, c, q)$

The symmetrically-reversible nature of those primitive transformations allows the automatic, lossless translation of queries between schemas. For instance, let us assume that a schema S is transformed into a schema S' through an *addRel* or an *addAtt* primitive transformation. Translating a query Q over S into an equivalent query Q' over S' simply consists in substituting occurrences of the deleted construct in Q by the query q

---

[5] By contrast, the GER model of the DB-MAIN approach forms a *unifying* data model from which each particular data model can be defined by *selecting* a particular subset of model constructs.

[6] This is only an example, the same approach is equally applicable to any other data model that can be expressed on top of the HDM model.

defined in the transformation. Sequences of primitive transformations are are handled by successively applying such substitutions in order to obtain the final translated query Q'. This query translation mechanism can be applied in order to obtain the *derivation* of each global schema construct from the set of local schemas. These derivations can, in turn, be substituted into any query expressed over the global schema in order to obtain an equivalent query distributed over the local database schemas (GAV approach). Conversely, the very same translation scheme can be applied to obtain the derivation of all local schema constructs from the global schema, based on which LAV query translation can be provided.

In addition to the above primitive transformations, BAV also consider four schema transformations that do not fully preserve information capacity. Two *contracting* transformations, *contractRel* and *contractAtt*, behave in the same way as *deleteRel* and *deleteAtt*, respectively, except that the associated query q may only *partially* reconstruct the extent of the deleted schema construct.

Similarly, *extending* transformations *extendRel* and *extendAtt* are similar to *addRel* and *addAtt* transformations, but they indicate that their accompanying query q may only *partially* build the extent of the new schema construct. In the worst case, the query q associated with a *contract* or an *extend* transformation may be *void*, which then means that there is no information available on how to derive, even partially, the extent of the deleted/new construct from the remaining/pre-existing schema constructs.

Based on similar derivation and translation mechanisms as those described above, BAV provides immediate support to *schema evolution*, be it the evolution of the *global schema* or the evolution of a *local schema*.

Global schema evolution. Let us suppose that n local schemas $S_1, ..., S_n$ have been integrated into global schema S_g, using the schema transformation sequence T. The evolution of S_g towards a new global schema S'_g can also be modeled as a sequence of primitive schema transformations. Let us consider a single primitive transformation step t. Then the new transformation from the local schemas to the S'_g becomes $T' = T; t$, obtained by suffixing t to T. Three cases are to be considered:

1. t is an *add* or a *delete* transformation, i.e., S_g and S'_g are *semantically equivalent*, and T' is already correct.
2. t is a *contract* transformation, i.e., information capacity has been *decreased* when transforming S_g into S'_g. This means that some local constructs may have no representation in S'_g.
3. t is an *extend* transformation with a *void* accompanying query, then one needs to examine the derivability of relationship between the new global construct and the local schemas. If it is derivable, t has to be replaced in T' by a more informative *extend* or *add* transformation.

Local schema evolution. A similar reasoning can be followed when a local schema S_i evolves to S'_i. The only difference is that it may also involve global schema change. If T is the initial transformation sequence from $S_1 \cup ... \cup S_i \cup ... \cup S_n$ to S_g, generating a new transformation sequence T' from $S_1 \cup ... \cup S'_i \cup ... \cup S_n$ to S_g can be done by prefixing the *reverse* of t to T, i.e., $T' = t^{-1}; T$.

1. If t is an *add* or a *delete* transformation, then S_i and S'_i are *semantically equivalent*, i.e., any information in S_g derived from S_i can also be derived from S'_i. So, T' is already correct.
2. If t is a *contract* transformation, i.e., information capacity has been *decreased* when transforming S_i into S'_i, i.e., it may be the case that S_g now contains some constructs about which no information is derivable from any local schema. In this case, those constructs must be removed from S_g, and T' must be adapted accordingly.
3. If t is an *extend* transformation, then one needs to examine the derivability of relationship between the new local construct and S_g. If it is derivable through a transformation t_d, the latter must be appended to T'. If it is not derivable, an *extend* transformation, adding the new construct to S_g, must be appended to T'.

Bidirectional Characteristics: Considering the classification dimensions of Section 2, the BAV method exposes the following properties:

– Some of the primitive schema transformations supported by BAV, in particular the *contract* operators, actually **decrease the information capacity** of the schema.
– The *extend* schema transformations defined above **can increase information capacity**, and still allow to automatically derive an updated relationship between the global and the local schemas. This mainly depends on the nature of the accompanying query q.
– Developed in the context of data integration, BAV corresponds to a **virtual target state** scenario, where the global schema is *virtual* and the local schemas represent distributed store models.
– BAV also fits with the **model co-evolution** approach: incremental transformations successively applied to the global or to the local schema propagate, automatically or semi-automatically, as incremental adaptations of the inter-schema mappings and, in some cases, of the global schema.

3.6 Discussion

Consider a simple case of a mapping M that translates a schema S with a single table Person(id, salary, type) into schema T with two tables Sales(id, salary) and Engineer(id, salary). Mapping M partitions the table Person according to the value in field type, sending salespeople and engineers to the proper target table. This mapping can be expressed in any of the five tools introduced above.

In **lenses**, one constructs M out of functions *get* and *put* as follows, using relational algebra as the definition language and defining the functions one relation at a time:

$$get_{Sales} = \pi_{id,salary}\sigma_{type='sales'}\text{Person}$$

$$get_{Engineer} = \pi_{id,salary}\sigma_{type='engineer'}\text{Person}$$

$$put_{Person} = (\text{Sales} \times \{'sales'\}) \cup (\text{Engineer} \times \{'engineer'\})$$

In **PRISM**, one constructs M out of a short sequence of SMOs[7]:

```
– partition table PERSON into SALES with type = 'sales', ENGINEER
– drop column type from SALES
– drop column type from ENGINEER
```

In a **channel**, one constructs M as $HMerge(\{Sales, Engineer\}, Person, type)$, noting that a channel is built in reverse order which is the reason for the operator being "merge" instead of "partition". $HMerge$ handles the transformation that fills in the type column during updates.

In **DB-MAIN**, the partitioning will be interpreted, at the conceptual level, as the creation of two additional entity types $Sales$ and $Engineer$, each defined as a subtype of $Person$ through an *is-a* relationship of type *partition* (*disjointness* and *totality*). Then the logical design process is updated so that the new is-a relationship is transformed via downward inheritance, and both attributes *type* are removed.

In **BAV**, one constructs M as a sequence of *addRel* and *deleteRel* operations as follows:

- $addRel(\langle\langle Sales, id\rangle\rangle, q_1)$
- $addAtt(\langle\langle Sales, salary\rangle\rangle, notnull, q_2)$
- $addRel(\langle\langle Engineer, id\rangle\rangle, q_3)$
- $addAtt(\langle\langle Engineer, salary\rangle\rangle, notnull, q_4)$
- $deleteAtt(\langle\langle Person, salary\rangle\rangle, notnull, q_5)$
- $deleteAtt(\langle\langle Person, type\rangle\rangle, notnull, q_6)$
- $deleteRel(\langle\langle Person, id\rangle\rangle, q_7)$

Accompanying queries q_1, q_2, q_3 and q_4 incrementally partition the data of relation *Person* into relations $Sales$ and $Engineer$. Accompanying queries q_5, q_6 and q_7 merge $Sales$ and $Engineer$ into *Person*.

This simple analysis shows that there is overlap between the tools mentioned here, but also begins to show how they differ. One consideration that becomes immediately apparent is the level of granularity in specification. For this specific example, one can use a channel to express mapping M in a single step. However, a simple change to the example — letting the tables SALES and ENGINEER retain the type column — would make $HMerge$ inappropriate to use, while allowing PRISM to now express the mapping with a single operator.

Another point to note is that, in PRISM, the `drop column` SMO does not specify what the value in the dropped column could or should be, and that the value is assumed to be a labeled null or Skolem function in the operator's inverse function [9], this is handled more clearly with the introduction of ICMOs in PRISM++ [10], but still require some manual intervention in the definition of inverses.

Contrast that feature with BAV, whose syntax for a *deleteAttr* or *deleteRel* operation requires that the user provide the means for replacing the value (one could use the *contract* forms with a *void* argument to indicate that the value need not be recovered). Further contrast those options with the $HMerge$ channel transformation, which

---

[7] In PRISM++ the same example would include ICMOs, which will force the type column to be set to default, thus freeing the SMO steps from any information-capacity responsibility. Inverting the ICMO would require manual intervention.

holistically provides both transformation directions and thus prescribes that the value for the missing column be filled in with the name of the table of origin.

These options demonstrate that these tools often trade atomicity for functionality or vice versa. The *HMerge* channel transformation is the least atomic of all of the tools; however, it is unclear whether the schema co-evolution and data update features of channels would be possible if the transformation were broken into smaller parts. BAV offers a more atomic approach, but requires the user to express more in the queries that form the arguments to the operators.

4   Research Opportunities

Now that we have analyzed each of these five incremental *bx*tools, a natural question to ask is, for each tool, what it would take for it to work seamlessly in the context of the other tools' core scenarios? More concretely:

- Can the definition of a lens be analyzed on extended in such a way that allows a lens to support model co-evolution, or declarative integrity constraints?
- Can channels be extended to work in a way that more flexibly increases or decreases information capacity?
- Can automatic specification tools (e.g., [25,35,36]) be used to write mappings in BAV without needing to manually write queries?

Any attempt to unify the capabilities of these tools takes one step towards the construction of more general-purpose tools. Of course, there is no guarantee that these tools can be unified, in which case concrete evidence as to *why* unification is not possible can lead to a well-founded taxonomy of incremental *bx* tools.

Assuming that unification is possible in full or in part, there are other opportunities to leverage the flexibility of incremental *bx* tools. For the rest of this section, we examine one such case: data exchange mappings.

Data Exchange literature in the database field has centered around source-to-target tuple-generating-dependencies (st-tgds), a subset of first-order predicate calculus, as the formalism of choice for expressing mappings [1]. Such mappings are by their very nature unidirectional, so several efforts have been made to examine how to invert them to provide a measure of bidirectionality (e.g., [2]). However, many mappings have no unique inverse, or an inverse that is highly lossy.

Both Data Exchange and symmetric lenses describe their motivating scenario similarly: as synchronizing two data sources. In addition, if one examines st-tgds as a mapping language relative to the criteria in Section 2, however, one can see that they are strikingly similar to lenses. A data exchange mapping with no existential quantifiers can be information capacity decreasing but not increasing, like a classical lens; adding in existential quantifiers allows it to increase capacity as well, like a symmetric lens. Data transfer and operations in a data exchange setting assume materialized states (though in theory, those states may include placeholders or labeled nulls, which complicate query answering). And the schema evolution scenario in data exchange is a source-invariant one, where model evolution is represented itself as a mapping, and that mapping is composed with the original in some way to construct the new mapping [1,34,37].

Seeing this parallelism between Data Exchange and Lenses may be fruitful, especially for the Data Exchange field. For instance, the following might be possible:

- In the same way that relational calculus expressions can be translated into relational algebra, establish a set of operators that form a basis set for st-tgds. Whether a single such basis set would exist is unknown, as different sets may have different desirable properties, but each such set would need a conversion algorithm from st-tgds.
- Design the basis set of operators so that, like BAV and DB-MAIN, it can be partitioned into those that increase, decrease, or preserve information capacity.
- For each operator in the basis set, author the operator as an edit (symmetric) lens to allow for both information capacity increases and decreases.
- For each operator in the basis set, extend the set of allowable update operations on either side of the lens to include arbitrary INSERT, UPDATE, and DELETE SQL statements as in channels.
- For each operator in the basis set, allow primitive schema evolution operations to propagate through it as in channels. Also, ensure that each primitive schema evolution is itself reflected in the operator basis set, so that as in PRISM, one can instead choose to append the evolution to the mapping rather than propagate it.
- For each operator that increases information capacity, allow the user to specify options as to how to populate that information capacity above and beyond the default (as in channels — consider a case where an operator adds a column, but the user chooses to populate that column with the current time).
- Whenever either model evolves as a result of model co-evolution, maintain an audit trail of those evolutions as PRISM SMOs to maintain version compatibility.

As a result, one may end up constructing a tool similar to that of Clio [14], where one can use a declarative language to specify data exchange mappings. However, one would as a result able to construct a mapping where either endpoint is updatable declaratively, either endpoint is evolvable (and those evolutions can either propagate or work in-place), model evolutions result in an audit trail that mitigates version dependencies, users would have a wide variety of options for filling in missing data, and the mapping would still have the same formal foundation that allows for deep analysis and potential for optimization. If one does not like using declarative languages for mappings, one could instead use an Extract-Transform-Load-style tool [33] where one constructs a mapping directly from composing the individual operations.

5   Conclusion

All *bx* tools differ in the particular level of genericity, usability, expressive power, automation, etc. they provide, and often are based on comparable yet different theoretical backgrounds. For instance, channels or SMO choose a narrower target scenario and provide theoretically sound yet very usable solutions, while DB-MAIN opted for a broader theoretical framework (e.g., providing very high flexibility in terms of the data model they support). We believe this field is starting to mature, and it is now time to start thinking about a unification effort, capable of leveraging the flexibility of certain approaches, and the usability of others.

As a potential first step, a proper conceptual mapping across theoretical backgrounds can be of great use for future research efforts by providing a stronger foundation where new research efforts can operate in a way that is more aware of the existing landscape.

A more ambitious goal would be to form a truly unified theoretical foundation to incremental *bx*.

Such a unification would not be *sufficient*, since it would obviously not magically translate as a single generic and efficient tool to solve all problems in all possible scenarios. Such a unification would not be *trivial* to accomplish, since it will require a huge collaborative effort, involving researchers from distinct communities, countries and scientific cultures. However, such a unification would be *desirable*, both for better addressing existing bidirectional scenarios and for tackling largely unexplored, yet important scenarios. In the first case, the BAV approach constitutes a perfect example of a new approach designed to combine the respective advantages of the LAV and GAV approaches, which were initially seen as incompatible competitors in data integration. In the second case, new challenging problems are currently emerging that will require a better understanding of the existing *bx* tools and of their differences (1) to avoid reinventing the wheel forever, (2) to choose the adequate (combination of) tool(s) to address a given (set of) scenario(s), and (3) to face the current limitations of each tool in the context its own scenario(s).

In short, such a unified framework would greatly benefit to the entire research community, simultaneously acting as a structured survey of the field, as a systematic way of identifying novel research directions, and as a means to foster collaboration between several researchers or teams. This paper is not only a first humble step towards this objective; it is also a call to arms.

References

1. Arenas, M., Barceló, P., Libkin, L., Murlak, F.: Relational and XML Data Exchange. Morgan and Claypool Publishers (2010)
2. Arenas, M., Peréz, J., Riveros, C.: The recovery of a schema mapping: bringing exchanged data back. In: PODS 2008 (2008)
3. Bohannon, A., Pierce, B.C., Vaughan, J.A.: Relational lenses: a language for updatable views. In: PODS 2006 (2006)
4. Boyd, M., McBrien, P.: Comparing and Transforming Between Data Models Via an Intermediate Hypergraph Data Model. In: Spaccapietra, S. (ed.) Journal on Data Semantics IV. LNCS, vol. 3730, pp. 69–109. Springer, Heidelberg (2005)
5. Cleve, A.: Program analysis and transformation for data-intensive system evolution. PhD thesis. University of Namur, Chapter 10 (2009)
6. Cleve, A., Brogneaux, A.-F., Hainaut, J.-L.: A Conceptual Approach to Database Applications Evolution. In: Parsons, J., Saeki, M., Shoval, P., Woo, C., Wand, Y. (eds.) ER 2010. LNCS, vol. 6412, pp. 132–145. Springer, Heidelberg (2010)
7. Cleve, A., Hainaut, J.-L.: Co-transformations in Database Applications Evolution. In: Lämmel, R., Saraiva, J., Visser, J. (eds.) GTTSE 2005. LNCS, vol. 4143, pp. 409–421. Springer, Heidelberg (2006)
8. Cleve, A., Henrard, J., Roland, D., Hainaut, J.-L.: Wrapper-based system evolution — application to CODASYL to relational migration. In: CSMR 2008, pp. 13–22. IEEE CS (2008)
9. Curino, C.A., Moon, H., Zaniolo, C.: Graceful Database Schema Evolution: the PRISM Workbench. In: VLDB 2008 (2008)
10. Curino, C.A., Moon, H., Deutsch, A., Zaniolo, C.: Update Rewriting and Integrity Constraint Maintenance in a Schema Evolution Support System: PRISM++. In: VLDB 2011 (2011)
11. Czarnecki, K., Foster, J.N., Hu, Z., Lämmel, R., Schürr, A., Terwilliger, J.F.: Bidirectional Transformations: A Cross-Discipline Perspective. In: Paige, R.F. (ed.) ICMT 2009. LNCS, vol. 5563, pp. 260–283. Springer, Heidelberg (2009)

12. Dayal, U., Bernstein, P.: On the Correct Translation of Update Operations on Relational Views. ACM Transactions on Database Systems 8(3) (September 1982)
13. Deutsch, A., Nash, A., Remmel, J.: The chase revisited. In: PODS 2008 (2008)
14. Fagin, R., Haas, L.M., Hernández, M., Miller, R.J., Popa, L., Velcgrakis, Y.: Clio: Schema Mapping Creation and Data Exchange. In: Borgida, A.T., Chaudhri, V.K., Giorgini, P., Yu, E.S. (eds.) Mylopoulos Festschrift. LNCS, vol. 5600, pp. 198–236. Springer, Heidelberg (2009)
15. Foster, J.N., Greenwald, M.B., Moore, J.T., Pierce, B.C., Schmitt, A.: Combinators for bidirectional tree transformations: A linguistic approach to the view-update problem. ACM Trans. Program. Lang. Syst. 29(3) (2007)
16. Foster, J.N., Pilkiewicz, A., Pierce, B.C.: Quotient Lenses. In: ICFP 2008 (2008)
17. Hainaut, J.-L.: A generic entity-relationship model. In: Proc. of the IFIP WG 8.1 Conference on Information System Concepts: an In-depth Analysis, pp. 109–138. North-Holland (1989)
18. Hainaut, J.-L.: The Transformational Approach to Database Engineering. In: Lämmel, R., Saraiva, J., Visser, J. (eds.) GTTSE 2005. LNCS, vol. 4143, pp. 95–143. Springer, Heidelberg (2006)
19. Hainaut, J.-L., Cleve, A., Hick, J.-M., Henrard, J.: Migration of Legacy Information Systems. Software Evolution (2008)
20. Hick, J.-M., Hainaut, J.-L.: Database application evolution: A transformational approach. Data & Knowledge Engineering 59, 534–558 (2006)
21. Hofmann, M., Pierce, B.C., Wagner, D.: Edit lenses. In: POPL 2012 (2012)
22. Hofmann, M., Pierce, B.C., Wagner, D.: Symmetric lenses. In: POPL 2011 (2011)
23. Hu, Z., Schürr, A., Stevens, P., Terwilliger, J.F.: Dagstuhl seminar on bidirectional transformations (BX). SIGMOD Record 40(1) (2011)
24. Hu, Z., Schürr, A., Stevens, P., Terwilliger, J.F.: Bidirectional Transformation "bx" (Dagstuhl Seminar 11031). Dagstuhl Reports 1(1) (2011)
25. Matsuda, K., Hu, Z., Nakano, K., Hamana, M., Takeichi, M.: Bidirectionalization transformation based on automatic derivation of view complement functions. In: ICFP 2007 (2007)
26. McBrien, P., Poulovassilis, A.: Data Integration by Bi-Directional Schema Transformation Rules. In: ICDE 2003 (2003)
27. McBrien, P., Poulovassilis, A.: Schema Evolution in Heterogeneous Database Architectures, A Schema Transformation Approach. In: Pidduck, A.B., Mylopoulos, J., Woo, C.C., Ozsu, M.T. (eds.) CAiSE 2002. LNCS, vol. 2348, pp. 484–499. Springer, Heidelberg (2002)
28. Moon, H.J., Curino, C.A., Deutsch, A., Hou, C.-Y., Zaniolo, C.: Managing and querying transaction-time databases under schema evolution. In: VLDB 2008 (2008)
29. Moon, H.J., Curino, C., Zaniolo, C.: Scalable Architecture and Query Optimization for Transaction-time DBs with Evolving Schemas. In: SIGMOD 2010 (2010)
30. Poulovassilis, A., McBrien, P.: A general formal framework for schema transformation. Data & Knowlegde Engineering 28(1), 47–71 (1998)
31. Terwilliger, J.F., Delcambre, L.M.L., Logan, J.: Querying Through a User Interface. Journal of Data and Knowledge Engineering (DKE) 63(3)
32. Terwilliger, J.F., Delcambre, L.M.L., Maier, D., Steinhauer, J., Britell, S.: Updatable and Evolvable Transforms for Virtual Databases. PVLDB 3(1) (VLDB 2010)
33. Vassiliadis, P., et al.: A generic and customizable framework for the design of ETL scenarios. Information Systems 30(7)
34. Velegrakis, Y., Miller, R.J., Popa, L.: Preserving Mapping Consistency Under Schema Changes. VLDB Journal 13(3) (2004)
35. Voigtländer, J.: Bidirectionalization for Free! In: POPL 2009 (2009)
36. Voigtländer, J., Hu, Z., Matsuda, K., Wang, M.: Combining syntactic and semantic bidirectionalization. In: ICFP 2010 (2010)
37. Yu, C., Popa, L.: Semantic Adaptation of Schema Mappings When Schemas Evolve. In: VLDB 2005 (2005)

Using Models of Partial Knowledge
to Test Model Transformations

Sagar Sen[1], Jean-Marie Mottu[2], Massimo Tisi[1], and Jordi Cabot[1]

[1] AtlanMod, École des Mines de Nantes - INRIA, France
{sagar.sen,massimo.tisi,jordi.cabot}@inria.fr
[2] Université de Nantes - LINA (UMR CNRS 6241)
jean-marie.mottu@univ-nantes.fr

Abstract. Testers often use partial knowledge to build test models. This knowledge comes from sources such as requirements, known faults, existing inputs, and execution traces. In Model-Driven Engineering, test inputs are models executed by model transformations. Modelers build them using partial knowledge while meticulously satisfying several well-formedness rules imposed by the modelling language. This manual process is tedious and language constraints can force users to create complex models even for representing simple knowledge. In this paper, we want to simplify the development of test models by presenting an integrated methodology and semi-automated tool that allow users to build only small partial test models directly representing their testing intent. We argue that partial models are more readable and maintainable and can be automatically completed to full input models while considering language constraints. We validate this approach by evaluating the size and fault-detecting effectiveness of partial models compared to traditionally-built test models. We show that they can detect the same bugs/faults with a greatly reduced development effort.

1   Introduction

Model transformations are core components that automate important steps in Model-Driven Engineering (MDE), such as refinement of input models, model simulation, refactoring for model improvement, aspect weaving into models, exogenous/endogenous transformations of models, and the classical generation of code from models. Models and transformations have a widespread development in academia and industry because they are *generic* artifacts to represent complex data structures, constraints, and code abstractions. However, there is little progress in techniques to test transformations [5]. Testing requires the specification of software artifacts called *test models* that aim to detect faults in model transformations. Specifying test models manually is a tedious task, complicated by the fact that they must conform to a modelling language's specification and numerous well-formedness rules. For instance, the specification of the UML contains numerous inter-related concepts and well-formedness rules for its models such as class diagrams, activity diagrams, and state machines. The issue becomes crucial when a tester needs to create several hundred test models for a model transformation.

The knowledge to create test models can come from various sources. Usually some tests have a direct correspondence with application requirements, others are conceived

Z. Hu and J. de Lara (Eds.): ICMT 2012, LNCS 7307, pp. 24–39, 2012.

by imagining corner cases that can cause an error in a transformation. Several methods exist to derive further testing knowledge. For instance, analyzing a model transformation can reveal locally used classes, properties and possibly some of their values or bounds on values called a *footprint* [16]. Similarly, analyzing a localized fault via techniques such as dynamic tainting [12] in a model transformation can reveal patterns in the input modelling language that evoked the fault. Other sources include existing models via model slicing [7] and execution traces of a model transformation [3]. However, most of this knowledge is incomplete or *partial* in the sense that it must be completed and certified as a valid input test model that conforms to the well-formedness rules and constraints of the modelling language. We face and rise to three challenges:

Challenge 1: How can we express partial knowledge in a modelling language? We call the artifact containing this knowledge a *partial model.*

Challenge 2: How can we automatically complete a partial model?

Challenge 3: Are these automatically completed models effective in detecting the same faults that a human-made model containing the partial knowledge detects?

In this paper, we provide a methodology to generate effective test models from partial models and a semi-automated supporting tool [25]. The methodology to generate complete test models from partial knowledge is divided into two phases. In the first phase, we need to specify partial model(s) as required by *Challenge 1*. We propose to represent a partial model as a model conforming to a relaxed version of the original input metamodel of the model transformation. The relaxed metamodel allows specification of elements in a modelling language without obligatory references, containments, or general constraints that were in the original metamodel. Our tool adopts the transformation in [24] to generate a relaxed metamodel suitable to specify a partial model. In the second phase, we automatically complete partial models by integrating our tool PRAMANA [26] as required by *Challenge 2*. PRAMANA transforms the input metamodel to a base constraint satisfaction problem A_b in Alloy [15]. In this paper, we re-write partial models as *predicates in Alloy* that are juxtaposed to A_b. We also specify finite bounds on the satisfaction problem called the *scope*. The scoping strategy can be modified depending on whether we would like to generate minimally sized or large test models. We solve the constraint satisfaction problem in Alloy using a SAT solver to generate one or more test models that complete the partial models and satisfy all well-formedness rules of the input modelling language. If the partial model conflicts with a modelling language constraint, which has higher priority, we do not generate a test model, but we give feedback to the tester about the partial model being invalid. Our approach is applicable to transformations written using model transformation languages (e.g., ATL [17], Kermeta [22]) or also general-purpose languages (e.g., Java).

Finally, we experimentally evaluate a case study, to tackle the last *Challenge 3*, but also to show the benefits of our methodology in drastically reducing the manual aspects of test specification in MDE. Our experimentation shows that a set of small partial models can detect all the faults that human-made complex models detect. We compare, by using *mutation analysis* for model transformations [11] [21], the bug-detecting effectiveness between a test set of completed partial models and human-made test models. In our experiments, we employ the representative case study of transforming simplified UML class diagram models to database (RDBMS) models called class2rdbms. Mutation analysis on this transformation reveals that a set of 14 concise partial models can detect the same bugs as 8 complex man-made models. The 8 complete models contain more than twice (231 vs. 109) the number of elements compared to partial models. It is also important to note that the complete models were constructed meticulously to satisfy well-formedness rules while the partial models contain loosely connected objects that are automatically linked by our tool. The partial models are noise-free in the sense that they illustrate the *testing intent* precisely without the clamor of well-formedness rules. This result suggests that concise expression or extraction of partial knowledge for testing is *sufficient, comparatively effective,* and *less tedious* than manually creating test models hence solving *Challenge 3* and promoting our approach.

The paper is organized as follows. In Section 2 we present the representative case study for transformation testing. In Section 3 we present our integrated methodology to generate complete test models from partial knowledge. In Section 4 we present our experimental setup and results. In Section 5 we discuss related work. We conclude in Section 6.

2   Case Study

In the paper, we consider the case study of transforming simplified UML Class Diagram models to RDBMS models called class2rdbms. We briefly describe class2rdbms in this section and discuss why it is a representative transformation to validate our approach.

For testing a model transformation, the user provides input models that conform to the input metamodel *MM* (and possibly transformation pre-condition *pre(MT)*). In Figure 1, we present the simplified UMLCD input metamodel for class2rdbms. The concepts and relationships in the input metamodel are stored as an Ecore model [10] (Figure 1 (a)). Part of all the invariants on the simplified UMLCD Ecore model, expressed in Object Constraint Language (OCL) [23], are shown in Figure 1 (b). The Ecore

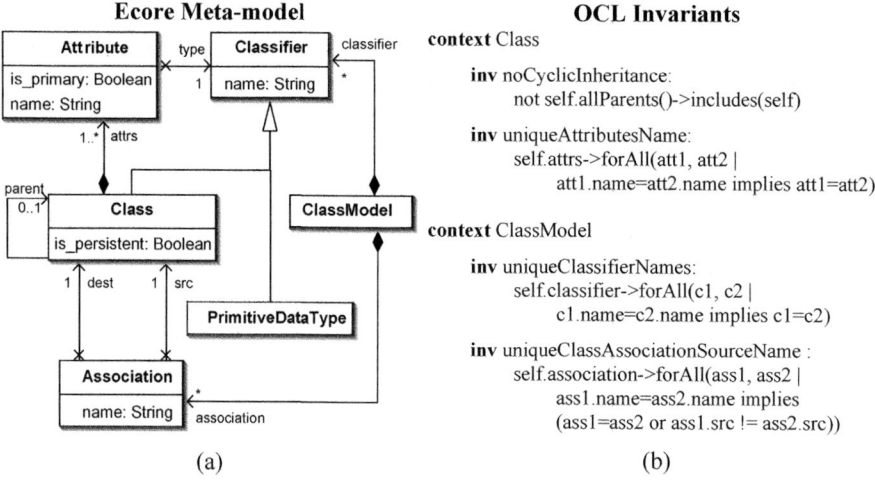

(a)   (b)

Fig. 1. (a) Simplified UML Class Diagram Ecore Metamodel (b) OCL constraints on the MM

model and the invariants together represent the true input metamodel for class2rdbms. The OCL and Ecore are industry standards used to develop metamodels and specify different invariants on them.

For this paper we use the Kermeta implementation of class2rdbms, provided in [21]. The transformation class2rdbms serves as a sufficient case study for several reasons. The transformation is the benchmark proposed in the MTIP workshop at the MoDELS 2005 conference [6] to experiment and validate model transformation language features. The input domain metamodel of simplified UMLCD covers all major metamodelling concepts such as inheritance, composition, finite and infinite multiplicities. The constraints on the simplified UMLCD metamodel contain both first-order and higher-order constraints. There also exists a constraint to test transitive closure properties on the input model such as there must be no cyclic inheritance. The transformation exercises most major model transformation operators such as navigation, creation, and filtering (described in more detail in [21]) enabling us to test essential model transformation features.

3   Methodology

We present a methodology to generate complete models from partial models, and the supporting tool [25] we developed. We describe the process in three phases: 1) Partial Model Specification, 2) Transformation to ALLOY, 3) Model Completion.

1) Partial Model Specification

A partial model is essentially a graph of elements such that: (1) The elements are instances of classes in the modelling language metamodel MM (2) The partial model does not need to conform to the language metamodel MM or its invariants expressed in a textual constraint language such as OCL. A complete model on the other hand contains all the objects of the partial model and additional objects or property value assignments in new/existing objects such that it conforms both to the metamodel and its invariants.

In the first phase, the user can specify partial models using an automatically-generated relaxed language. Given an input metamodel MM we generate a relaxed metamodel MM_r using a relaxation transformation as shown in Figure 2 (a). The transformation is adopted from Ramos et al. [24]. The relaxed metamodel MM_r allows the specification of partial models that need not satisfy a number of constraints enforced by the original metamodel MM. In Figure 2 (b), we show the relaxed metamodel derived from the simple class diagram metamodel shown in Figure 1 (a). The relaxed metamodel allows the specification of partial models in a modelling language. For instance, in Figure 4(a), we show a partial model specified using MM_r. It is important to clarify that PObject acts as a linking object to another metamodel that allows writing of patterns called *model snippets* on the relaxed metamodel. for a pattern-matching framework.This pattern matching metamodel from Ramos et al.[24] is not used in this paper and is hence not shown. It is interesting to note that the partial model specified using MM_r allows specification of objects required for testing. For instance, we specify Class and Association without the need to specify their containment in ClassModel. Similarly, a full specification of all property values is not required. The only properties specified in the partial model

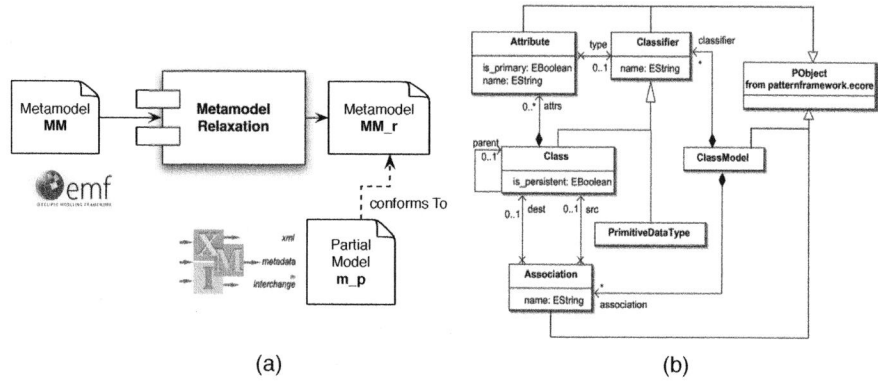

Fig. 2. (a) Metamodel relaxation to help specifying a partial model (b) Relaxed metamodel MM_r

are Class.is_persistent, Association.src, Association.dest, and Class.parent. On the contrary, a model that conforms to MM, must satisfy the metamodel MM's constraints and its well-formedness rules such as no cyclic inheritance such as in Figure 1(a), (b). Using OCL [23] constraints is another way to specify partial models. However, it is more complex for the tester to write an OCL constraint equivalent to a partial model: it is based on the complete MM to navigate the model from its root to the constrained concepts. For instance, whereas a partial model constrains one class to have its attribute is_attribute = true is made only with this class with only this attribute initialized, the same constraint written in OCL is:

```
input.classifier.select(c|c.oclIsType(Class)).exists(cs|cs.attrs.exists(a|a.is_primary = true))
```

In our approach, the concepts in the partial model are those that are available in MM but no one is mandatory.

2) Transformation to ALLOY

In the second phase, we transform the original metamodel MM to Alloy [15]. We integrate the tool PRAMANA presented in Sen et. al. [26], for the transformation in [25]. PRAMANA first transforms a metamodel expressed in the Ecore format using the transformation rules presented in [26] to ALLOY. Basically, classes in the input metamodel are transformed to ALLOY signatures and implicit constraints such as inheritance, opposite properties, and multiplicity constraints are transformed to ALLOY facts. Second, PRAMANA does not fully address the issue of transforming invariants and pre-conditions expressed on metamodels in the industry standard OCL to ALLOY. The automatic transformation of OCL to ALLOY presents a number of challenges that are discussed in [1]. We do not claim that all OCL constraints can be manually/automatically transformed to ALLOY for our approach to be applicable in the most general case. The reason being that OCL and ALLOY were designed with different goals. OCL is used mainly to query a model and check if certain invariants are satisfied.

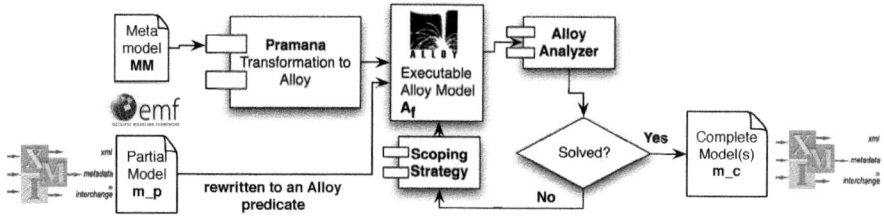

Fig. 3. Generation of Complete Model(s) from a Partial Model

ALLOY facts and predicates on the other hand enforce constraints on a model. The core of ALLOY is declarative and is based on first-order relational logic with quantifiers while OCL includes higher-order logic and has imperative constructs to call operations and messages making some parts of OCL overly expressive for the purpose of finite-domain constraint solving. In our case study, we have been successful in *manually transforming* all constraints on the simplified UMLCD metamodel to ALLOY from their original OCL specifications. Can we fully automate this step? This question remains to be answered in our approach. However, the manual transformation is a one time mental effort for a fixed set of constraints.

The ALLOY model generated in the previous phase has to be extended with the information coming from the partial test models. A partial model, such as in Figure 4 (a), is *manually re-written* to an ALLOY predicate as shown in Figure 4 (b). For the translation we navigate all objects of a certain type and put them together as an ALLOY predicate. The predicate states that there exists three objects c1,c2,c3 of type Class such that they are unequal, and only c1.is_persistent =True. The Class objects c2 and c3 inherit from c1. There exists also an Association object a1 such that its source a1.src is Class object c2 and destination a1.dest is c3. The name properties of the Class objects c1,c2,c3 and Association object a1 are not specified. They also do not contain a primary attribute which is mandatory for the transformation **class2rdbms**. The partial model objects also do not have to be contained in a ClassModel object. This process can be automated to generate a concise and effective ALLOY predicate. For instance, in [28], the authors have automated the transformation of partial models specified in a model editor. It can also be improved to consider negative application conditions (false-positives) or bounds on maximum/minimum of objects that can be specified in a partial model.

3) Model Completion

The final phase in our methodology is that of solving the translated ALLOY predicate in the executable ALLOY model A_f to obtain one or more complete models.

Model completion requires finite values such as the upper bound on the number of objects of the classes in the *MM*, or the exact number of objects for each class, or a mixture of upper bounds and exact number of objects for different classes. These values are called the *scope*. We call the approach to specify the scope the *scoping strategy*. The default scoping strategy is the upper bound defined by the number of objects of

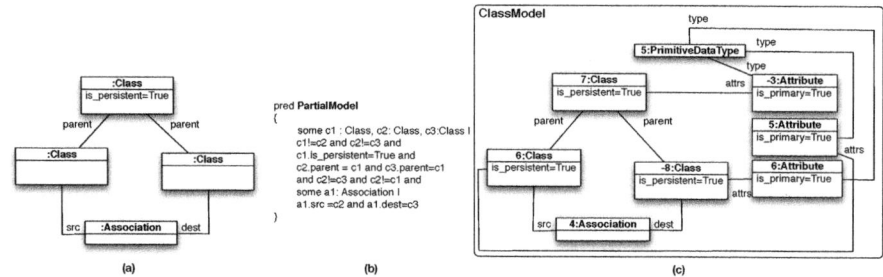

Fig. 4. (a) Partial Model P3 (b) Re-written Alloy Predicate (c) Complete Model from P3

each type in the partial model. For objects that are not part of the partial model we manually fine tune the scope to a minimum value such that a complete model can be generated. The scope of the top-level container class (ClassModel in our case study) is often exactly one. The scoping strategy is generated in the form of a ALLOY *run command* that is finally inserted in the executable ALLOY model A_f as shown in Figure 3. For example, the run command generated and fine-tuned for the partial model in Figure 4 (a) is:

```
run partialModel for exactly 1 ClassModel, 4 int, 3 Class, 3 Attribute, 1 PrimitiveDataType, 1 Association
```

The above run command is *manually fine-tuned* for Attribute and PrimitiveDataType. Fine-tuning is necessary in our case study since all Class objects require at least one primary Attribute object.

The ALLOY model is transformed to a set of expressions in relational calculus by the ALLOY analyzer. These expressions are then transformed to Conjunctive Normal Form (CNF) [29] using KodKod. Finally, the CNF is solved using a SAT solver [19]. The low-level SAT solutions are transformed back to XMI models that conform the initial metamodel. The resulting XMI models are validated by loading them into an EMF model editor. The editor ensures both the satisfaction with respect to an Ecore metamodel and OCL constraint checker. The use of an industrial tool helps ensure that the input models contain objects of valid classes, and conform to all metamodel and OCL constraints

There is always the option to either automatically/manually fine tune the scoping strategy to generate big or concise complete models. For instance, in Figure 4 (c), we show a concise complete model that we generate for the partial model in Figure 4 (a). One or more non-isomorphic solutions can be generated by ALLOY by adding a symmetry breaking constraint for each new solution. The symmetry breaking constraints are generated and added automatically by Alloy whenever a new solution is requested. This in-turn allows us to generate several hundred different models that are non-isomorphic if possible. **All elements of the partial model are preserved in the complete test models. Minimal number of objects are added so that all well-formedness rules and other constraints are satisfied.** In Table 1, we present the degree of automation achieved in our tool.

Table 1. Degree of automation in our tool

Aspect of Tool	Degree of Automation
Metamodel to Alloy in PRAMANA	automatic
OCL to Alloy	manual
Metamodel to relaxed metamodel	automatic
Partial Model Specification	manual
Partial Model to Alloy Predicate	currently manual/can be automated [28]
Solution Scoping	default and manually tunable
Solving Alloy Model from API in PRAMANA	automatic
XMI model instance from solution in PRAMANA	automatic

4 Experiments

In this section we perform experiments to address two questions:

Q1 Can tests derived from partial models detect the same faults that human-made models detect?

Q2 Are partial models more concise than equivalently powerful human-made models?

The inputs to our experiments are (a) the simplified UMLCD input metamodel from class2rdbms, (b) a set of well-formedness rules and class2rdbms pre-conditions in OCL. The experimentation proceeds as follows:

1. A set of random faults are injected in the class2rdbms model transformation, using a mutation tool;
2. A first modeler, aware of the previous mutations, develops a set of models that kills all the injected faults;
3. A second modeler, aware of the mutations, develops a set of partial models to kill all the injected faults, following our approach;
4. We measure the size of test models and compare the effectiveness of the two test sets.

The two modelers are both experienced in testing and modeling. For our final evaluation, we use a Macbook Pro Intel dual-core processor 2.7 GHz, 8 GB RAM to generate complete models and perform the analysis.

4.1 Injecting Faults in the Model Transformation

Our experimental evaluation is based on the principles of mutation analysis [11]. Mutation analysis involves creating a set of faulty versions or *mutants* of a program. A test set must distinguish the correct program output from all the output of its mutants. In practice, faults are modelled as a set of mutation operators where each operator represents a class of faults. A mutation operator is applied to the program under test to create each mutant injecting a single fault. A mutant is killed when at least one test model

Table 2. Partition of the class2rdbms mutants depending on the mutation operator applied

Mut. Oper.	CFCA	CFCD	CFCP	CACD	CACA	RSMA	RSMD	ROCC	RSCC	Total
# of Mutants	19	18	38	11	9	72	12	12	9	200

detects the pre-injected fault. It is detected when program output and mutant output are different. A test set is relatively adequate if it kills all mutants of the original program. A mutation score is associated to the test set to measure its effectiveness in terms of percentage of the killed/revealed mutants.

To inject faults in our transformation, we use the mutation operators for model transformations presented by Mottu et al. [21]. These mutation operators are based on three abstract operations linked to the basic treatments in a model transformation: the navigation of the models through the relations between the classes, the filtering of collections of objects, the creation and the modification of the elements of the output model. Using this basis, Mottu et al. defined the following mutation operators:

Relation to the same class change (RSCC): The navigation of one association toward a class is replaced with the navigation of another association to the same class.

Relation to another class change (ROCC): The navigation of an association toward a class is replaced with the navigation of another association to another class.

Relation sequence modification with deletion (RSMD): This operator removes the last step off from a navigation which successively navigates several relations.

Relation sequence modification with addition (RSMA): This operator does the opposite of RSMD, adding the navigation of a relation to an existing navigation.

Collection filtering change with perturbation (CFCP): The filtering criterion, which could be on a property or the type of the classes filtered, is disturbed.

Collection filtering change with deletion (CFCD): This operator deletes a filter on a collection; the mutant operation returns the collection it was supposed to filter.

Collection filtering change with addition (CFCA): This operator does the opposite of CFCD. It uses a collection and processes an additional filtering on it.

Class compatible creation replacement (CCCR): The creation of an object is replaced by the creation of an instance of another class of the same inheritance tree.

Classes association creation deletion (CACD): This operator deletes the creation of an association between two instances.

Classes association creation addition (CACA): This operator adds a useless creation of a relation between two instances.

We apply all these operators on the class2rdbms model transformation. We identify in the transformation code all the possible matches of the patterns described by each operator. For each match we generate a new mutant. This way we produce two hundred mutants from the class2rdbms model transformation with the partition indicated in Table 2.

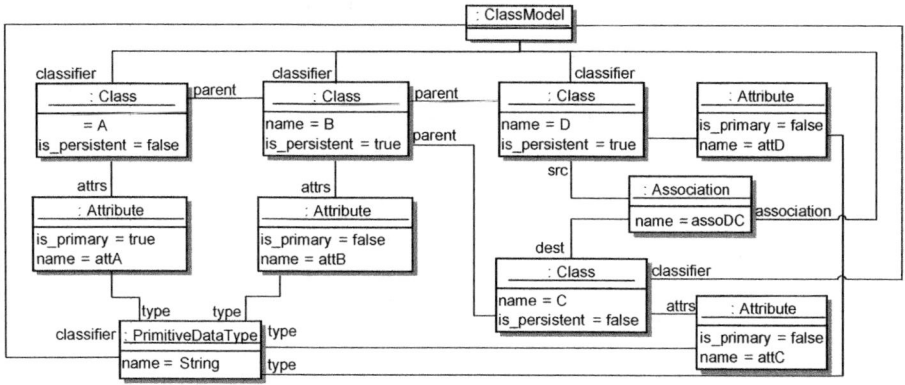

Fig. 5. Human-made Test Model M1

In general not all injected mutants become faults, since the semantics of some of them is equivalent to the correct program, and therefore they can never be detected. The controlled experiments presented in this paper use mutants presented in our previous work [21] where we have clearly identified mutants and equivalent mutants.

4.2 Building Manual Tests

A modeler manually designed a set of test models, by analyzing the 200 faults we injected in class2rdbms as described earlier. The objective was to obtain a test set able to kill all the mutants we generated. The resulting test set is composed by 8 models conforming to the fully constrained UMLCD modelling domain (metamodel + well-formedness rules + pre-conditions for class2rdbms). Together, the hand-made test models count 231 elements. We used mutation analysis to verify that these 8 human-made models can kill all 194 mutants (excluding 6 equivalent mutants identified).

4.3 Building Partial Models

For the experimentation, we could have built the partial models from scratch (as it normally happens) but we prefer to extract partial models from the manual models derived in the previous phase. We need it to have a direct correspondence between traditional models and partial models, that is useful in the evaluation. This does not influence the global experimentation results: from scratch or from a traditional model, the second modeler wants to kill one mutant. With both methods, only the mutant is targeted. Additionally with our experimental approach, partial models contain elements comparable with traditional models.

We incrementally derive our partial models from the 8 human-made models. Given a reference human-made model (as shown in Figure 5) we extract a partial model (in Figure 4 (a)). A partial model is extracted from a human-made model by keeping only the useful elements to kill one target mutant. We use our methodology in Section 3 to automatically complete the partial model to get 10 test models. Second, we apply

Table 3. Size Comparison Between Partial Models and Human-made Models

partial model	P1	P2	P3	P4	P5	P6	P7	P8	P9	P10	P11	P12	P13	P14	M1	M2	M3	M4	M5	M6	M7	M8
from reference model	M1	M1	M1	M3	M5	M6	M3	M4	M7	M3	M4	M2	M1	M8								
#ClassModel	0	0	0	0	0	0	0	0	0	0	0	0	0	0	1	1	1	1	1	1	1	1
#Class	1	3	3	2	2	2	2	2	3	3	4	1	1	3	4	3	3	3	2	3	2	3
#name attr.	0	0	0	0	0	0	0	0	0	0	0	0	0	0	4	3	3	3	2	3	2	3
#is_persistent attr.	1	3	1	2	0	0	2	0	0	3	1	0	1	1	4	3	3	3	2	3	2	3
#parent relation	0	1	2	0	1	1	1	0	0	1	1	0	0	1	3	1	2	1	1	1	0	1
#Association	0	0	1	1	0	2	1	2	0	0	2	0	0	1	1	1	1	2	0	3	0	1
#name attr	0	0	1	1	0	2	0	0	0	0	0	0	0	1	1	1	1	2	0	3	0	1
#src relation	0	0	1	1	0	2	1	2	0	0	2	0	0	1	1	1	1	2	0	3	0	1
#dest relation	0	0	0	0	0	0	1	2	0	0	2	0	0	1	1	1	1	2	0	3	0	1
#Attribute	1	0	0	0	2	0	0	0	2	1	0	1	1	0	4	3	3	3	2	3	3	2
#is_primary attr.	1	0	0	0	0	0	0	0	2	1	0	1	1	0	4	3	3	3	2	3	3	2
#name attr.	0	0	0	0	2	0	0	0	0	0	0	0	0	0	4	3	3	3	2	3	3	2
#type relation	0	0	0	0	0	0	0	0	2	1	0	1	1	0	4	3	3	3	2	3	3	2
#PrimitiveDataType	0	0	0	0	0	0	0	0	0	0	0	1	1	0	1	1	1	1	1	1	1	1
#name attr	0	0	0	0	0	0	0	0	0	0	0	0	0	0	1	1	1	1	1	1	1	1
Total #concept	4	7	9	7	7	9	8	8	9	10	12	5	6	8	38	29	30	33	18	37	21	25
Total #concept per set	of partial models:					109									of models:				231			

mutation analysis to obtain a score for the 10 complete test models if the partial model is solvable. If the mutation score is not 100% we extract more partial models from the set of complete models to kill live mutants. We repeat this process until test models generated from partial models give a 100% mutation score.

For instance, to kill a mutant we extract a partial model P3 in Figure 4 (a) from the human-made model M1 in Figure 5. The human-made model contains many concepts (classes, classes' attributes and relationships), it satisfies the invariants of the meta-model and the pre-conditions of the model transformation under test class2rdbms. The class on top of the partial model matches class B of the reference model, the class on the left matches class D (both are persistent), the class on the right matches class C (name, is_persistent, attrs are not concerned with this matching), and the association of the partial model matches association assoDC. Then the partial model P3 matches the model M1. We notice that the completed model illustrated in Figure 4 (c) doesn't match the model M1 (both subclasses are persistent for instance).

We apply this iterative process to obtain 14 partial models that when completed give a 100% mutation score. In Table 3, we list the partial models, the reference human-made models they were extracted from, and the objects they contained. We notice that partial models do not need specification of all attribute values. For instance, the partial model P3 (Figure 4) has three classes but is_persistent attribute is set to true for only one of them (true or false or nothing in different partial models). Moreover, thanks to the relaxed metamodel, we do not have to instantiate the root container class ClassModel for the partial model. In the right hand side of Table 3, we describe the elements in the 8 human-made reference models. We do not list the abstract classes *classifier* since it is never instantiated and the property *attrs* which is redundant for all models.

4.4   Test Sets Comparison

The set of 14 partial models is made of 109 elements while the set of reference models contain 231 elements which is more than twice. We express less information in the partial models, considering only testing information based on potential faults simulated

Table 4. Summary of Mutation Analysis

Human-made model	Mutation Score of the model	Partial models matching the human-made model	Mutation Score of the completed partial models	Mutant killed by model also killed by completed partial models	Additional mutants killed by the completed partial models (% total #mutants)
M1	57.2%	P2, P3, P13	87.1%	100%	29.9%
M2	57.2%	P1, P12	28.9%	50.5%	0%
M3	49%	P1, P4, P7, P10, P12	74.2%	89.3%	39.7%
M4	57.2%	P1, P8, P11, P12	85.6%	98.2%	29.4%
M5	35.6%	P1, P5, P12	39.7%	100%	4.1%
M6	61.3%	P1, P2, P4, P6, P8, P12, P13, P14	83%	100%	21.6%
M7	47.4%	P1, P9, P12	56.2%	100%	8.8%
M8	58.2%	P1, P12, P14	75.8%	100%	17.5%
8 models	100%	14 partial models	100%	100%	0%

in the mutants. We also need to specify fewer meta-classes and properties in partial models. Our tool generates complete test models form partial models while satisfying the metamodel specification, well-formedness rules, pre-conditions, and a finite scope. In case a partial model violates a pre-condition the tester is notified such that he/she can modify the partial model to satisfy the constraints. The concise size of a partial model facilitates this process. From these results we deduce that creating partial models is simpler compared to creating complete human-made models hence addressing **Q2**.

In a second experiment, in order to study the resilience in quality (in terms of mutation score) irrespective of solver technology (Alloy and SAT in our case) in our methodology, we generate 100 complete models for 14 partial models giving rise to 1400 test models.

A sample of a completed model is illustrated in Figure 4 (c) for a partial model in Figure 4 (a). The time taken to generate a complete test model was almost instantaneous. Setting up a problem in CNF for a SAT solver takes about 400 ms for our case study. The scope to generate complete models is controlled such that the models are structurally minimal. The integer scope was set to $\pm 2^{10}$, and the number of objects per class was adjusted such that it either equals the number of objects in the partial model or the scope is incremented such that a consistent solution is generated. For instance, all Class objects require a primary Attribute object hence the number of Class and Attribute objects were equal or the number of Attribute objects was higher if required by the partial model.

We compute the mutation score of the human-made models and compare it to the score obtained by completing partial models. The results of mutation analysis are summarized in Table 4. The time taken for mutation analysis using 1400 models to detect 200 mutants was about 6 hours and 1 minute.

In Table 4, Column 1 lists the human-made models and column 3 lists the partial models that match parts of the human-made models. In Column 2, we present the mutation score obtained for a human-made model lets say *Mx*. In Column 4, we present the mutation score obtained by the set of completed partial models that match *Mx*.

For instance, M1 kills 57.2% of all mutants, while its set of completed partial models from P2, P3, and P16 (3*100 completed models) kills 87.1% mutants. The column 5 lists the percentage of the mutants killed by reference human-made model that are also killed by the completed models. In column 6, we also present the mutants additionally killed by the completed partial models that were not killed by their reference human-made model. The decomposition in different lines comparing each human-made model with its partial models illustrates that (i) a partial model can match several models and that (ii) a set of several partial models matching one model is more efficient than it. The most important line is the last one where the results of the entire set of human-made models and the entire set of completed partial models are the same : both 100 %. A general conclusion is that both human-made models and completed partial models give a 100% mutation score hence addressing question **Q1**. An in-depth analysis reveals that for 6 human-made models among 8, more than 98% of the mutants killed by the reference model are also killed by the completed partial models. The completed models from P1, P12, P14 kill 89.3% of the mutants killed by M3. Only completed models from P1, P12 have a lower score of 28.9% compared to the score of M2 viz. 57.2%.

Despite generating different solutions for the same partial model our mutation scores remain consistent and comparable to human-made models. We illustrated that a set of partial models has the same efficiency in revealing bugs than the set of human-made models they match. It is not necessary to write complicated human-made models since partial models have the same efficiency.

5 Related Work

The work in this paper integrates some contributions coming from previous work. In [26] Sen et al. have introduced a tool CARTIER (now called PRAMANA) based on the constraint solving system ALLOY to resolve the issue of generating models such that constraints over both objects and properties are satisfied simultaneously. PRAMANA does not integrate a way to translate existing models to ALLOY since it's thought for model synthesis, and does not have any support for partial models. In this work we are able to apply PRAMANA to model completion by feeding it with a suitable translation of the partial models. The idea of generating test models from partial knowledge developed from our previous work in [28]. In [28], we present the idea of generating suggestions to complete models in a domain-specific model editor. We qualify our approach using our previous work [21], where we extend mutation analysis to MDE by developing mutation operators for model transformation languages.

We explore two main areas of related work : specifying partial models and test model synthesis.

The notion of a partial model in MDE has been previously developed for various objectives. In [14], the authors present the notion of *model fragments* that are essentially partial models with test coverage criteria. Similarly, in [24] the authors propose the notion of *model snippets*. Model snippets are partial models that are specified on a relaxed version of the original metamodel. In this paper, we use the notion of model snippets to define partial models. In [20], the authors propose partial models as a way to represent uncertainty and variation in an already available complete model. They use ALLOY to generate variable complete models.

The second area of related work is about model generation or using partial knowledge to synthesize models for applications such as testing. Model generation is more general and complex than generating integers, floats, strings, lists, or other standard data structures such as those dealt with in the Korat tool of Chandra et al. [8]. Korat is faster than ALLOY in generating data structures such as binary trees, lists, and heap arrays from the Java Collections Framework but it does not consider the general case of models which are arbitrarily constrained graphs of objects. The constraints on models makes model generation a different problem than generating test suites for context-free grammar-based software [18] which do not contain domain-specific constraints. Models are complex graphs that must conform to an input meta-model specification, a transformation pre-condition and additional knowledge such as a partial model to help detect bugs. In [9], the authors present an automated generation technique for models that conform only to the class diagram of a metamodel specification. A similar methodology using graph transformation rules is presented in [13]. Generated models in both these approaches do not satisfy the constraints on the metamodel. In [27], we present a method to generate models given partial models by transforming the metamodel and partial model to a Constraint Logic Programming (CLP). We solve the resulting CLP to give model(s) that conform to the input domain. However, the approach does not add new objects to the model. We assume that the number and types of models in the partial model is sufficient for obtaining complete models. The constraints in this system are limited to first-order horn clause logic. Previous work exists in mapping UML to AL-LOY for a similar purpose. The tool UML2Alloy [2] takes as input UML class models with OCL constraints. The authors present a set of mappings between OCL collection operations and their ALLOY equivalents.

6 Conclusion

Manually creating test models that conform to heterogeneous sources of knowledge such as a metamodel specification, well-formedness rules, and testing information is a tedious and error-prone operation. Moreover the developed test models are often unreadable, since their complexity obfuscates the testing intent.

In this paper, we present methodology based on models from partial knowledge to simplify the development of effective test models. The emphasis in this paper is to push towards the development of partial models by analyzing requirements, existing test models [7], the transformation under test [16], or fault locations [12]. We provide a semi-automated tool [25] to support the development of partial models and completing them by automatic solving. We show that the specification of partial models is concise, they can detect the same bugs that a human-made model can detect, and they precise capture testing intent. We also perform experiments show that partial models are effective irrespective of the underlying solver (which is Alloy in this case). Different non-isomorphic complete models obtained by solving a single partial model consistently detect the bug the partial model was initially targeted to kill.

We believe our work reinforces a human-in-the-loop testing strategy that sits in between two prominent schools of thought: manual test case specification and automated test generation. Partial models are an effective way to insert human testing knowledge or sophisticated analysis in a purely automated generation methodology. The approach presented in the paper contains steps that are automated and those that are not. For instance, the transformation of a partial model to ALLOY could be automated to a certain degree. The scoping strategy can be improved using various heuristics to synthesize a diverse set of complete models. We would also like to explore strategies to combine mutually consistent partial models to generate a smaller set of complete test models. Finally, we would like to see the creation of industry-strength tools that allow convenient ways to specify partial models with fully automated complete test model synthesis.

References

1. Anastasakis, K., Bordbar, B., Georg, G., Ray, I.: UML2Alloy: A Challenging Model Transformation. In: Engels, G., Opdyke, B., Schmidt, D.C., Weil, F. (eds.) MODELS 2007. LNCS, vol. 4735, pp. 436–450. Springer, Heidelberg (2007)
2. Anastasakis, K., Bordbar, B., Georg, G., Ray, I.: On Challenges of Model Transformation from UML to Alloy. Software and Systems Modeling 9(1), 69–86 (2008); Special Issue on MoDELS 2007
3. Aranega, V., Mottu, J.-M., Etien, A., Dekeyser, J.-L.: Traceability mechanism for error localization in model transformation
4. Bardohl, R., Taentzer, G., Minas, M., Schurr, A.: Handbook of Graph Grammars and Computing by Graph transformation, vII: Applications, Languages and Tools. World Scientific (1999)
5. Baudry, B., Ghosh, S., Fleurey, F., France, R., Le Traon, Y., Mottu, J.-M.: Barriers to Systematic Model Transformation Testing. Communications of the ACM 53(6) (2010)
6. Bézivin, J., Schürr, A., Tratt, L.: Model Transformations in Practice Workshop. In: Bruel, J.-M. (ed.) MODELS 2005. LNCS, vol. 3844, pp. 120–127. Springer, Heidelberg (2006)
7. Blouin, A., Combemale, B., Baudry, B., Beaudoux, O.: Modeling Model Slicers. In: Whittle, J., Clark, T., Kühne, T. (eds.) MODELS 2011. LNCS, vol. 6981, pp. 62–76. Springer, Heidelberg (2011)
8. Boyapati, C., Khurshid, S., Marinov, D.: Korat: automated testing based on java predicates. In: Proceedings of the 2002 ACM SIGSOFT International Symposium on Software Testing and Analysis (2002)
9. Brottier, E., Fleurey, F., Steel, J., Baudry, B., Traon, Y.L.: Metamodel-based test generation for model transformations: an algorithm and a tool. In: Proceedings of ISSRE 2006, Raleigh, NC, USA (2006)
10. Budinsky, F.: Eclipse Modeling Framework. The Eclipse Series. Addison-Wesley (2004)
11. DeMillo, R., Lipton, R., Sayward, F.: Hints on test data selection: Help for the practicing programmer. IEEE Computer 11(4), 34–41 (1978)
12. Dhoolia, P., Mani, S., Sinha, V.S., Sinha, S.: Debugging Model-Transformation Failures Using Dynamic Tainting. In: D'Hondt, T. (ed.) ECOOP 2010. LNCS, vol. 6183, pp. 26–51. Springer, Heidelberg (2010)
13. Ehrig, K., Küster, J.M., Taentzer, G., Winkelmann, J.: Generating Instance Models from Meta Models. In: Gorrieri, R., Wehrheim, H. (eds.) FMOODS 2006. LNCS, vol. 4037, pp. 156–170. Springer, Heidelberg (2006)
14. Fleurey, F., Baudry, B., Muller, P.-A., Traon, Y.L.: Towards dependable model transformations: Qualifying input test data. Software and Systems Modelling (2007) (accepted)
15. Jackson, D.: (2008), http://alloy.mit.edu
16. Jeanneret, C., Glinz, M., Baudry, B.: Estimating footprints of model operations. In: International Conference on Software Engineering (ICSE 2011), Honolulu, USA. IEEE (May 2011)

17. Jouault, F., Kurtev, I.: On the Architectural Alignment of ATL and QVT. In: Proceedings of ACM Symposium on Applied Computing (SAC 2006), Dijon, FRA (April 2006)
18. Hennessy, M., Power, J.: An analysis of rule coverage as a criterion in generating minimal test suites for grammar-based software. In: Proc. of the 20th IEEE/ACM ASE, NY, USA (2005)
19. Mahajan, Y.S., Fu, Z., Malik, S.: Zchaff2004: An Efficient SAT Solver. In: Hoos, H.H., Mitchell, D.G. (eds.) SAT 2004. LNCS, vol. 3542, pp. 360–375. Springer, Heidelberg (2005)
20. Famelis, M., Ben-David, S., Chechik, M., Salay, R.: Partial models: A position paper. In: MoDeVVa Workshop (co-located with MODELS 2011) (2011)
21. Mottu, J.-M., Baudry, B., Traon, Y.L.: Mutation analysis testing for model transformations. In: Proceedings of ECMDA 2006, Bilbao, Spain (July 2006)
22. Muller, P.-A., Fleurey, F., Jézéquel, J.-M.: Weaving Executability into Object-Oriented Meta-languages. In: Briand, L.C., Williams, C. (eds.) MODELS 2005. LNCS, vol. 3713, pp. 264–278. Springer, Heidelberg (2005)
23. OMG. The Object Constraint Language Specification 2.0, OMG: ad/03-01-07 (2007)
24. Ramos, R., Barais, O., Jézéquel, J.-M.: Matching Model-Snippets. In: Engels, G., Opdyke, B., Schmidt, D.C., Weil, F. (eds.) MODELS 2007. LNCS, vol. 4735, pp. 121–135. Springer, Heidelberg (2007)
25. Sen, S.: Partial model completion tool (2011), https://sites.google.com/site/partialmodeltool/
26. Sen, S., Baudry, B., Mottu, J.-M.: On combining multi-formalism knowledge to select test models for model transformation testing. In: IEEE International Conference on Software Testing, Lillehammer, Norway (April 2008)
27. Sen, S., Baudry, B., Precup, D.: Partial model completion in model driven engineering using constraint logic programming. In: International Conference on the Applications of Declarative Programming (2007)
28. Sen, S., Baudry, B., Vangheluwe, H.: Towards Domain-specific Model Editors with Automatic Model Completion. SIMULATION 86(2), 109–126 (2010)
29. Torlak, E., Jackson, D.: Kodkod: A relational model finder. In: Tools and Algorithms for Construction and Analysis of Systems, Braga, Portugal (March 2007)

Specification-Driven Test Generation for Model Transformations

Esther Guerra

Universidad Autónoma de Madrid, Spain
Esther.Guerra@uam.es

Abstract. Testing model transformations poses several challenges, among them the automatic generation of appropriate input test models and the specification of oracle functions. Most approaches to the generation of input models ensure a certain level of source meta-model coverage, whereas the oracle functions are frequently defined using query or graph languages. Both tasks are usually performed independently regardless their common purpose, and sometimes there is a gap between the properties exhibited by the generated input models and those demanded to the transformations (as given by the oracles).

Recently, we proposed a formal specification language for the declarative formulation of transformation properties (invariants, pre- and postconditions) from which we generated partial oracle functions that facilitate testing of the transformations. Here we extend the usage of our specification language for the automated generation of input test models by constraint solving. The testing process becomes more *intentional* because the generated models ensure a certain coverage of the interesting properties of the transformation. Moreover, we use the same specification to consistently derive both the input test models and the oracle functions.

1 Introduction

Model transformations are the pillars of Model-Driven Engineering (MDE), and therefore they should be developed using sound engineering principles to ensure their correctness [12]. However, most model transformation technologies are nowadays centered on supporting the implementation phase, and few efforts are directed to the specification of requirements, design or testing of transformations. As a consequence, transformations are frequently hacked, not engineered, being hard to maintain, incorrect or buggy.

In order to alleviate this situation, we proposed in the past *trans*ML, a family of modelling languages for the *engineering* of transformations using an MDE approach [12]. *trans*ML provides support for the gathering of requirements, its formal specification, the architectural, high- and low-level design, as well as the specification of test scripts, which themselves are also models. An engine called *mtUnit* is able to execute these test suites in an automated way.

*trans*ML includes a visual language with formal semantics called PAMoMo (Pattern-based Model-to-Model Specification Language) [11] for the contract-based specification of transformation requirements. In this way, the designer may

Z. Hu and J. de Lara (Eds.): ICMT 2012, LNCS 7307, pp. 40–55, 2012.

specify requirements of the input models of a transformation (preconditions), expected properties of the output models (postconditions), as well as properties that any pair of input/output models should satisfy (invariants). Similar to software requirement specification languages like Z or Alloy, PaMoMo's formal semantics enables reasoning at the level of requirements, while being independent of the particular transformation language used for the implementation.

In [11], we explored the use of PaMoMo for testing. In particular, we showed how to automatically derive OCL partial oracle functions from PaMoMo specifications, and used these oracles to assert whether a particular implementation satisfied a specification. Still, the transformation tester had the burden to: (a) produce a reasonable set of input test models, (b) build a *mtUnit* script to exercise the transformation with the different input models, and (c) select the partial oracle functions produced from the specification to assert whether the tests passed or failed. In particular, the manual creation of input models is tedious and time-consuming, and it does not guarantee an appropriate coverage of all requirements in the specification.

In this paper, we tackle these problems by deriving, from the transformation specification, not only the oracle functions, but also a set of input test models ensuring a certain level of coverage of the properties in the specification. These input models are calculated using constraint solving techniques. Besides, a dedicated *mtUnit* test suite is generated for the automated testing of the transformation implementation using the generated input models and oracle functions.

While there are several approaches for the automated testing of transformations, ours is unique because our test models aim at testing the requirements and properties of interest as given in a specification. Current approaches either focus on producing input test models ensuring a certain coverage of the input meta-model [7,21], or do not consider specification-based testing. Hence, our approach is directed to test the *intention* of the transformation. Moreover, the use of the same specification to consistently derive both the input models for testing and the oracle functions is also novel.

Paper organization: Section 2 reviews existing approaches to model transformation testing. Then, Section 3 sketches our proposal. Section 4 introduces our specification language PaMoMo, whereas Section 5 describes our approach to derive input test models with a certain level of specification coverage. We present tool support in Section 6 and discuss some conclusions in Section 7.

2   State of the Art

There are three main challenges in model transformation testing [2]: the generation of input test models, the definition of test adequacy (or coverage) criteria, and the construction of oracle functions.

Most works dealing with the generation of input test models consider only the features of the input meta-model but not properties of the transformation (i.e. they support black-box testing). For instance, in [7,21], the authors perform automatic generation of input test models based on the input meta-model and some coverage criteria (e.g. partitioning of attribute values and number of classes). In [10], the generation of input test models must be hand-coded using an imperative language with features for randomly choosing attribute values and association ends. There are a few white-box testing approaches, like [15] where the authors propose using all possible overlapping models of each pair of rules in a transformation as input models for testing.

Regarding the third challenge, we distinguish between complete and partial oracle functions. The former are defined by having the output models at hand. For instance, the test cases for the C-SAW transformation languages [16] consist of a source model and its expected output model. Partial oracle functions express contracts that the input and output models of a transformation should fulfil. Most proposals to partial oracle functions use OCL to specify the contracts [6,10,18]. The approaches in [8,9] follow a similar philosophy to the xUnit framework, and the oracle functions can be specified as OCL/EOL assertions. Finally, some approaches permit the specification of partial oracle functions as graph patterns or model fragments [1]. None of these approaches provide a mechanism to assert the adequacy of the specified tests and automate their generation.

In conclusion, we observe that some transformation testing approaches provide automated test execution [8,9], but do not support the generation of input models, and the oracle needs to be specified manually. Other works focus on the automatic generation of input models [7,21], but without considering transformation properties. Finally, the works proposing contracts for specifying transformations do not use the contracts for input test generation. In this paper, we will present our approach to specification-based transformation testing which automates the generation of the input test models, the oracle function and executable test scripts from the same transformation specification.

It is worth noting that the idea of synthesizing both input test data and oracle functions from a specification has been successfully applied to general software testing, if we look at the broader scope of model-based testing. For instance, in [3], the authors generate both artifacts for automated testing of Java programs based on Java predicates from which all possible non-isomorphic inputs (up to a certain size bound) are efficiently generated. This yields complete coverage of the input state space. In our case, we aim at generating test models exhibiting relevant properties; complete meta-model coverage (i.e. generating all meta-model instances of a certain size) does not guarantee this, and may lead to the so-called state explosion problem. The model-based testing approach in [23] uses symbolic execution to generate unit tests ensuring a certain path coverage, i.e., it supports white-box testing. In our case, we follow a black-box testing approach.

3   A Framework for Specification-Driven Testing

Fig. 1 shows the working scheme of our approach. First, the designer specifies the requirements (i.e. the pre/postconditions and invariants) of the transformation using our language PAMOMO. The developer can use this specification as a guide to implement the transformation using his favourite language (e.g. ATL [13],

ETL [14], etc.). Starting from the specification, the transformation tester can automatically generate a complete test suite which can be directly used to test the transformation implementation. This test suite comprises: (i) an oracle function that encodes the invariants and postconditions in the specification as assertions [11]; (ii) a set of input test models which enables the testing of all requirements in the specification according to certain coverage criteria; and (iii) a test script that automates the execution of the transformation for each test model, checks the conformance of the result using the oracle function, and reports any detected error using our *mtUnit* engine [12].

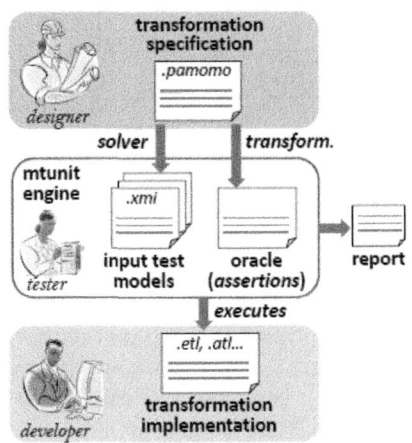

Fig. 1. Framework

4 A Specification Language for Model Transformations

PAMOMO is a formal, pattern-based, declarative, bidirectional specification language to describe, in an implementation-independent way, correctness requirements of the transformations and of their input and output models [11]. These requirements may correspond to *preconditions* that the input models should fulfil, *postconditions* that the output models should fulfil (beyond meta-model constraints), as well as *invariants* of the transformation (i.e. requirements that the output model resulting from a particular input model should satisfy).

Preconditions, postconditions and invariants are represented as graph patterns, which can be positive to specify expected model fragments, or negative to specify forbidden ones. They can have attached a logical formula stating extra conditions, typically (but not solely) constraining the attribute values in the graph pattern. In this paper and in our prototype tool, these formulas are written in OCL. Optionally, patterns can define one enabling condition and any number of disabling conditions, to reduce the scope of the pattern to the locations where the enabling condition is met, and the disabling conditions are not.

Formally, an invariant $I = (C, C_{en}, \{C_{dis}\})$ is made of a main constraint C, an enabling condition C_{en} (which may be empty), and a set $\{C_{dis}\}$ of disabling conditions. The main constraint and conditions $C_x = \langle G_s, G_t, \alpha \rangle$ are made of two graphs typed by the source and target meta-models of the transformation, and a formula α over their elements. A positive invariant holds on a pair of source and target models if: (i) for each occurrence Occ of the source graph of the main constraint C plus the enabling condition C_{en}, (ii) if there is no occurrence of the disabling conditions $\{C_{dis}\}$ in the context of Occ, (iii) then there is an occurrence of the target graph of C in the context of Occ. If the invariant is negative, then

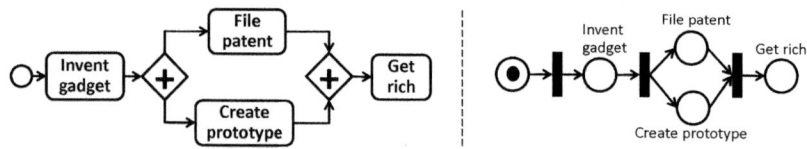

Fig. 2. BPMN model (left) and equivalent Petri net (right)

we should not find an occurrence of the target graph of C in step (iii). Pre- and postconditions have the same structure as invariants, but the target (source) graph in preconditions (postconditions) is always empty. Their interpretation is also different. A pre/postcondition holds if, for each occurrence of the enabling condition, there is an occurrence of the main constraint for which no occurrence of the disabling conditions is found. See [11] for a complete description of the formal semantics of PaMoMo.

As a running example, we use PaMoMo to specify a transformation from the Business Process Modeling Notation (BPMN) [4] to Petri nets. The goal is to analyse BPMN models to detect deadlocks, incorrect termination conditions or tasks that can never be completed. The left of Fig. 2 shows a BPMN model. It specifies a flow initiated in a start event (the circle), and consisting in the completion of different tasks (rounded rectangles). The diamonds in the model are called parallel gateways, and split the execution in several parallel branches (first gateway) which are later synchronized (second gateway). From this BPMN model, our transformation should create a Petri net like the one to its right.

The left of Fig. 3 shows some transformation preconditions, expressing requirements that any input model should fulfil beyond its meta-model constraints[1]. For instance, our transformation expects models with one start event from which only one sequence flow goes out. This is formalised by the positive precondition *OneStartEvent* (i.e. there must exist one start event with one outgoing flow in the input model) and the negative precondition *MultipleStartEvents* (there cannot be several start events). These conditions are not demanded by the BPMN meta-model, which allows models with any number of start events, each one of them with multiple outgoing flows, but are required by our transformation. The figure shows another precondition, *PathsForGateway*, with an enabling condition. It demands that each gateway (enabling condition) defines at least one input and one output flows (main constraint). The precondition contains the abstract class *Gateway*, becoming applicable to all concrete gateway types inheriting from it.

Postconditions express requirements of the output models, beyond their meta-model constraints. The right of Fig. 3 shows some postconditions for the generated Petri nets, like the absence of unconnected places (*UnconnectedPlaces*), the existence of input and output places for all transitions (*ConnectedTransitions*), and the existence of a single place with one token and without input transitions (*InitialPlace* and *InitialMarking*).

---

[1] In this section, we use a graphical concrete syntax for the specification. In Section 6, we will show an alternative textual syntax that is supported by our prototype tool.

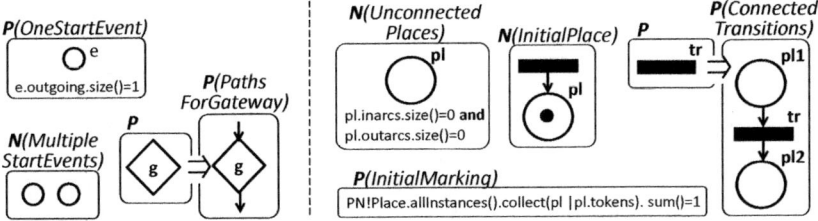

Fig. 3. Preconditions (left) and postconditions (right) of the transformation

Finally, Fig. 4 shows some invariants describing the transformation of tasks and gateways. Tasks must be transformed into equally named places (*Task1*). Since tasks can only have one outgoing flow, the corresponding places cannot be connected to two output transitions (*Task2*). Each parallel gateway should be transformed into a transition (*ParallelGateway1*), and the places for all incoming tasks to the gateway should be input to the transition (*ParallelGateway2*). Invariant *ParallelGateway3* states that if a parallel gateway does not have a task *t2* as input (disabling condition), then the place for *t2* cannot be connected to the transition (as the invariant is negative). There are similar invariants for the tasks going out from a parallel gateway. The two remaining invariants state that exclusive (also called choice) gateways should be transformed into an intermediate place, plus one transition for each outgoing branch.

The use of a formal specification language like PAMOMO to specify transformation properties has the following advantages: (i) it enables reasoning on the transformation requirements before their implementation, as well as detecting contradictions in the requirements early in the project; (ii) it provides a high-level notation to specify pre/postconditions and invariants of the transformations; and (iii) it is possible to automate the generation of an oracle function from the specification and use it for automated testing [11]. However, the challenge of generating appropriate input test models remains, as these have to be built by hand, which is a tedious and error-prone task. Moreover, it is difficult to ensure that the input test set will enable the testing of all relevant properties in the specification. To solve this problem, next we present an approach to generate input test models ensuring the coverage of a specification.

5   Specification-Driven Generation of Input Test Models

Our approach to specification-driven testing consists of the following steps: (1) translation of the properties in the specification into a suitable format for model finding, (2) selection of a level of specification coverage, resulting in a particular strategy to build expressions that demand the satisfaction (or not) of a number of properties in the generated models, (3) use of a constraint solver to find

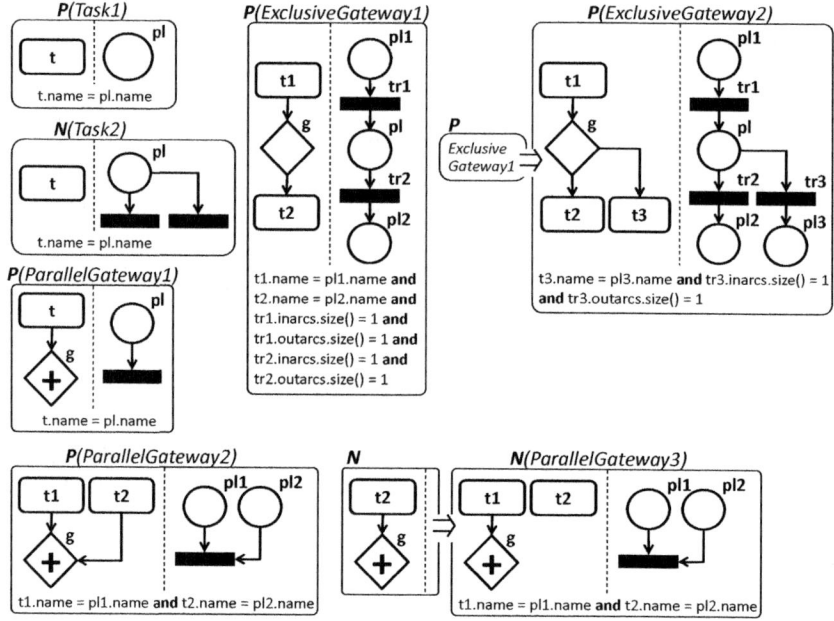

Fig. 4. Invariants of the transformation

models satisfying concrete combinations of properties and the input meta-model integrity constraints, and finally, (4) identification of the assertions that should be checked after testing the transformation with a particular input model. In the remaining of this section we present in detail this procedure.

5.1   Translation of Properties in the Specification

As a first step, we translate the specification into a language that allows automating the generation of models. In particular, we use OCL as target language because we already had support to generate OCL assertions from our specifications [11], there are available solvers that find models satisfying a set of OCL constraints [5], and we do not need to parse the OCL formulas in the properties of the specification to a different language. Nonetheless, this is our particular option and the framework could be used with a different target language whenever a translation from our specification language is provided.

Although a specification includes preconditions, postconditions and invariants, only preconditions and invariants contain useful information for the input model generation. Postconditions refer to properties of the output models and are only used to generate oracle functions, but not input models.

An invariant expresses a property of the form: *if certain source pattern appears in the input model, then certain target pattern should be present (or not) in the output model.* Thus, it is interesting to generate input models containing instances of the source pattern, to test whether transforming these models actually yields output models containing the target pattern. For this purpose, from each PaMoMo invariant we generate an OCL expression which characterises the source pattern of the invariant. Listing 1 shows a scheme of the generated expression. It iterates on the objects of the source graph of the main constraint (lines 1–3), and checks that there is no occurrence of the source graph of any disabling condition (lines 4–7, this code is generated for each disabling condition). The function conditions corresponds to an OCL expression checking the conditions that the traversed objects should fulfil, namely the existence of the links specified in the invariant (o_i.link->includes(o_j)), inequalities for the objects with same type ($o_i<>o_j$), and all terms in the invariant formula over elements of the input domain only. The enabling condition of the invariants is ignored as we do not want models where the invariant is satisfied vacuously due to the absence of the enabling condition in the models. Moreover, if the invariant is negative, the generated expression is the same (i.e. it is not preceded by the not particle) because the source part of the invariant is still positive (*if X appears then...*).

```
o1.type::allInstances()->exists(o1 | ...
  oi.type::allInstances()->exists(oi |
    conditions(o1,...,oi)
    < and not
      oj.type::allInstances()->exists(oj | ...
        ok.type::allInstances()->exists(ok |
          conditions(o1,...,oi,oj,...,ok) >* ) ...)
```

Listing 1. OCL for invariants

```
< not >?
< o1.type::allInstances()->forAll(o1 | ...
  oi.type::allInstances()->forAll(oi |
    conditions(o1,...,oi)
    implies >?
    oj.type::allInstances()->exists(oj | ...
      ok.type::allInstances()->exists(ok |
        conditions(o1,...,oi,oj,...,ok)
        < and not
          ol.type::allInstances()->exists(ol | ...
            om.type::allInstances()->exists(om |
              conditions(o1,...,oi,oj,...,ok,ol,...,om) >* ...)
```

Listing 2. OCL for preconditions

As an example, from invariant *ParallelGateway3* we generate the expression:

```
Task::allInstances()->exists(t1 |
  Task::allInstances()->exists(t2 |
    ParallelGateway::allInstances()->exists(g |
      t1.outgoing->includes(g) and t1<>t2 and not t2.outgoing->includes(g) )))
```

Frequently, specifications include invariants with same source and different target. For instance, *Task1* and *Task2* have both a task as source, whereas the former specifies how to translate a task correctly, the latter identifies an incorrect translation. In this case, generating an input model containing a task enables the testing of both invariants. Thus, from the set of generated OCL expressions, we eliminate redundant source conditions (i.e. equal source in the main constraint and disabling conditions). We do not eliminate subsumptions to allow for the testing of models with different size and context conditions.

Finally, preconditions specify requirements of the input models of a transformation. A transformation is not demanded to work properly for input models that do not satisfy these preconditions. The validity of the input models is hardly ever done by the transformation, but by an external procedure, or otherwise it is ensured by the transformation application context. Thus, we take the convention that all generated input models must fulfil all preconditions in the specification. For this purpose, we generate an OCL constraint from each precondition, and enforce their satisfaction in all generated input models by adding them to the expressions used to generate them (see next subsection). The scheme of the generated OCL code is shown in Listing 2. The expression looks for all occurrences of the enabling condition (lines 2–4), and demands that for each one of them there is an occurrence of the main constraint (lines 6–8) satisfying the disabling conditions (lines 9–10). If the precondition has no enabling condition the resulting expression is the same as the one for invariants, and if it is negative, the generated expression is preceded by not.

5.2   Input Model Generation for Different Coverage Criteria

The model generation process is performed in two steps. First, we compose an OCL expression for each input model to be generated, identifying the properties that this model should fulfil. These expressions are built according to certain specification coverage criteria. Then, we feed each expression, together with the input meta-model and the OCL code generated from the preconditions, to a constraint solver. The solver will try to find a valid input model satisfying the given OCL expression, preconditions and meta-model integrity constraints. For a particular expression, the solver may not find a model in the given scope, or due to some inconsistency in the specification. In such a case, we can either widen the search scope, or do not generate a model for that particular expression.

We identify seven levels of specification coverage for the generated test set, with increasing degrees of exhaustivity: *property, closed property, t-way, closed t-way, combinatorial, closed combinatorial* and *exhaustive*. The property, t-way and exhaustive levels generate models enabling the testing of a number of invariants in the specification by combining their source models. The remaining levels generate also models that do not contain occurrences of certain invariants. In the following, we present in detail each one of them.

Property coverage. This is the least exhaustive level of coverage, appropriate when the invariants in the specification are independent. It generates as many input models as invariants in the specification, each one including at least one occurrence of the source of an invariant. The rationale is to use each generated model to test one property of the transformation, given by one invariant in the specification. For this purpose, given a specification with $I = \{I_1, ..., I_n\}$ invariants (with different source), we generate n expressions of the form $ocl(I_i)$. Each expression demands the existence of an occurrence of the source of invariant I_i. As an example, Table 1 shows in the first column the expressions generated from a specification with three invariants, where each i_x term represents the OCL code generated from the invariant x.

Closed property coverage. This criterion extends the previous one by generating additional models that do not contain occurrences of the source of some invariant in the specification. The goal is checking whether the transformation under test handles properly the absence of certain patterns in the input models. These limit cases, usually due to underspecifications, frequently lead to errors in the final implementations, yielding malformed output models. Thus, given a specification with $I = \{I_1, ..., I_n\}$ invariants, we also generate n additional expressions of the form $not\ ocl(I_i)$. The second column of Table 1 shows the generated expressions assuming three invariants.

Interestingly, any model that does not contain the source of an invariant will satisfy the invariant vacuously, as an invariant states the consequences of having some pattern in the source model, but not the consequences of its absence. Nonetheless, the input models generated in this way are still interesting because their transformation have to yield valid target models satisfying the rest of invariants and postconditions in the specification as well as the target meta-model integrity constraints.

Finally, this coverage criterion is also indicated for specifications that use a closed world assumption (i.e. any property not included in the specification is false) by generating models which potentially may not belong to the input language according to the specification. Currently, PaMoMo does not support a closed world semantics.

t-way coverage. Most faults in software systems are due to the interactions of several factors or properties. Based on this observation, t-wise testing [22] consists of the generation of test cases for all possible combinations of t properties in the system under test. *Pairwise* testing is a particular case of this kind of testing for $t = 2$ (i.e. the generation of test cases for pairs of properties) which yields smaller test suites than exhaustive generation yet being able to find many errors. In our case, we are interested in detecting errors coming from an incorrect implementation of the combination of several requirements in a specification. These errors are frequent when each requirement is implemented as a rule or relation that interacts with other rules in the transformation, e.g. through explicit invocation.

In this case, given a specification with $I = \{I_1, ..., I_n\}$ invariants, we generate an expression of the form $ocl(I_j)\ and\ ...\ and\ ocl\ (I_k)$ for each t-tuple of invariants in the specification, demanding the existence of an occurrence of the source part of each invariant in the tuple. In the limit, 1-way testing

Table 1. Expressions generated from a specification with 3 invariants. The terms i_1, i_2 and i_3 represent the OCL code generated from the invariants in the specification.

property	closed property	2-way	closed 2-way	combinatorial	closed combinatorial	exhaustive (for i_1, i_2)
i_1	i_1	i_1 and i_2	i_1 and i_2	i_1	i_1	-
i_2	i_2	i_1 and i_3	i_1 and i_3	i_2	i_2	i_1
i_3	i_3	i_2 and i_3	i_2 and i_3	i_3	i_3	i_2
	not i_1		not i_1	i_1 and i_2	i_1 and i_2	not i_1
	not i_2		not i_2	i_1 and i_3	i_1 and i_3	not i_2
	not i_3		not i_3	i_2 and i_3	i_2 and i_3	i_1 and i_2
				i_1 and i_2 and i_3	i_1 and i_2 and i_3	i_1 and not i_2
					not i_1	not i_1 and i_2
					not i_2	not i_1 and not i_2
					not i_3	

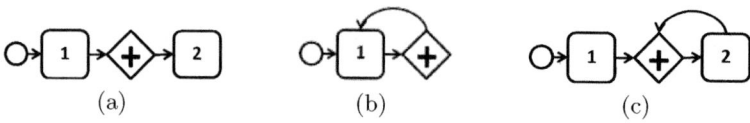

(a)   (b)   (c)

Fig. 5. Generated models for the OCL expressions: (a) *ocl(Task1) and ocl(Parallel-Gateway3)*; (b) *ocl(ParallelGateway1) and not ocl(ParallelGateway3)*; (c) *ocl(Parallel-Gateway1) and not ocl(ParallelGateway3)*, with failure due to disabling condition

is equivalent to property coverage. Table 1 shows the expressions generated for pairwise (i.e. 2-way) testing.

As an example, Fig. 5(a) shows a model generated for pairwise testing, considering the properties *Task1* and *ParallelGateway3*. The model contains two tasks, the first one is input to the gateway, and the second one not (as required by the disabling condition of the second invariant). The solver introduces a start event which does not appear in any of the invariants, as it is required by precondition *OneStartEvent*. Moreover, the tasks in the two invariants are not required to be different in the generated model, hence we obtain a model with two tasks instead of three.

In the MDE community, pairwise testing is being successfully used for software product line testing [19,20], considering pairs of features in a feature model. In our case there are additional challenges, because our specifications do not explicitly encode dependencies between their requirements, and the model generation procedure has to consider the constraints given by the input meta-model and preconditions.

Closed *t*-way coverage. As discussed previously, sometimes it is desirable to test also that the input models that do not contain occurrences of the source of invariants are handled correctly. Hence, in this criteria we generate the same models as in *t*-way coverage, as well as models generated from expressions of the form *not ocl(I_i)*, as Table 1 shows for *t=2*.

Combinatorial coverage. It generates all models for *1*-way, *2*-way, ... *t*-way coverage, where *t* is the number of invariants in the specification. Thus, here we consider all combinations of properties, including all of them simultaneously (*t*-way case). A total of $2^n - 1$ models are generated (see Table 1).

Closed combinatorial coverage. It generates the same models as in combinatorial coverage, and a model from each negated invariant (see Table 1).

Exhaustive coverage. This is the most exhaustive level of coverage, generating models for all combinations of the occurrence or absence of the source of the invariants in a specification, or their obliteration. For this purpose it generates different OCL expressions where the existence of the source of each invariant can be either mandatory (*ocl(I_i)*), forbidden (*not ocl(I_j)*) or ignored (i.e. the invariant is not taken into account). This yields a number of 3^n potential models. The last column of Table 1 shows the OCL expressions for a specification with two invariants.

As an example, Fig. 5(b) shows a model generated for the OCL expression *ocl(ParallelGateway1) and not ocl(ParallelGateway3)*. In particular, invariant *ParallelGateway3* is not satisfied in the model as there is no occurrence of its main constraint (i.e. there are not two different tasks).

For a more exhaustive coverage we can enforce the absence of a property in the generated models in several ways. Up to now, this was achieved by negating the source of the invariant ($not \ ocl(I_j)$). However, there are different ways in which we can "disable" the testing of a particular invariant: either because there is no occurrence of the source of its main constraint, or because there are occurrences of the main constraint but these do not satisfy some disabling condition. Thus, we generate an OCL expression for each way to disable the property (the source of the main constraint of the invariant is not found, or it is found but it does not fulfil some disabling condition). Fig. 5(c) shows a model used to test invariant *ParallelGateway1* and the absence of *ParallelGateway3*, the latter due to the occurrence of its disabling condition (both tasks are input to the gateway).

Altogether, this coverage uses a brute-force approach to the generation of test models. Unfortunately, many of the generated OCL expressions are unsatisfiable because they contain contradicting requirements. For instance, the expression $ocl(ParallelGateway2) \ and \ not \ ocl(ParallelGateway1)$ has no solution because it looks for an input model with two tasks connected to a gateway (first invariant), and simultaneously forbids having tasks connected to gateways (negation of the second invariant). The problem is that the negated invariant is included in the required one. Given that the generation of models using constraint solving is time-consuming, it is advisable to discard unsatisfiable expressions prior to model generation. Thus, if the source of one invariant is included in another one, no expression requiring the invariant with bigger source and negating the other should be considered.

Finally, it is for us an open question whether such a deep degree of exhaustivity is worth for certain kinds of specifications, or whether it is more effective to use less exhaustive types of coverage as the previous ones, enriched with heuristics that allow for the generation of bigger sets of test input models (for instance, generating several models from the same OCL expression, or demanding more than one occurrence of the invariants).

Regardless the chosen level of coverage, there are some configurable aspects (or heuristics) in the model generation process, which may affect the size and number of generated models. For example, when looking for models aimed at testing several invariants with non-empty intersection, different levels of overlapping between them can be considered, ranging from non-overlapping (the source of the invariants is taken to be disjoint) to a maximal overlap. Second, for specifications with a high number of requirements or for exhaustive testing, we can minimise the size of the generated test set by skipping the generation of a model for a particular combination of properties if this combination is already present in a model previously generated.

Finally, as the reader may have noticed, the solver may yield the same model for the resolution of two different OCL expressions. For instance, if the input meta-model for our running example requires exclusive gateways to have at least two output tasks, then the solver will always try to complete the source model in invariant *ExclusiveGateway1* with a new task connected to the gateway, i.e., it will try to find a model like the one in invariant *ExclusiveGateway2*. Thus, the expressions *ocl(ExclusiveGateway1)* and *ocl(ExclusiveGateway2)* are likely to produce the same input model. At this point, we can simply remove one of the generated input models from the test set and continue processing the next OCL expression, as the model enables the testing of both invariants.

5.3   Linking Input Models and Oracles

As a final step, we automatically generate an *mtUnit* script – another language in our *transML* family of languages [12] – to automate the testing of the transformation using the generated models. The script includes a test case for each invariant and postcondition in the specification, defining the input models to be used in the test case, and the oracle function checking the particular invariant or postcondition. By default, all generated models are added to all test cases. However, if a model was generated from an expression that negated an invariant, then it is not added to the test case for that invariant, as we already know that such a model satisfies the invariant vacuously. This allows for a more efficient testing process, as a given oracle function will not be checked in any model for which we already know that the oracle function will always hold.

Altogether, the generation of input models, oracle functions and scripts is done automatically from the same specification. This has the advantage that the transformation tester does not need to build them separately by hand, identify the oracle functions to be used for each input model, and build a script to execute the test. More importantly, being generated from the same specification, both the models and the oracle functions will work together to validate the same properties of interest: the models will enable the testing of these properties, and the oracle functions will check their satisfaction.

The right of Fig. 6 shows an excerpt of the *mtUnit* script generated from our specification example, using the *property* coverage level. Lines 4–17 in the upper window contain the definition of the test case generated from the invariant *ParallelGateway1*. For space constraints, the figure only shows two of the input models for this test case (lines 5–6). Below, the figure partially shows the result of running the test.

6   Tool Support

The presented framework is supported by an Eclipse, EMF-based prototype tool which allows building PAMOMO specifications using a textual editor, and automates the generation of input models and *mtUnit* test scripts for them. The left of Fig. 6 shows part of our specification example using the textual editor, in particular the definition of the invariants *ParallelGateway1* (lines 4–12) and *ParallelGateway3* (lines 14–31). The generation of the test suite and input models from this specification is *push-button*. In our case, it yields the *mtUnit* file that is partially shown to the right of the figure, in the upper right window. The first two lines declare the file with the transformation to be tested (either ATL or ETL) and the source and target meta-models. Lines 4–17 correspond to the test case for the *ParallelGateway1* invariant (for brevity we only show two of its input test models in lines 5–6). Executing the test suite will run the transformation for each input model and report whether the result verifies the assertions in the different test cases (see lower right window in Fig. 6).

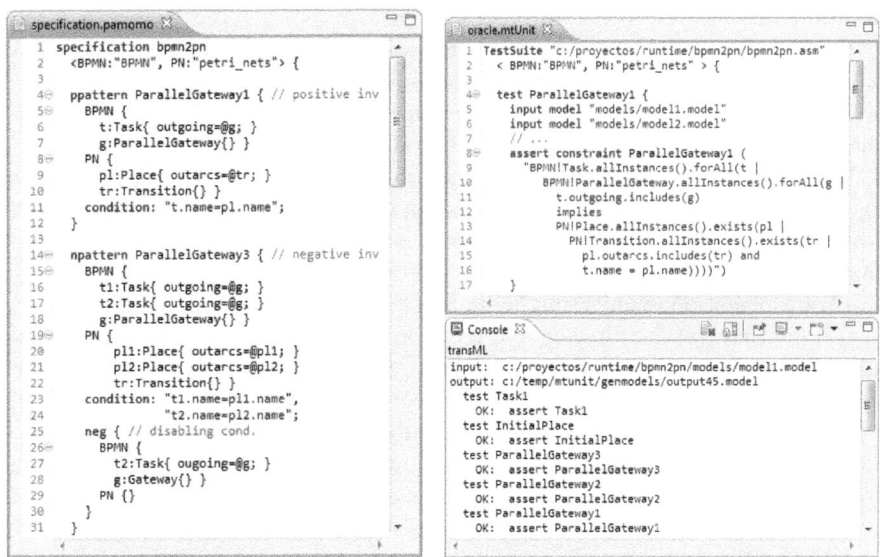

Fig. 6. Tool support for PAMOMO specifications (left) and testing (right)

In the back-end, we are using the UMLtoCSP constraint solver [5] for model finding. UMLtoCSP receives an Ecore meta-model and a file with OCL invariants, and generates a ".dot" file with a model that satisfies the meta-model integrity constraints and the OCL invariants. Then, we parse this file into an EMF-conformant representation for its use in *mtUnit*. Currently, we do not provide support for model generation heuristics like different overlapping degrees or detection of redundant models.

7   Discussion and Lines of Future Work

As discussed in Section 2, most black-box testing approaches use meta-model coverage criteria to ensure that the generated input models will include, altogether, instances of all classes and associations in the meta-model, and extreme values for the attributes. However, it is difficult to ensure that the generated models will include certain structures enabling the testing of relevant transformation properties, whereas unimportant class instances or model fragments may appear repeatedly in the generated models.

In contrast, the presented specification-driven approach aims at testing the intention of the transformation, and ensures that the generated models will allow testing transformation properties of interest. In this sense, the quality of the generated test set highly relies on how complete a specification is. If a specification only covers part of the transformation requirements, then the generated models may not enable the testing of the underspecified parts. For instance, our running example does not include invariants over the *EndEvent* BPMN class, and therefore the generated test models may not include instances of this type, leaving its transformation untested. Thus, we foresee complementing our techniques with additional coverage criteria, also meta-model based.

Finally, the models we generate with our technique tend to be small. This has the advantage that the test models remain intentional: they are generated for testing a particular combination of transformation invariants, which will be checked by the oracle function more efficiently.

We are currently conducting some experiments of our approach with promising results. For the specification in this paper, we have implemented an ATL transformation of 120 lines of code, and performed pairwise testing. To test the effectiveness of the generated test set, we manually created 40 mutants of this transformation by injecting faults that followed the systematic classification in [17] (i.e. navigation, filtering and creation mutations), and then used the test set on the transformation mutants. The test discovered the faults in 28 out of the 40 mutants, which gives a mutation score (or effectiveness) of 70% (28/40).

Starting from the results in this and subsequent experiments, in the future we plan to investigate the effectiveness of our generated input models to detect transformation failures. This is called *vigilance*, which is the degree in which contracts can detect faults in the running system. A relevant question is the level of detail required in contracts to find a significant number of failures and obtain high vigilance. Another interesting issue is whether the size of the generated input test models has an influence on the effectiveness of the test set. In order to obtain "bigger" test models, we are considering (a) the possibility of including extra constraints, stating that e.g., models should have a certain number of instances of each class, and (b) extending the coverage criteria to allow several instances of the same invariant. Regarding tool support, the most critical factor is the constraint solver, which is time-costly, and therefore we are currently working towards a domain-specific constraint solver for models.

Acknowledgements. Work funded by the Spanish Ministry of Economy and Competitivity (TIN2011-24139) and by the R&D programme of Madrid Region (S2009/TIC-1650).

References

1. Balogh, A., Bergmann, G., Csertán, G., Gönczy, L., Horváth, Á., Majzik, I., Pataricza, A., Polgár, B., Ráth, I., Varró, D., Varró, G.: Workflow-Driven Tool Integration Using Model Transformations. In: Engels, G., Lewerentz, C., Schäfer, W., Schürr, A., Westfechtel, B. (eds.) Nagl Festschrift. LNCS, vol. 5765, pp. 224–248. Springer, Heidelberg (2010)
2. Baudry, B., Ghosh, S., Fleurey, F., France, R.B., Traon, Y.L., Mottu, J.-M.: Barriers to systematic model transformation testing. CACM 53(6), 139–143 (2010)

3. Boyapati, C., Khurshid, S., Marinov, D.: Korat: automated testing based on Java predicates. In: ISSTA 2002, pp. 123–133 (2002)
4. BPMN, http://www.bpmn.org/
5. Cabot, J., Clarisó, R., Riera, D.: UMLtoCSP: a tool for the formal verification of UML/OCL models using constraint programming. In: ASE 2007, pp. 547–548 (2007)
6. Cariou, E., Marvie, R., Seinturier, L., Duchien, L.: OCL for the specification of model transformation contracts. In: ECEASST, vol. 12, pp. 69–83 (2004)
7. Fleurey, F., Baudry, B., Muller, P.-A., Traon, Y.: Qualifying input test data for model transformations. SOSYM 8, 185–203 (2009)
8. García-Domínguez, A., Kolovos, D.S., Rose, L.M., Paige, R.F., Medina-Bulo, I.: EUnit: A Unit Testing Framework for Model Management Tasks. In: Whittle, J., Clark, T., Kühne, T. (eds.) MODELS 2011. LNCS, vol. 6981, pp. 395–409. Springer, Heidelberg (2011)
9. Giner, P., Pelechano, V.: Test-Driven Development of Model Transformations. In: Schürr, A., Selic, B. (eds.) MODELS 2009. LNCS, vol. 5795, pp. 748–752. Springer, Heidelberg (2009)
10. Gogolla, M., Vallecillo, A.: *Tract*able Model Transformation Testing. In: France, R.B., Kuester, J.M., Bordbar, B., Paige, R.F. (eds.) ECMFA 2011. LNCS, vol. 6698, pp. 221–235. Springer, Heidelberg (2011)
11. Guerra, E., de Lara, J., Kolovos, D., Paige, R.: A visual specification language for model-to-model transformations. In: VL/HCC 2010, pp. 119–126 (2010)
12. Guerra, E., de Lara, J., Kolovos, D., Paige, R., dos Santos, O.: Engineering model transformations with transML. Software and Systems Modeling (in press, 2012)
13. Jouault, F., Allilaire, F., Bézivin, J., Kurtev, I.: ATL: A model transformation tool. Science of Computer Programming 72(1-2), 31–39 (2008)
14. Kolovos, D.S., Paige, R.F., Polack, F.: The Epsilon Transformation Language. In: Vallecillo, A., Gray, J., Pierantonio, A. (eds.) ICMT 2008. LNCS, vol. 5063, pp. 46–60. Springer, Heidelberg (2008)
15. Küster, J.M., Abd-El-Razik, M.: Validation of Model Transformations – First Experiences Using a White Box Approach. In: Kühne, T. (ed.) MODELS 2006. LNCS, vol. 4364, pp. 193–204. Springer, Heidelberg (2007)
16. Lin, Y., Zhang, J., Gray, J.: A framework for testing model transformations. In: Model-driven Soft. Devel. - Research and Practice in Sof. Eng. Springer (2005)
17. Mottu, J.-M., Baudry, B., Le Traon, Y.: Mutation Analysis Testing for Model Transformations. In: Rensink, A., Warmer, J. (eds.) ECMDA-FA 2006. LNCS, vol. 4066, pp. 376–390. Springer, Heidelberg (2006)
18. Mottu, J.-M., Baudry, B., Le Traon, Y.: Reusable MDA Components: A Testing-for-Trust Approach. In: Wang, J., Whittle, J., Harel, D., Reggio, G. (eds.) MoDELS 2006. LNCS, vol. 4199, pp. 589–603. Springer, Heidelberg (2006)
19. Oster, S., Zorcic, I., Markert, F., Lochau, M.: MoSo-PoLiTe: tool support for pairwise and model-based software product line testing. In: VaMoS 2011. ACM International Conference Proceedings Series, pp. 79–82. ACM (2011)
20. Perrouin, G., Oster, S., Sen, S., Klein, J., Baudry, B., Traon, Y.: Pairwise testing for software product lines: a comparison of two approaches. Soft. Qual. J. (2011) (in press)
21. Sen, S., Baudry, B., Mottu, J.-M.: Automatic Model Generation Strategies for Model Transformation Testing. In: Paige, R.F. (ed.) ICMT 2009. LNCS, vol. 5563, pp. 148–164. Springer, Heidelberg (2009)
22. Sherwood, G.B., Martirosyan, S.S., Colbourn, C.: Covering arrays of higher strength from permutation vectors. J. of Combinat. Designs 14(3), 202–213 (2005)
23. Tillmann, N., Schulte, W.: Unit tests reloaded: Parameterized unit testing with symbolic execution. IEEE Software 23(4), 38–47 (2006)

Typing Model Transformations Using Tracts

Antonio Vallecillo[1] and Martin Gogolla[2]

[1] GISUM/Atenea Research Group, Universidad de Málaga, Spain
[2] Database Systems Group, University of Bremen, Germany
av@lcc.uma.es, gogolla@informatik.uni-bremen.de

Abstract. As the complexity of MDE artefacts grows, there is an increasing need to rely on precise and abstract mechanisms that allow system architects to reason about the systems they design, and to test their individual components. In particular, assigning types to models and model transformations is needed for realizing many key MDE activities. This paper presents a light-weight approach to type model transformations using *tracts*. Advantages and limitations of the proposal are discussed, as well as the applicability of the proposal in several settings.

1 Introduction

Types are essential in Model-Driven Engineering (MDE) for understanding, managing and manipulating all artefacts involved in the analysis, design, development, operation and evolution of software systems. In particular, assigning types to models and to model transformations (i.e., *typing* them) is required for characterizing, in a precise and abstract manner, the operations we can perform on them, their valid inputs and outputs, and how they behave. Types are also very useful for ensuring their error-free composition, their safe replaceability by newer versions or by other artefacts, and for checking that a given instance or implementation is correct—by checking that it conforms to the appropriate type.

Typing models is something that the MDE community has already addressed. In a nutshell, the type of a model is essentially its metamodel (modulo its internal packaging structure) [1]. Then, the notion of model subtyping (i.e., safe replaceability) becomes easy to define [2, 3] and to check by tools [2].

However, the situation is not so bright for model transformations, mainly because of their dual nature: they can be considered to be both models and operations. Thus, the community must come up with new ideas and approaches for transformation typing. As models they can be naturally typed by the metamodel of their modeling language (e.g., QVT or ATL). However, typing them as operations is not easy. In general, specifying the type of any software artefact that exhibits behaviour (be it a function, operation, object, component, or a model transformation) is far from a trivial task, specially when its behaviour is rather complex. Furthermore, manipulating and reasoning about these kinds of behavioural types tend to be rather cumbersome and computationally expensive: normally these types try to capture the full behaviour of the artefact of interest independently from any context of use and require heavyweight reasoning techniques and tools, such as theorem provers.

Z. Hu and J. de Lara (Eds.): ICMT 2012, LNCS 7307, pp. 56–71, 2012.

In this paper we discuss central ideas building the cornerstones of a light-weight and modular approach to model transformation typing, using *Tracts*. Tracts were introduced in [4] as a specification and black-box testing mechanism for model transformations. Thus every model transformation can be specified by means of a set of tracts, each one covering a particular scenario or *context of use*—which is defined in terms of particular input and output models and how they should be related by the transformation. In this way, tracts allow partitioning the full input space of the transformation into smaller, more focused behavioural units, and to define specific tests for them. This approach to typing provides a form of "Duck typing" [5]. Basically, what we do with the tracts is to identify the scenarios of interest to the user of the transformation (each one defined by a tract) and check whether the transformation behaves as expected in these scenarios. Another characteristic of our proposal is that we not require complete proofs, just to check that the transformation works for the tract test suites, hence providing a *light-weight* form of verification.

The organization of this paper is as follows. After this introduction, Section 2 presents model transformations, discussing the problem of model transformation typing, the issues of current approaches, and a frame which we use for formalizing model transformations. Then, Section 3 briefly presents Tracts, introducing its main characteristics and constituent elements. Section 4 describes our ideas underlying a light-weight approach to model transformation typing using sets of tracts, and discusses the kinds of analysis that are possible with our proposal, how to conduct them, as well as its current advantages and limitations. Finally, Section 5 compares our work to other related proposals and Section 6 draws the final conclusions and outlines some lines for future work.

2   Typing in MDE

2.1   Typing Models

Model types are useful in many MDE activities. For example, model types are needed for describing the signature (i.e., input and output parameters) of model operations and services, which in MDE are defined in terms of their metamodels. Thus, to perform an operation or a transformation on a model (conforming to a metamodel) we need to check first if it is a valid input for the operation. This situation is even more justified if modeling tools need to be connected, or for chaining several model transformations together. For connecting them, it is essential to check the type substitutability between the output of a service and the input of another, in such a way that type safety is guaranteed.

In our context, the type of a model is essentially its metamodel (modulo its internal packaging structure) [3]. Then we can consider that every metamodel M defines a collection of models $\mathcal{R}_M$ with the models that conform to M.

Let M and M' be metamodels (which can be considered as *types* for the sets of models that conform to them). We say that M' *extends* M (denoted by $M' <: M$) iff $\mathcal{R}_M \subseteq \mathcal{R}_{M'}$. In other words, $M' <: M$ implies that all models that conform to M also conform to M'. This is the equivalent operation to object subtyping in object-oriented type systems [6], and therefore it also implies safe replaceability. This operator is also similar to the *matching* operator ($<\#$) defined in [7, 3], although matching is in general weaker than subtyping.

For example, consider a metamodel SM to describe simple states machines, and another metamodel Composite-SM that adds to SM the possibility of allowing composite states [3]. Metamodel Composite-SM extends SM because every simple state machine can be considered as a particular case of composite state machine. Similarly, we could say that data type Int extends Nat, i.e., Int $<:$ Nat because every value of Nat is a valid value of Int.

Intuitively, $M' <: M$ if: M' contains all classes and relationships in M; all attributes of M classes are present in the corresponding M' classes; and M' imposes the same or even stronger constraints to M elements than those that M imposes (including cardinality constraints). For a more complete definition we refer the reader to [2]. Note as well that the $<:$ operator defines a *partial order*: It is Reflexive ($M <: M$), Transitive ($M' <: M \wedge M'' <: M' \Rightarrow M'' <: M$) and Asymmetric ($M' <: M \wedge M <: M' \Rightarrow M = M'$). It is partial because not any two metamodels can be related by this relation (e.g., the metamodels of simple state machines and of sequence charts).

2.2   Typing Model Transformations

Model transformation (MT) type systems are helpful in many situations. For example, they can be used to check that a transformation can be chained (or composed) with others, check that their behaviour is correct (w.r.t. its type), rule out transformations that would produce models that are not proper instance of their metamodels, identify useless transformations (e.g., that navigate never existing paths in a model), etc.

MT typing is not as easy to define as model typing because of the intrinsic "behavioural" aspects of model transformations, i.e., the way they transform model elements from the source metamodel into model elements of the target metamodel. We need to realize that model transformations comprise two different aspects: structure and behaviour. The former aspect defines the structural relation that should hold between source and target models, whilst the latter specifies how the specific source model elements are transformed into target model elements.

This intrinsic duality needs to be especially taken into account when reasoning about MT subtyping (or extensibility), which in our context has to do with *safe replaceability*. Replaceability refers to the ability of a software entity to substitute another, in such way that the change is transparent to external clients [8]. In the realm of model transformations, we say that transformation T' extends another model transformation T (and write $T' <: T$) if T' behaves as T with all valid input models of T. In other words, we will say that $T' <: T$ iff T' can safely replace T without being noticed by the clients of T.

We can identify at least three kinds of types for model transformations. In the first place we have the *language type*. The fact that model transformations are also models [9] provides one (naive) approximation to the problem of typing transformations, by which the type of a MT is the metamodel of the language in which it is written (e.g, QVT or ATL). For example, the metamodel of QVT defines the set of all transformations written in that language.

Secondly, we have the *structural type*, defined by the fact that model transformations can be considered as operations, and therefore can be typed by the types of their input and output metamodels, as proposed in [10]. Then, if $T : M \to N$ is a transformation, its structural type is $M \to N$. With this, MT *structural subtyping* becomes similar to traditional subtyping of functions defined in type theory, which relies only on the contravariance of argument types and covariance of the return type.

More precisely, let $T : M \to N$ and $T' : M' \to N'$ be two transformations. We say that T' extends (or can structurally substitute) T ($T' <:_s T$) iff ($M' <: M) \land (N <: N')$. But again, this approach to typing model transformations is not sufficient [6]. It is like typing functions by their input and output parameters, or typing operations by their signatures. For instance, functions $\sqrt{x} : \mathtt{Nat} \to \mathtt{Nat}$ and $x^2 : \mathtt{Nat} \to \mathtt{Nat}$ become indistinguishable if we use this approach.

This is the reason for having to consider the *behavioural type* of a transformation T for defining it properly. In the most general case such a type needs to define how every valid input model is transformed into a valid output model. This can be specified in terms of a set $\mathfrak{S}$ of (*source*) constraints that defines the valid input models for T, a set $\mathfrak{T}$ of (*target*) constraints that defines the valid output models, and a set $\mathfrak{R}$ of (*source-target*) constraints that describe how individual source and target models should be related.

In other words, $\mathfrak{S}$ defines the preconditions that must hold for all input models of the transformation; $\mathfrak{T}$ defines the postconditions that must hold for the output models that the transformation produces; and $\mathfrak{R}$ defines conditions that should hold relating the individual source and target models. To express this, if $C[\![m]\!]$ means that a model m satisfies a constraint C (which is nothing but a logic predicate), then the *behavioural type* of a model transformation $T : M \to N$ is a triplet $(\mathfrak{S}, \mathfrak{T}, \mathfrak{R})$ such that: $\forall m \in \mathcal{R}_M \bullet \mathfrak{S}[\![m]\!] \Rightarrow \mathfrak{T}[\![T(m)]\!] \land \mathfrak{R}[\![(m, T(m))]\!]$.

To formally express behavioural subtyping, let $T : M \to N$ and $T' : M' \to N'$ two model transformations, for which $T' <:_s T$ (structural subtyping should be a requirement for behavioural subtyping), and let $(\mathfrak{S}, \mathfrak{T}, \mathfrak{R})$ and $(\mathfrak{S}', \mathfrak{T}', \mathfrak{R}')$ be the specification of the behavioural types of T and T', respectively. Then, T' can behaviourally substitute T ($T' <: T$) iff $(\mathfrak{S} \Rightarrow \mathfrak{S}') \land (\mathfrak{T}' \Rightarrow \mathfrak{T}) \land (\mathfrak{R}' \Rightarrow \mathfrak{R})$.

This is similar to Liskov's substitutability principle [6], which states that if S is a subtype of T ($S <: T$) , then objects of type T in a program may be replaced with objects of type S without altering any of the desirable properties of that program (e.g., correctness). Liskov's principle imposes some standard requirements on signatures (adopted later in contract-based design [11]): contravariance of method arguments in the subtype; covariance of return types in the subtype; preconditions cannot be strengthened in a subtype; postconditions cannot be weakened in a subtype; and finally, invariants of the supertype must be preserved in a subtype.

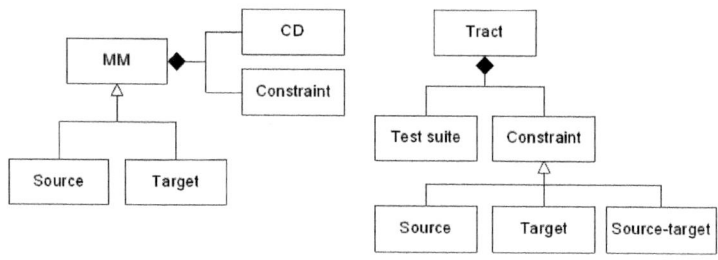

Fig. 1. Concepts in a Tract [4]

Although this definition of Model Transformation subtyping is appropriate from a theoretical point of view, it is not easy to check in practice without the aid of heavyweight tools such as theorem provers. Nevertheless, the given formulation admits one interesting way to check whether a model transformation T' can replace another one T, for which we know the behavioural type $(\mathfrak{S}, \mathfrak{T}, \mathfrak{R})$. It is enough to check that T' conforms to that type, i.e., $\forall\, m \in \mathcal{R}_M \bullet \mathfrak{S}[\![m]\!] \Rightarrow \mathfrak{T}[\![T'(m)]\!] \wedge \mathfrak{R}[\![(m, T'(m))]\!]$.

Using this notion of behavioural subtyping for model transformations we can search for a required transformation in a repository of transformations (such as `http://www.eclipse.org/m2m/atl/atlTransformations/`), or check that a given transformation can easily replace (or implement) another one, or that a given implementation of a model transformation conforms to its type (i.e., is correct w.r.t. its expected usage).

3   Tracts

One of the problems of the previous specification of MT behavioural type lies in its complexity. The specifications of an MT type can become monstrously large as far as the transformation is not trivial (even far more complex than the transformation itself). The reasons are, among others, the lack of modularity, having to deal with too many details at the same time, and excessive size. Because the type specifications try to capture all the model transformation behaviour in one huge set of constraints, they become hard to write, debug and maintain. In addition, checking the conformance of MT implementations and conducting other tests over these specifications become quite cumbersome, complex, and computationally prohibitive tasks.

In order to deal with the problems, we propose the use of *tracts*. They provide modular pieces of specification, each one focusing on a particular scenario. They have the structure of a behavioural type, plus a test suite that allows operationalizing the conformance tests. We do not provide the full behavioural specification of a model transformation, but just a set of tracts that defines how the transformation should behave in certain particular scenarios (or use cases) which are the ones of interest to the user. We do not care how the transformation

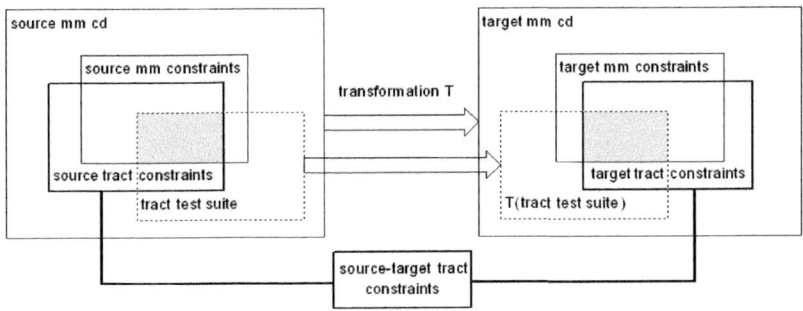

Fig. 2. Building Blocks of a Tract [4]

works in the rest of the cases. In this respect, this approach to typing is a form of "Duck typing": "If it looks like a duck, swims like a duck, and quacks like a duck, then it probably is a duck" [5].

In a nutshell, a tract defines a set of constraints on the *source* and *target* metamodels, a set of *source-target* constraints, and a tract *test suite*, i.e., a collection of source models satisfying the source constraints. Such constraints serve as "contracts" (in the sense of contract-based design [11]) for the transformation in some particular scenarios, and are expressed by means of OCL invariants. Tracts are composed by conjunction, similarly to the modular specification of an operation using several pre- and postconditions, each one defining a specific situation or use case of the operation.

Assume a source model m being an element of the test suite and satisfying the metamodel source and the tract source constraints is given. Then, the tract essentially requires that the result $T(m)$ of applying transformation T satisfies the target metamodel and the target tract constraints and the pair $(m, T(m))$ satisfies the source-target tract constraints. The source-target tract constraints are crucial insofar that they can establish a correspondence between a source element and a target element in a declarative way by means of a formula. In technical terms, a source tract constraint is basically an OCL expression with free variables over source elements, a target tract constraint has free variables over target elements, and a source-target tract constraint possesses free variables over source and target elements.

In Fig. 2 we have displayed the central ingredients of our approach for transformation testing: a source and target metamodel, the transformation T under study, and a transformation contract, for short tract, which consists of a tract test suite and tract constraints. The test suite and its transformation result are shown with dashed lines and the different tract constraints with thick lines. Five different kinds of constraints are present: the source and target class diagrams are restricted by source and target metamodels constraints, and the tract imposes source, target, and source-target tract constraints. Such constraints are expressed by means of OCL invariants. The context of these invariants is a class representing a transformation contract, a so-called tract class. An example of a tract class called 2S1T-Tract is shown later in this section.

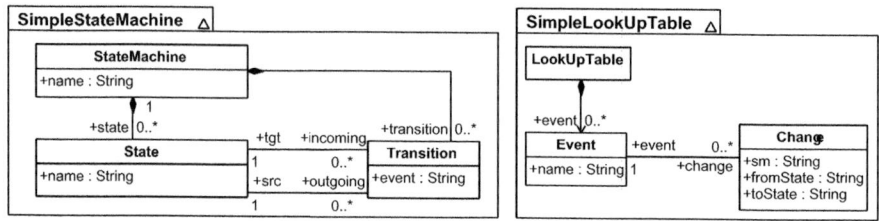

Fig. 3. Source and Target Metamodels of transformation SM2T

In Fig. 2, the rectangles indicate possible overlap (resp. disjointness) of source and target models. Basically, the tract — consisting of the test suite and the three kinds of constraints — checks for the correctness of the transformation in the sense that correct source models from the test suite are transformed into correct target models, i.e., our approach checks that in Fig. 2 the grey source section is transformed into the grey target section. In general, there can be more than one tract for a single transformation because particular source models are constructed in the test suite which then induce particular tract constraints. We show the dashed rectangles with the test suites not necessarily inside source/target tract constraint rectangles in order to allow, e.g., the definition of negative tests for the transformation.

To test a transformation T against a tract t, the input test suite models can be automatically generated using languages like ASSL [12], and then transformed into their corresponding target models. These models can also be automatically checked with the USE tool [13] against the constraints defined for the transformation. The checking process can be automated, allowing the model transformation tester to process a large number of models in a mechanical way.

Although this approach to testing does not guarantee full correctness, it provides very interesting benefits. In particular, it can be useful for identifying bugs in a cost-effective manner. Moreover, it allows dealing with industrial-size transformations without having to transform them into any other formalism or to abstract away from any of its features. Furthermore, tracts provide a modular approach to specification and testing, enabling the partition of the full input space of the transformation into smaller, more focused behavioural units, and to define precise specifications for them. These are important advantages over other approaches that prove full correctness but at a higher computational cost.

For illustration purposes, let us consider a model transformation SM2T between simple state machines and a lookup table that lists the events and their associated transitions [3]. The source and target metamodels of this transformation are shown in Figure 3. In this case, we want only one lookup table to be built, whose entries are all the events of all the state machines in the source model. In addition to the (multiplicity) constraints shown in these class diagrams, we need to add uniqueness on names of the state machines, and uniqueness of names of states within the same state machine:

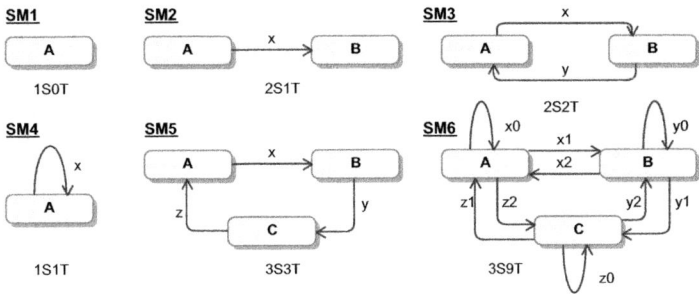

Fig. 4. Test suites samples for the 6 tracts defined for model transformation SM2T

```
context StateMachine inv uniqueNames:
  self.state->isUnique(name) and
  StateMachine.allInstances->isUnique(name)
```

To specify the SM2T transformation we can define the following six tracts, whose test suite models are illustrated in Figure 4 (literals SM1...SM6 represent the names of the state machines):

- 1S0T: state machines with single states and no transitions.
- 2S1T: state machines with two states and one transition between them.
- 2S2T: state machines with two states and two transition between them.
- 1S1T: state machines with single states and one transition.
- 3S3T: state machines with three states and three transitions, forming a cycle.
- 3S9T: state machines with three states and 9 transitions (see Figure 4).

Let us show here one of these tracts, 2S1T, for illustration purposes. The rest follow similar patters. In the first place, the *tract source constraint* that specifies the source models is defined by OCL invariant SCR_2S1T:

```
context 2S1T-Tract
inv SCR_2S1T:
  StateMachine.allInstances->forAll (sm |
    (sm.state->size() = 2) and (sm.transition->size() = 1)
    (sm.transition.src <> sm.transition.tgt)
```

We need to decide what the transformation should do when these models are used as input models. There is no restriction on the kinds of entries that can be produced in the lookup table, but we need to state that only one lookup table is produced. This is expressed by the following OCL constraint:

```
context 2S1T-Tract
inv TRG_2S1T: LookUpTable.allInstances->size() = 1
```

Regarding the *source-target constraints*, given that every state machine has only one transition, there should be one change in the lookup table for every state machine, and the attributes should match with the events and states related by the corresponding transition in the state machine. This is expressed by the following source-target constraint:

```
context 2S1T-Tract
inv SRC_TRG_2S1T:
  StateMachine.allInstances->size() = LookUpTable.change->size() and
  LookUpTable.change->forAll (c |
    StateMachine.allInstances->one(sm | (sm.name = c.sm) and
      (sm.transition.src->collect(name) = c.fromState.asSet()) and
      (sm.transition.tgt->collect(name) = c.toState.asSet()) and
      (sm.transition.event = c.event.name) )
```

Finally, the test suite for this tract is defined by an ASSL procedure that generates the input models (not shown here for space reasons).

4   Model Transformation Typing Using Tracts

Let us explain how (sub-)typing works for tracts. A tract is responsible for specifying how to transform a source model into a target model.

In Fig. 5 we see that TractG transforms metamodel SourceG into metamodel TargetG. 'G' and 'S' stand for 'general' (resp. 'specific'). SourceS is a specialization of SourceG in the the sense that it extends SourceG by adding new elements (classes, attributes, associations) and possibly more restricting constraints.

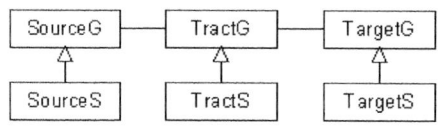

Analogously this is the case for TargetS. TractS is a specialization of TractG and inherits from TractG its connecting associations. Constraints must guarantee that tract TractS connects SourceS and

Fig. 5. Tract subtyping

TargetS elements. Both, TractG and TractS are established with a test suite generating a set of SourceG models (resp. a set of SourceS models).

4.1   Tract Typing by Example

Fig. 6 shows an example for tract subtyping, using a different case study. The first source metamodel is the plain Entity-Relationship (ER) model with entities, relationships and attributes only. An ER model is identified by an object of class ErSchema. The second source metamodel is a specialization of the Entity-Relationship model which adds cardinality constraints for the relationship ends. Objects of class ErSchemaC are associated with ER models which additionally possess cardinality constraints.

The first target metamodel is the relational data model allowing primary keys to be specified for relational schemas. Objects of class RelDBSchema identify relational database schemas with primary keys. The second target metamodel describes relational database schemas with primary keys and additional foreign keys. The upper part of the diagram shows the principal structure with respective source and target as well as general and special elements. The lower part shows the details. Please note that the four source and target metamodels have a common part, namely the class Attribute.

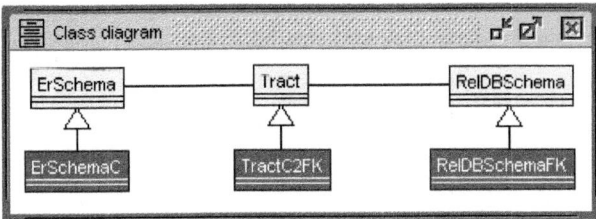

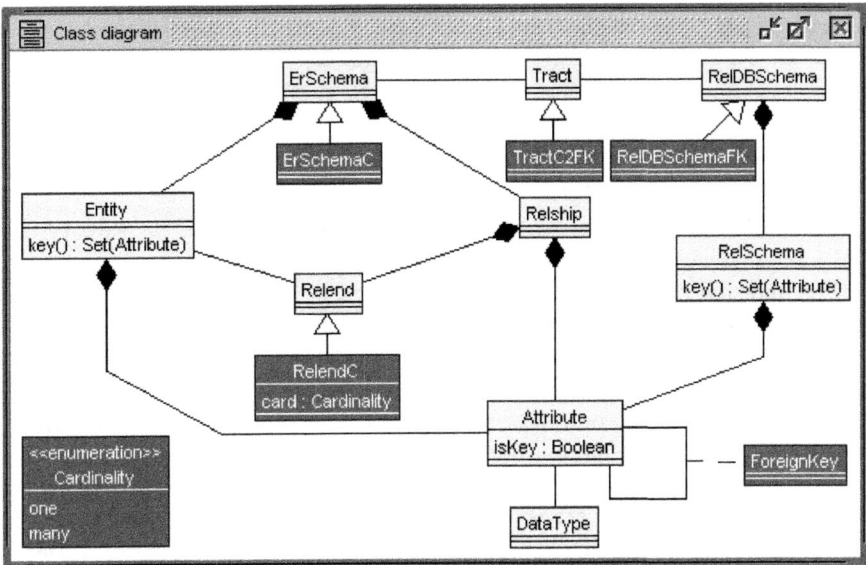

Fig. 6. An example of tract subtyping

It would also be possible to have disjoint source and target models by introducing classes `ErAttribute` and `ErDataType` for the ER model as well as `RelAttribute` and `RelDataType` for the relational model. The association class `ForeignKey` belongs exclusively to the relational database metamodel with foreign keys. This could be made explicit by establishing a component relationship, a black diamond, from class `RelDBSchemaFK` to `ForeignKey`. The central class `Tract` specifies the transformation contract and has access, through associations, to both the source and target metamodel. Tract subtyping is expressed through the fact that class `TractC2FK` is a subtype of class `Tract`.

The scenario `Town-liesIn-Country` depicted in Fig. 7 shows informally what will be represented further down as a formal instantiation of the metamodels. Three transformations are shown. The first one `ER_2_Rel` transforms a plain ER schema (without cardinalities) into a relational database schema with primary keys only. The second one `ERC_2_Rel` goes from an ER schema with cardinalities into a relational database schema with only primary keys. The third transformation `ERC_2_RelFK` takes the ER schema with cardinalities and yields a relational database schema with primary keys and foreign keys. Please note that the three

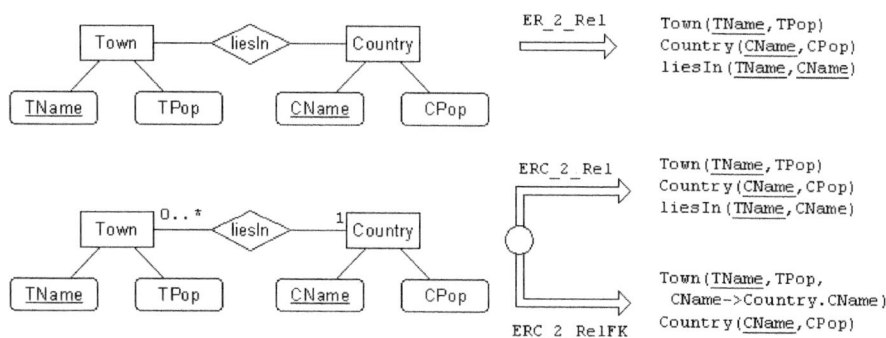

Fig. 7. Town-liesIn-Country scenario

relational database schemas can be distinguished by their use of primary keys and foreign keys.

The informal scenario Town-liesIn-Country is formally presented in Fig. 8 with object diagrams instantiating the metamodel class diagrams. The most interesting parts which handle the primary and foreign keys are pictured in a white-on-black style. Please pay attention to the typing of the source, target, and tract objects which are different in each of the three cases and which formally reflect the chosen names of the transformations (trafo_GG, trafo_SG, trafo_SS).

As shown in Fig. 9, in the ER and relational database metamodel example we see three different transformations: trafo_GG, trafo_SS, and trafo_SG. trafo_GG and trafo_SS are the transformations directly obtained from the respective tracts. Another transformation is trafo_SG, which takes SourceS models, builds TargetG models and checks them against the TargetG constraints. As shown in the right lower part, the example transformations trafo_SS and trafo_SG are subtypes of trafo_GG.

4.2   Working with Tract Types

As mentioned above, the type of a model transformation T is specified in terms of a set of tracts $\{t_1, t_2, \ldots, t_n\}$. This section briefly discusses the kinds of analysis that can be conducted with tracts, as well as the pros and cons of our proposal.

Correctness of a MT Implementation. The first thing we can do is to check whether a given transformation behaves as expected, i.e., its implementation is correct w.r.t. a specification. In our approach, this is just checking that a given transformation conforms to a type. For example, a developer can come up with an ATL [14] model transformation that implements the SM2T specification, and we need to test whether such MT is correct. This was the original intention of Tracts, and such a testing process is fully described in [4].

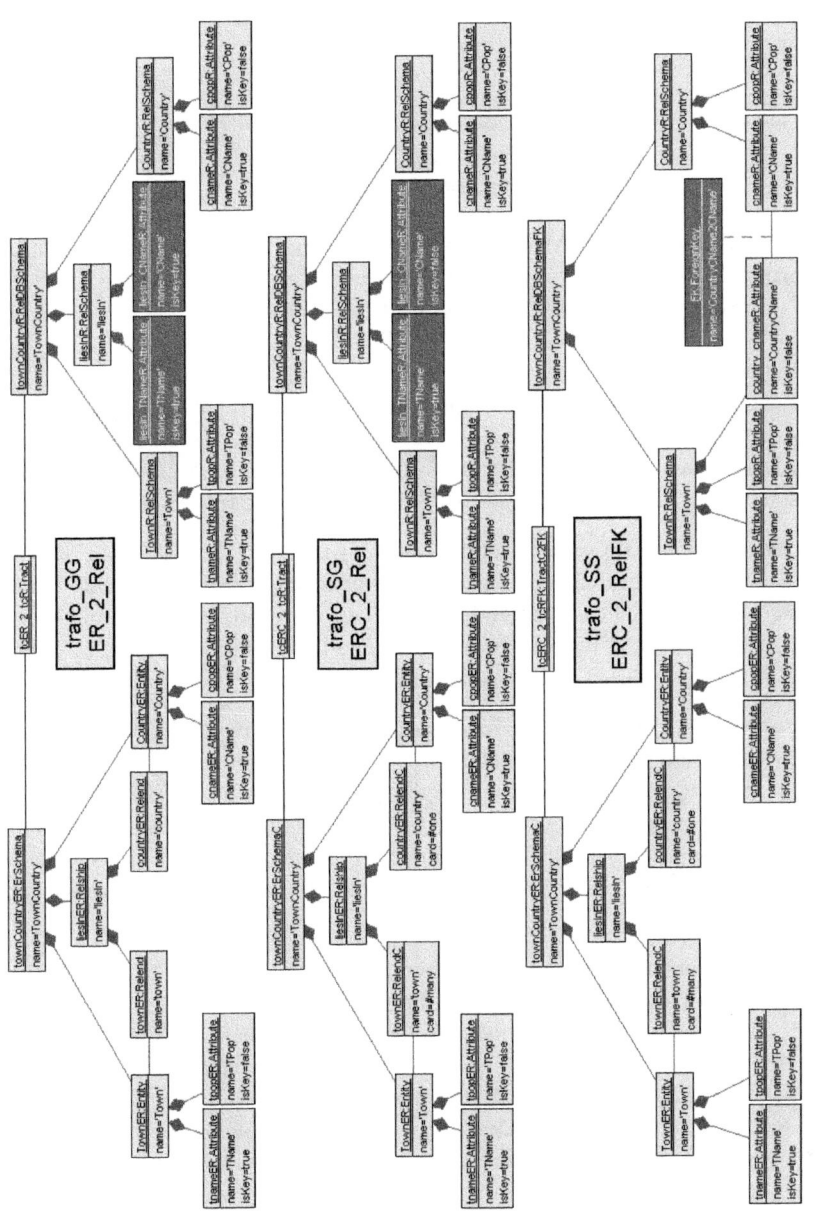

Fig. 8. Town-liesIn-Country object diagram

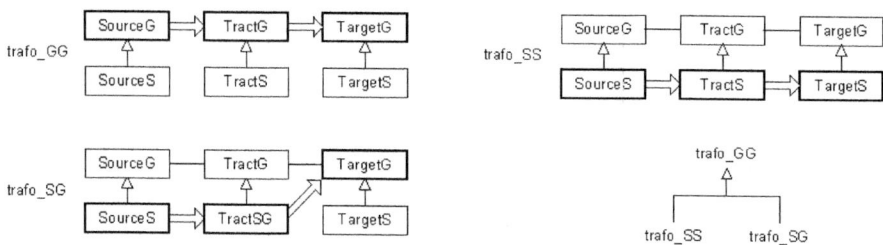

Fig. 9. Relationship between Example Transformations

Safe Substitutability of Model Transformations. Now, given another model transformation T', how do we decide if T' can safely substitute T ($T' <: T$)? In our approach, it is a matter of testing that T' satisfies all T tracts, which can be checked in an automated way [4]. This is a two step process: first of all a number of input models is automatically generated and then for each of these models we can check whether the transformation fulfills the associated tract. We will not get 100% assurance that $T' <: T$ for all possible models, but we will be able to know that at least it will work in all scenarios that we have identified as relevant for us with the tracts, and for the test suites of interest.

Incrementality of Transformation Development. The ERC_2_RelFK example uses an incremental methodology for transformation development. Source and target metamodels are extended by subtyping through small increments which are accompanied by corresponding tracts including test suites. The tract test suites can give direct feedback on the correctness of the increment.

Declarative vs. Imperative Tracts. Tracts may have a descriptive nature when only the relationship between source and target elements is characterized. Tracts may also be described in an operational way when the tract includes operations that map source elements to target elements. Operational tracts may be understood as implementations of descriptive ones and their correctness can be checked against the descriptive tract by employing the descriptive test suite for the operational tract.

Pros and Cons. In general, we have found that typing model transformations using tracts provides interesting advantages, such as modularity, usability, and cost-effectiveness, but at the cost of sacrificing completeness and full verification. Furthermore, having a high-level specification of what the transformation should do at the tract level (independently of how it actually implements it) becomes beneficial because both descriptions provide two complementary views (specifications) of the behaviour of the transformation. Then, during the checking process the tract specifications and the code help testing each other: we believe in an incremental and iterative approach to model transformation testing, where tracts are progressively specified and the transformation checked against them. The errors found during the testing process are carefully analyzed and either the tract or the transformation refined accordingly.

5   Related Work

There are several kinds of contributions that can be related to our work. In the first place we have the works that define contracts for model transformations, using different notations. One of the earlier works [15] introduces the concept of "transformation contract" in a similar way to ours—although without incorporating test suites. However, the authors propose to specify contracts by means of OCL operations, which causes many technical problems for writing contracts—as the own authors discuss in their paper. Besides, they do not discuss any practical way of using their contract specifications for model testing. The work in [16] also proposes OCL for defining transformation contracts, although in their paper they just provide a general view of what they think that could be done with model transformation contracts, but without delving into the details about how to achieve it. Finally, Poernomo [17] defines a similar proposal in spirit, but using constructive type theory instead of first-order logic.

Another proposal defines the type of a model transformation in terms of its input and output metamodels [10]. However, as mentioned in the introduction, such a *structural type* is not enough for capturing all relevant aspects: behaviour should also be taken into account. The work in [18] utilizes transformation types in an XML context.

The proposal presented in [19] is also of interest. The authors show how to derive some invariant-based verification properties that should be preserved by the transformation (which are similar to our tracts) by analysing the internal rules that compose a transformation. Although they follow a white-box approach to model transformation testing, it could probably be combined with ours if their approach could help us identify some more tracts for a transformation written in any of the languages they deal with (TGG and QVT). In this sense, we fully agree with one of the reviewer's suggestions about the interest to investigate the implications of the similarity between the tracts and these languages' transformation rules: when using tracts with these kinds of model transformations, is it a matter of typing using the existing rules, or is it necessary to have separate rules for implementation and specification, and if so, how structurally distinct should they be?

Furthermore, our idea of modularizing the specifications into smaller units could be transferred to other techniques apart from Tracts, e.g., to the invariant-based verification properties presented in [19] or to existing contract-based approaches [15, 16].

Another group of works (see, e.g., [20–24]) also use a white-box approach to model-transformation specification and testing, aiming at fully validating the behaviour of the transformation (including other properties such as confluence of the rules, termination, etc.) using formal methods and their associated toolkits—which include, e.g., Alloy, Maude, or graph rewriting techniques. Although more powerful than our approach from a theoretical perspective, their computational complexity generally makes them inappropriate for testing large model transformations. In addition, the drawback of a white-box approach is that it is tightly coupled to the transformation language and thus it would need to be adapted or completely redefined for another transformation language [25].

6 Conclusions

As MDE is becoming more widely applied and adopted, larger transformations are being developed, with thousands of lines of code. This makes them error-prone and brittle, becoming hard to understand, develop, debug, maintain and reuse. In fact, model transformations, like any other Software Engineering arte-fact, must be systematically designed and implemented [26]. The need to have effective mechanisms for specifying and properly testing them is now critical.

In this paper we have developed central ideas using *Tracts* for Model Transformation typing, and discussed benefits and limitations of this approach.

There are several lines of work that we plan to address next. For instance, we would like to study how to choose the tracts that compose the type of a model transformation, to ensure enough coverage and completeness. In this respect, we plan to investigate how to improve our proposal by incorporating some of the existing works on the effective generation of input test cases. We also plan to improve the current tool support for tracts, incorporating the creation and maintenance of libraries of tracts. Finally, larger case studies will be carried out in order to stress the applicability of our approach and to obtain more extensive feedback.

Acknowledgements. We would like to thank the anonymous reviewers for their constructive and insightful comments, and to Loli Burgueño and Fernando López for developing the tool support. This work is supported by Research Projects TIN2008-03107 and P07-TIC-03184.

References

1. Steel, J., Jézéquel, J.-M.: Model Typing for Improving Reuse in Model-Driven Engineering. In: Briand, L.C., Williams, C. (eds.) MoDELS 2005. LNCS, vol. 3713, pp. 84–96. Springer, Heidelberg (2005)
2. Romero, J.R., Rivera, J.E., Durán, F., Vallecillo, A.: Formal and tool support for model driven engineering with Maude. Journal of Object Technology 6(9), 187–207 (2007)
3. Steel, J., Jézéquel, J.M.: On model typing. Software and Systems Modeling 6(4), 401–413 (2007)
4. Gogolla, M., Vallecillo, A.: Tractable Model Transformation Testing. In: France, R.B., Kuester, J.M., Bordbar, B., Paige, R.F. (eds.) ECMFA 2011. LNCS, vol. 6698, pp. 221–235. Springer, Heidelberg (2011)
5. Heim, M.: Exploring Indiana Highways: Trip Trivia. Exploring America's Highway. Travel Organization Network (2007), http://en.wikipedia.org/wiki/Duck_test
6. Liskov, B., Wing, J.M.: A behavioral notion of subtyping. ACM Trans. Program. Lang. Syst. 16(6), 1811–1841 (1994)
7. Bruce, K.B., Vanderwaart, J.: Semantics-driven language design: Statically type-safe virtual types in object-oriented languages. ENTCS 20(1), 1–26 (2004)
8. Nierstrasz, O.: Regular types for active objects. In: Nierstrasz, O., Tsichritzis, D. (eds.) Object-Oriented Software Composition, pp. 99–121. Prentice-Hall (1995)

9. Bézivin, J.: On the unification power of models. Software and Systems Modeling (SoSyM) 4(2), 171–188 (2005)
10. Vignaga, A., Jouault, F., Bastarrica, M.C., Brunelière, H.: Typing in Model Management. In: Paige, R.F. (ed.) ICMT 2009. LNCS, vol. 5563, pp. 197–212. Springer, Heidelberg (2009)
11. Meyer, B.: Applying design by contract. IEEE Computer 25(10), 40–51 (1992)
12. Gogolla, M., Bohling, J., Richters, M.: Validating UML and OCL Models in USE by Automatic Snapshot Generation. Software and Systems Modeling 4(4), 386–398 (2005)
13. Gogolla, M., Büttner, F., Richters, M.: USE: A UML-based specification environment for validating UML and OCL. Science of Computer Programming 69, 27–34 (2007)
14. Jouault, F., Allilaire, F., Bézivin, J., Kurtev, I.: ATL: A model transformation tool. Science of Computer Programming 72(1-2), 31–39 (2008)
15. Cariou, E., Marvie, R., Seinturier, L., Duchien, L.: OCL for the specification of model transformation contracts. In: Proc. of the OCL and Model Driven Engineering Workshop (2004)
16. Baudry, B., Dinh-Trong, T., Mottu, J.M., Simmonds, D., France, R., Ghosh, S., Fleurey, F., Traon, Y.L.: Model transformation testing challenges. In: Proc. of IMDD-MDT 2006 (2006)
17. Poernomo, I.H.: Proofs-as-Model-Transformations. In: Vallecillo, A., Gray, J., Pierantonio, A. (eds.) ICMT 2008. LNCS, vol. 5063, pp. 214–228. Springer, Heidelberg (2008)
18. Matsuda, K., Hu, Z., Takeichi, M.: Type-based specialization of xml transformations. In: Proc. of PEPM 2009, pp. 61–72. ACM (2009)
19. Cabot, J., Clarisó, R., Guerra, E., de Lara, J.: Verification and validation of declarative model-to-model transformations through invariants. Journal of Systems and Software 83(2), 283–302 (2010)
20. Baresi, L., Ehrig, K., Heckel, R.: Verification of Model Transformations: A Case Study with BPEL. In: Montanari, U., Sannella, D., Bruni, R. (eds.) TGC 2006. LNCS, vol. 4661, pp. 183–199. Springer, Heidelberg (2007)
21. Ehrig, H., Ehrig, K., de Lara, J., Taentzer, G., Varró, D., Varró-Gyapay, S.: Termination Criteria for Model Transformation. In: Cerioli, M. (ed.) FASE 2005. LNCS, vol. 3442, pp. 49–63. Springer, Heidelberg (2005)
22. Küster, J.M.: Definition and validation of model transformations. Software and Systems Modeling 5(3), 233–259 (2006)
23. Anastasakis, K., Bordbar, B., Küster, J.M.: Analysis of model transformations via Alloy. In: Proc. of MODEVVA (2007),
    http://www.cs.bham.ac.uk/~bxb/Papres/Modevva07.pdf
24. Troya, J., Vallecillo, A.: A rewriting logic semantics for ATL. Journal of Object Technology 10, 5:1–29 (2011)
25. Baudry, B., Ghosh, S., Fleurey, F., France, R., Traon, Y.L., Mottu, J.M.: Barriers to systematic model transformation testing. Communications of the ACM 53(6), 139–143 (2010)
26. Guerra, E., de Lara, J., Kolovos, D.S., Paige, R.F., Santos, O.: Engineering model transformations with transML. Software and Systems Modeling (2011) (in press)

Reusable and Correct Endogenous Model Transformations

Suzana Andova, Mark G.J. van den Brand, and Luc Engelen

Department of Mathematics and Computer Science
Eindhoven University of Technology
P.O. Box 513, 5600 MB, Eindhoven, The Netherlands
{S.Andova,M.G.J.v.d.Brand,L.J.P.Engelen}@tue.nl

Abstract. Correctness of model transformations is a prerequisite for generating correct implementations from models. Given refining model transformations that preserve desirable properties, models can be transformed into correct-by-construction implementations. However, proving that model transformations preserve properties is far from trivial. Therefore, we aim for simple correctness proofs by designing model transformations that are as fine-grained as possible. Furthermore, we advocate the reuse of model transformations to reduce the number of proofs. For a simple domain-specific language, SLCO, we define a formal framework to reason about the correctness, reusability, and composition of the fine-grained model transformations used to transform a given model to three target languages: NQC, Promela and POOSL. The correctness criterion induces that the original model and the resulting model obtained after a proper sequence of transformations have the same observable behavior.

1 Introduction

Domain-specific modeling languages (DSMLs) and model transformations are the key concepts in model driven engineering [16]. A DSML enables domain experts, through appropriate notations and abstractions, to develop models using concepts in their own domain. DSML models can be transformed into models in other languages using model transformations.

Model transformations can be used, for example, to transform DSML models to languages suited for validation, execution, testing, and visualization. In such cases, they should not hamper the quality or change the behavior of the source model; the requirements and properties modeled initially have to be propagated to the target implementation through (sequences of) transformations. In general, the target implementation is more complex, due to added implementation details. The increased complexity makes analysis of the quality more difficult, more time consuming, and in some cases even impossible [2]. The source model, however, is relatively small, so its properties can be inspected and validated. To check preservation of quality, one can analyze all or some of the intermediate models, but this means that a large portion of the analysis has to be duplicated. In addition, this procedure has to be repeated for every new model, even if only

Z. Hu and J. de Lara (Eds.): ICMT 2012, LNCS 7307, pp. 72–88, 2012.
© Springer-Verlag Berlin Heidelberg 2012

small changes to the model have been made. The efficient and general solution to this problem is to prove property preservation per transformation, and to localize the proof only on the changes induced by the transformation. In this paper, we present an approach that provides such a solution.

Our approach is demonstrated on a small but non-trivial DSML. The *Simple Language of Communicating Objects* (SLCO) [1] is a DSML for the specification of systems consisting of objects that operate in parallel and communicate with each other. In SLCO, such systems can be specified on various levels of abstraction. From a given SLCO model on a high level of abstraction, different compositions of fine-grained model transformations are used to generate NQC [5] models for execution on Lego Mindstorms controllers, POOSL [17] models for simulation, and Promela models for formal verification using the model checker SPIN [12]. The part of SLCO used for the specification of high-level models and the three target languages have different properties, and therefore, several semantic gaps need to be bridged [3]. Each of these gaps is bridged by one or more model transformations that add implementation details to the original SLCO model, resulting in a refined SLCO model that is closer to one of the target languages. To improve the reusability of these transformations, and to deal with only one language for the majority of the correctness proofs, we only use endogenous transformations for the refinement of models, instead of exogenous ones. To be able to use endogenous transformations for refinement, we extended SLCO with constructs to specify systems on a lower level of abstraction too.

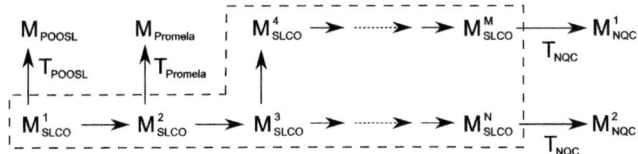

Fig. 1. Sequences of fine-grained transformations for three target languages

Figure 1 depicts a number of composed model transformations that transform an SLCO model to various target languages. The arrows inside the dashed shape depict endogenous transformations that transform SLCO models into more refined SLCO models. Each of the endogenous transformations leads to a model with observationally equivalent behavior. The arrows across the border of the dashed shape depict exogenous transformations. Because the semantic gaps between SLCO and the target languages are bridged completely by the endogenous transformations, these exogenous transformations are straightforward translations of SLCO constructs into equivalent constructs in the target languages. By developing endogenous transformations that are as fine-grained as possible, we improve their reusability within transformation sequences, aim for straightforward correctness proofs, and reduce the number of these proofs.

In this paper, we discuss how the transformations of SLCO models to the three target languages are decomposed into sequences of fine-grained transformations, and the way these are composed and reused within sequences. Furthermore, we describe a formal framework for SLCO used to reason about the correctness of model transformations. First, the formal Structural Operational Semantics (SOS) [14] of SLCO is defined, which generates a Labeled Transition System (LTS) representation of the dynamics of an SLCO model. Second, for each transformation, a (behavioral) equivalence relation is established between the behavior of an SLCO model before and after transformation. Finally, we prove that the transformations preserve the corresponding equivalence relation.

The benefits of this approach are manyfold. The correctness of the aforementioned model transformations is proven in general, not only for particular model instances. The use of SOS and the behavioral equivalence relations allows us to focus only on the part of a model affected by a transformation when reasoning about the correctness of this transformation. Here, additional benefits of fine-grained transformations [2] are evident, since they allow for rather straightforward proofs. Furthermore, the constraints on input models required for some of the transformations, detected earlier during experimental work, can now be formally shown necessary for the correctness of the transformations, which shows that the sequences of transformations used to generate code are well composed.

Structure of the paper. In Section 2, the SLCO language is briefly presented. In Section 3, model transformations and the way they are reused are described. The correctness criterion and the proof of one of the transformations are given in Section 4, as well as the formal semantics of SLCO, required for the proof. Section 5 discusses the related work and Section 6 concludes the paper.

2   The Simple Language of Communicating Objects

The *Simple Language of Communicating Objects* (SLCO) provides constructs for specifying systems consisting of objects that operate in parallel and communicate with each other. In this section, we describe the basic notions and syntax of SLCO. The formal semantics of SLCO are discussed in Section 4.1.

An SLCO model consists of classes, objects, and channels, as indicated by the syntax definition in Figure 2. Objects are instances of classes and communicate with each other via channels. A class describes the structure and behavior of its instances. Ports and variables define the structure, and state machines describe the behavior of the objects that are defined as instances of the class. Ports are used to connect channels to objects. Channels connect the objects of a model and describe how they communicate. They can be bidirectional or unidirectional, and used for synchronous or asynchronous communication. Synchronous channels are lossless, whereas asynchronous channels are either lossless or lossy.

A state machine consists of variables, a finite set of states, and transitions between states. Each state machine has a designated initial state and a set of final states. A transition has a source, a target state, and possibly a statement. A statement is either a boolean expression, an assignment, a signal sending statement, a delay, or a conditional signal reception. A boolean expression represents a statement that blocks the execution of the transition until the expression evaluates to **true**. A transition with a delay is enabled after a specified amount of

```
model   ::= mn class* obj* chan*
class   ::= cn var* pn* sm*
obj     ::= on " : " cn
chan    ::= uchan  |  bichan
uchan   ::= chn "(" type* ")" chtype "from" on "." pn "to" on' "." pn'
bichan  ::= chn "(" type* ")" chtype "between" on "." pn "and" on' "." pn'
type    ::= "Boolean"  |  "Integer"  |  "String"
chtype  ::= "sync"  |  "async lossless"  |  "async lossy"
var     ::= "Boolean" bvn [ " = " bc]  |  "Integer" ivn [ " = " ic]
          | "String" svn [ " = " sc]
```

Fig. 2. Syntax of SLCO models

time has passed since entering its source state. A transition with a conditional signal reception is enabled if a signal is received via the indicated port and the boolean condition of the statement holds. It is allowed to refer to arguments of the signal just being received in the condition of a conditional signal reception. For convenience, a condition that is always **true** can be omitted. Besides its local variables, a state machine may use the variables of the class it belongs to as global variables. The syntax of state machines and their constituents is given in Figure 3, where be, ie, and se denote boolean, integer, and string expressions, respectively, and e denotes arbitrary expressions.

```
sm      ::= smn var* states trans*
states  ::= "initial" sn sn* ["final" sn+]
trans   ::= tn "from" sn "to" sn' [stat]
stat    ::= send  |  assign  |  rec  |  del  |  be
send    ::= "send" sgn "(" e* ")" "to" pn
assign  ::= bvn " := " be  |  ivn " : = " ie  |  svn " : = " se
rec     ::= "receive" sgn "(" vn* "|" be ")" "from" pn
del     ::= "after" nc "ms"
```

Fig. 3. Syntax of SLCO state machines

Figures 4 and 5 show a simple example of an SLCO model. In the example, for the sake of readability, we use the graphical representation of SLCO models [4]. Figure 4 shows the communication diagram of the model. It shows the two objects in the model: p, an instance of class P, and q, an instance of Q. Object p has ports $P1$, $P2$, and $P3$, connecting it to channels $c1$, $c2$, and $c3$, and object q has ports $Q1$, $Q2$, and $Q3$, connecting it to the same channels. Thus, objects p and q can communicate over three channels: a synchronous channel $c1$, an asynchronous, lossy channel $c2$, and an asynchronous, lossless channel $c3$.

Figure 5 describes the behavior of the objects. The left state machine specifies the behavior of object p and the right one specifies the behavior of object q. The figure specifies the following communication between p and q, which gives the

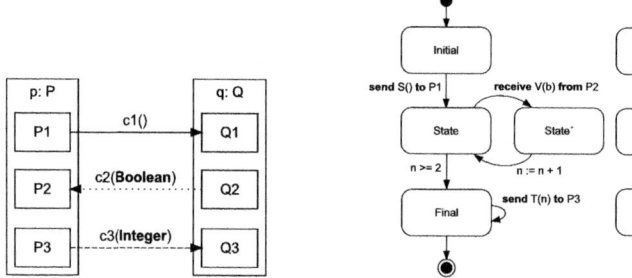

Fig. 4. Objects, ports, and channels in SLCO

Fig. 5. Two SLCO state machines

behavior of the model in total. The two objects first communicate synchronously over channel $c1$, after which q repeatedly sends signals to p over the lossy channel $c2$. When p has received at least two of these signals, it can send signals over channel $c3$, or terminate. After receiving such a signal, q terminates as well.

3   SLCO Model Transformations

DMSLs allow designers to reason at a high level of abstraction, and therefore, DSML models do not include many implementation details. The main goal of model transformations is to add more details to the model, thus bringing it closer to its implementation. To generate code (e.g. an NQC executable) from an SLCO model, a number of fine-grained, endogenous model transformations have been designed and implemented. By design, each model transformation transforms only a specific small part of the input model, because small transformations can be easily applied, composed, implemented, and analyzed. We have composed sequences of transformations for several target languages. Our correctness criterion guarantees that every intermediate model, including the last model in the sequence, has the same properties as the source model. Furthermore, the transformations are designed and composed such that the very last SLCO model in the chain contains all implementation details.

3.1   Reusability of SLCO Model Transformations

Table 1 lists eleven of thirteen endogenous model transformations that are defined to refine SLCO models. The other two transformations deal with time and are not discussed in this paper. Earlier work describes some of the model transformations in more detail [2]. There are two ways in which these transformations can be reused. First, a model transformation can be applied multiple times within the same sequence of transformations, as indicated in the second column of Table 1. In practise, the most reused model transformations are the *Clone Classes* transformation, which can be used to clone certain classes, and the *Remove Classes* transformation, which can be used to remove all classes that have no instances. They ensure that models adhere to the constraints imposed by most of the other transformations, and are therefore crucial for the successful composition of transformations. Second, a model transformation can be applied in multiple sequences leading to different target languages, as indicated in the third column. This type of reuse is less common, but supporting other target languages with similar semantic gaps would automatically lead to more reuse. The fourth column is discussed in Section 4. It shows the number of proof obligations that must be handled to prove the correctness of each transformation.

Table 1. Endogenous model transformations

Transformation Name	Reused within sequences	Reused for different target languages	Number of proof obligations
Bidirectional to Unidirectional	no	yes	1
Clone Classes	yes	yes	1
Exclusive Channels	yes	no	1
Identify Channels	no	no	1
Lossless to Lossy	no	no	74
Merge Channels	no	no	1
Merge Objects	yes	no	13
Names to Arguments	no	no	1
Remove Classes	yes	yes	1
Strings to Integer	yes	no	1
Synchronous to Asynchronous	no	no	4 and 34

It is not possible to consider all transformations from Table 1. Instead, we select the two variants of the *Synchronous to Asynchronous* transformation. The difference in complexity of these two transformations illustrates that the more generic transformations are, the more they need involved protocols for handling the introduced changes. One should search for the strongest possible constraints on the input models for such a transformation [2]. These constraints shall be realized in separate transformation steps that precede the more complex one in the transformation chain. This way, many unnecessary details are moved away from the core part of the transformation. The simple variant of the *Synchronous to Asynchronous* transformation also described below, although simple, is still complex enough to illustrate all the details of our approach.

3.2   Synchronous to Asynchronous Transformation

Synchronous communication is a typical example of a construct at a high level of abstraction that is often present in formal modeling languages. General-purpose programming languages, however, do not offer this concept. Synchronization should be realized with asynchronous interaction by correctly defined model transformations built around a properly defined communication protocol. While SLCO allows for synchronous communication, the communication between processes in NQC can be only asynchronous.

We defined two different transformations, T_{s2a}^S and T_{s2a}^G, to replace a synchronous by an asynchronous channel. As expected, this change requires and triggers further changes of the related classes, state machines, and transitions, to keep the observable behavior intact. T_{s2a}^S applies to restricted models, but is simple and does not greatly increase the complexity of the produced model. In contrast to T_{s2a}^S, T_{s2a}^G can be applied to any SLCO model, but as a more complex protocol is introduced by the transformation, it adds more complexity to the produced model, and this matters in practice. Both transformations require the following constraints to hold for their input: 1) the objects that communicate via the synchronous channel are the only instances of their classes, and 2) only a single pair of state machines from the two classes communicate over the channel. We stress, however, that this does not limit their applicability. By means of the *Exclusive Channels* and *Clone Classes* transformations, any SLCO model can be transformed into a model that meets these constraints. Thus, instead of having more complicated transformations that first change models to meet these constraints, and then replace synchronous communication by asynchronous communication, we opt for sequences of fine-grained transformations that have the same effect. The fact that the constraints hold can be used in the correctness proof of T_{s2a}^S and T_{s2a}^G, which greatly simplifies these proofs.

For the rest of the section, assume that in the model m, the synchronous channel $ch_s = chn()$ **sync from** $on_1.pn_1$ **to** $on_2.pn_2$ is to be transformed into an asynchronous one. Let object o_i with name on_i be an instance of class cl_i with name cn_i in model m, for $i = 1, 2$. We also assume, as explained above, that o_i is the only instance of cl_i, and that state machine sm_i is the only state machine in cl_i that uses channel ch_s, for $i = 1, 2$. Furthermore, we use $tr_s = tn_s$ **from** ss_1 **to** ss_2 **send** $sgn()$ **to** pn_1 to denote a transition of sm_1 of cl_1 that sends signals over ch_s, and $tr_r = tn_r$ **from** sr_1 **to** sr_2 **receive** $sgn()$ **from** pn_2 to denote a transition of sm_2 of cl_2 that receives signals over ch_s. Due to the uniqueness of the channel name and the previously mentioned assumption, the transformation of the channel ch_s induces a transformation of the classes cl_1 and cl_2 only. We show only the transformation of signals without arguments, but an extension to general signals is straightforward.

Simple Transformation. Transformation T_{s2a}^S modifies state machines by replacing some of their transitions. No essential changes are made to the other structures of a model. It is only applicable if, for every transition tr_s, there is no other transition with the same source state. For every transition tr_s of sm_1, and for every transition tr_r in sm_2, we define

$$T_{s2a}^S(tr_s, pn_1) = \langle\; ss_{nw},\; tn_s^1 \textbf{ from } ss_1 \textbf{ to } ss_{nw} \textbf{ send } ssgn() \textbf{ to } pn_1$$
$$tn_s^2 \textbf{ from } ss_{nw} \textbf{ to } ss_2 \textbf{ receive } asgn() \textbf{ from } pn_1 \rangle$$
$$T_{s2a}^S(tr_r, pn_2) = \langle\; sr_{nw},\; tn_r^1 \textbf{ from } sr_1 \textbf{ to } sr_{nw} \textbf{ receive } ssgn() \textbf{ from } pn_2$$
$$tn_r^2 \textbf{ from } sr_{nw} \textbf{ to } sr_2 \textbf{ send } asgn() \textbf{ to } pn_2 \rangle,$$

where ss_{nw} and sr_{nw} are fresh state names, tn_s^1, tn_s^2, tn_r^1, and tn_r^2 are fresh transition names, $ssgn \equiv$ "$s\_$" $+ sgn$, and $asgn \equiv$ "$a\_$" $+ sgn$. In the

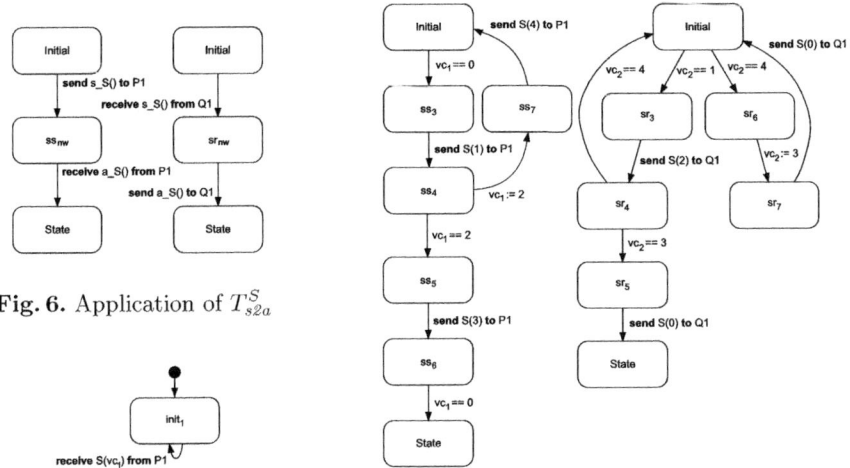

Fig. 6. Application of T_{s2a}^S

Fig. 7. A reader state machine

Fig. 8. Application of T_{s2a}^G

transformed model, the new states are added to the appropriate state machines, and the transitions tr_s and tr_r are replaced by the newly generated transitions. Applying T_{s2a}^S to the transitions concerning signal S of Figure 5 leads to the states and transitions of Figure 6.

General Transformation. Transformation T_{s2a}^G is more general than T_{s2a}^S, due to which it adds more complexity to the produced model. In this case, also classes are transformed, and new state machines are created. The restrictions we had on T_{s2a}^S are removed, thus T_{s2a}^G can be applied to any tr_s of sm_1 and any tr_r of sm_2 as defined above. Transformation T_{s2a}^G on transitions is defined as

$$
\begin{aligned}
T_{s2a}^G(tr_s, pn_1, vc_1) = \langle ss_3\ ss_4\ ss_5\ ss_6\ ss_7, \quad & ts_1 \text{ from } ss_1 \text{ to } ss_3\ vc_1 == 0 \\
ts_2 \text{ from } ss_3 \text{ to } ss_4 \text{ send } sgn(1) \text{ to } pn_1 \quad & ts_3 \text{ from } ss_4 \text{ to } ss_5\ vc_1 == 2 \\
ts_4 \text{ from } ss_5 \text{ to } ss_6 \text{ send } sgn(3) \text{ to } pn_1 \quad & ts_5 \text{ from } ss_6 \text{ to } ss_2\ vc_1 == 0 \\
ts_6 \text{ from } ss_7 \text{ to } ss_1 \text{ send } sgn(4) \text{ to } pn_1 \quad & ts_7 \text{ from } ss_4 \text{ to } ss_7\ vc_1 := 2 \rangle \\
T_{s2a}^G(tr_r, pn_2, vc_2) = \langle sr_3\ sr_4\ sr_5\ sr_6\ sr_7, \quad & tr_1 \text{ from } sr_1 \text{ to } sr_3\ vc_2 == 1 \\
tr_2 \text{ from } sr_3 \text{ to } sr_4 \text{ send } sgn(2) \text{ to } pn_2 \quad & tr_3 \text{ from } sr_4 \text{ to } sr_5\ vc_2 == 3 \\
tr_4 \text{ from } sr_5 \text{ to } sr_2 \text{ send } sgn(0) \text{ to } pn_2 \quad & tr_5 \text{ from } sr_4 \text{ to } sr_1\ vc_2 == 4 \\
tr_6 \text{ from } sr_7 \text{ to } sr_1 \text{ send } sgn(0) \text{ to } pn_2 \quad & tr_7 \text{ from } sr_6 \text{ to } sr_7\ vc_2 := 3 \\
tr_8 \text{ from } sr_1 \text{ to } sr_6\ vc_2 == 4 \rangle, &
\end{aligned}
$$

where ss_j and sr_j, and ts_j and tr_k are fresh state and transition names, for $j = 3, \dots, 7$ and $k = 3, \dots, 8$. Variables vc_1 and vc_2 are discussed below. Applying T_{s2a}^G to the transitions concerning signal S of Figure 5 leads to the states and transitions of Figure 8.

In the transformed model, the new states are added to the appropriate state machines, and transitions tr_s and tr_r are replaced by the new transitions. Additionally, a fresh integer variable vc_i and a state machine $reader_i$ are added to the classes cl_i, for $i = 1, 2$. Let Tr_i^G be the sets of all tr_s and tr_r-like transitions of sm_i, for $i = 1, 2$, Sgn_1 the set of all signal names occurring in the sending statements of transitions in Tr_1^G, and Sgn_2 the set of all signal names occurring in the reception statements of transitions in Tr_2^G. State machine $reader_i$ is a result of applying function Rsm, defined as

$$Rsm(pn_i, vc_i, Sgn_i) = reader_i \text{ initial } init_i$$
$$[tsgn_i \text{ from } init_i \text{ to } init_i \text{ receive } sgn(vc_i) \text{ from } pn_i \mid sgn \in Sgn_i],$$

where $reader_i$ is a fresh state machine name, $init_i$ is a fresh state name, and $tsgn \equiv \text{“}t\_\text{”} + sgn$. As defined, Rsm has a transition for every sgn from Sgn_i. Figure 7 shows this state machine for object p of Figure 5.

4   Correctness of Model Transformations

To reason about the correctness of model transformations, several dimensions have to be put together: 1) a description of the behavior of models, 2) definitions of model transformations, 3) the possibility to check the correctness criteria, for instance by comparing the behavior of models, and 4) the possibility to reason at the general language level rather than at the level of model instances. A sufficiently expressive and flexible formalism has to be used to cover all these aspects. We decided to use Labeled Transition Systems (LTSs) [14] as the underlying formalism to reason about the SLCO model transformations for several reasons. The LTS formalism is well-established and often used to describe system dynamics. Different equivalence relations between LTSs have been defined and used for comparison of behaviors. Since the SLCO transformations are defined as functions on SLCO models, we make use of such an equivalence relation later in this section for comparison of original and transformed models. Furthermore, in our earlier work, an executable prototype of the semantics of SLCO [4] is used to describe the behavior of models as LTSs. As the present work is largely inspired by and is a generalization of this prototype, it was quite natural to use the same formalism for the formal semantics of SLCO and utilize the gained experience.

To generate an LTS representation of the behavior of a given model, we first formally define the dynamic semantics of SLCO in the form of structural operational semantics [14]. Then, reasoning at the level of LTS-representations of the behaviors, we define our criterion for correctness of model transformations. The SLCO formal semantics is necessary for the correctness proofs.

In the sequel, we use the following conventions and notation. If lt is a list, we abuse the notation and write lt for the set of elements of lt. Thus, we write $elt \in lt$ for "there is an element elt in list lt". A number of functions are defined and used. $f[a/x]$ denotes the update of function f by $f(x) = a$. We also use the short-hand notation $f[a^*/x^*]$ for the updated values for the arguments in the list x^* with the corresponding values from the list a^*. If a given expression e evaluates to constant expressions ce with respect to evaluation functions V_1 and V_2, we write $\langle e, V_1, V_2 \rangle = ce$, and $\langle e^*, V_1, V_2 \rangle = ce^*$ for a list of expressions e^*. $\mathbb{D}$ denotes the domain of all integer, boolean, and string values.

4.1   Operational Semantics of SLCO

The formal semantics of SLCO is given in the form of Structural Operational Semantics (SOS) [14]. For an arbitrary model m, the SOS generates the complete behavior of the model in the form of an LTS. An LTS is a tuple $(S, \Lambda, \rightarrow, i)$, where S is a set of configuration, Λ is a set of labels, $\rightarrow \subseteq S \times \Lambda \times S$ is a ternary relation of labeled transitions, and $i \in S$ is the initial configuration. In our case, for an SLCO model $m = mn\ obj^*\ class^*\ chan^*$, with Var the set of all variables occurring in m, SM the set of all state machines of m, and $States$ the set of all states of these state machines, the *configurations* of the LTS generated for m are tuples $\langle m, S_{OSMS}, V_{OS}, V_{OSMS}, B \rangle$, where $S_{OSMS} : obj^* \mapsto (SM \mapsto States)$ is a function that indicates current states of the state machines of the objects in m, $V_{OS} : obj^* \mapsto (Var \mapsto \mathbb{D})$ is an evaluation function that assigns values to the (global) variables of the objects in m, $V_{OSMS} : obj^* \mapsto (SM \mapsto (Var \mapsto \mathbb{D}))$ is an evaluation function that assigns concrete values to the (local) variables of the state machines of the objects of model m, and $B : chan^* \times obj^* \times obj^* \mapsto (Signals \times CE^*) \cup \{\mathbf{nil}\}$ represents the content of the one-place buffers corresponding to the asynchronous channels in m. The content of a buffer can be $\mathbf{nil}$, denoting an empty buffer, or a tuple consisting of a signal name and a list of constant expressions. Two buffers are associated to each bidirectional asynchronous channel, one for each direction.

For the LTS of m, $LTS(m)$ in short, the *initial configuration* conforms to the following constraints: 1) all the buffers assigned to asynchronous channels are initialized to $\mathbf{nil}$, 2) all the state machines are in their initial state, and 3) all variables are initialized to values respecting their types.

A *transition label* can be ϵ, meaning that no statement is executed by this transition, $vn := ce$, representing an assignment, $sgn(ce*)$, representing synchronous communication, or **receiving** $sgn(ce*)$ or **sending** $sgn(ce*)$, for a signal name $sgn \in Signals$, representing asynchronous communication. The transitions of $LTS(m)$ are obtained as the least relation deduced from the SOS rules. Due to the space limitations, we give only a subset of the rules used in the correctness proof of the selected transformation given in this section.

The overall behavior of a model is defined by the behavior of its composite elements, the objects and the channels. The activities that the objects can execute and the way they interact via the channels determine the dynamics of the model, as captured by the SOS rules in Figures 11 and 12. The contributed activities of an object are deduced from the specification of its class, shown in Figure 11. At the most elementary level of the structure, the activities within a class are derived from the transitions specified for each state machine of the class, as shown in Figures 9 and 10.

Figure 9 shows some of the SOS rules which turn a single SLCO symbolic transition into a behavioral LTS transition. The auxiliary symbolic evaluation functions V_O and V_{SM}, which assign values to variables, are concretely interpreted within a model and its concrete evaluation functions. For a conditional signal reception (the second rule), the corresponding boolean expression has to be evaluated first, by the possibly updated evaluation functions, and only if it

evaluates to **true**, the transition step can be taken. The rule in Figure 10 shows
how the enabled transitions, derived by the SOS rules for transitions (Figure 9),
are lifted up to the level of a class. It also shows implicitly that all state machines
of a considered class are inspected. The symbolic function S_{SMS} keeps track of
the current states of the state machines, while V_{SMS} maps (local) variables to
their value for the state machines of the object, and V_O maps the (global) variables at the level of the object to their values.

$$\frac{\langle e^*, V_O, V_{SM} \rangle = ce^*}{\langle tn \text{ from } sn \text{ to } sn' \text{ effect send } sgn(e^*) \text{ to } pn, sn, V_O, V_{SM} \rangle \xrightarrow{\text{send } sgn(ce^*) \text{ to } pn}_{TR} \langle sn', V_O, V_{SM} \rangle}$$

$$\frac{\langle be, V_O', V_{SM}' \rangle = \textbf{true}, \quad V_O' = V_O[ce^*/vn^*], \quad V_{SM}' = V_{SM}[ce^*/vn^*]}{\langle tn \text{ from } sn \text{ to } sn' \text{ trigger receive } sgn(vn^*|be) \text{ from } pn, sn, V_O, V_{SM} \rangle \xrightarrow{\text{receive } sgn(ce^*) \text{ from } pn}_{TR} \langle sn', V_O', V_{SM}' \rangle}$$

Fig. 9. A subset of all deduction rules for transitions

$$\frac{sm = smn \; var^* \; states \; trans^*, \; tr \in trans^*, \; cl = cn \; var^* \; port^* \; sm_{cl}^*, \; sm \in sm_{cl}^*,}{\langle tr, S_{SMS}(sm), V_O, V_{SMS}(sm) \rangle \xrightarrow{l}_{TR} \langle sn', V_O', V_{SM}' \rangle, \; S_{SMS}' = S_{SMS}[sn'/sm], \; V_{SMS}' = V_{SMS}[V_{SM}'/sm]}{\langle cl, S_{SMS}, V_O, V_{SMS} \rangle \xrightarrow{l}_{CL} \langle S_{SMS}', V_O', V_{SMS}' \rangle}$$

Fig. 10. Deduction rule for classes

Objects behave as specified by their class. In a composition, objects participate and interact as described by the SOS rules for compositions of objects,
some of which are given in Figure 11. Every non-synchronizing transition of
one of the objects enabled for execution in the current configuration is executed
by the composition of the objects, and the functions are updated accordingly;
only the functions of the object this transition belongs to are updated. A non-
synchronizing transition receiving signals over an asynchronous lossless channel
is captured by the second rule. The first rule describes synchronization of two
objects via a synchronous channel.

4.2   Correctness of the T_{s2a}^S Transformation

The operational semantics of SLCO generates an LTS representation of the
model dynamics, for a given model and its initialization. Thus, to reason about
the correctness of and property preservation by a model transformation, we need
to compare the behaviors of two models, one before and one after the transformation, represented as LTSs.

$$cl_1, cl_2 \in class^*, \quad o_1, o_2 \in obj^*, \quad o_1 \equiv on_1 : cl_1, \quad o_2 \equiv on_2 : cl_2$$

$$chn(type^*) \text{ sync from } on_1.pn_1 \text{ to } on_2.pn_2 \in chan^*$$

$$\langle cl_1, S_{OSMS}(o_1), V_{OS}(o_1), V_{OSMS}(o_1) \rangle \xrightarrow{\text{send } sgn(ce^*) \text{ to } pn_1}_{CL} \langle S'_{SMS}, V'_O, V'_{SMS} \rangle,$$

$$\langle cl_2, S_{OSMS}(o_2), V_{OS}(o_2), V_{OSMS}(o_2) \rangle \xrightarrow{\text{receive } sgn(ce^*) \text{ from } pn_2}_{CL} \langle S''_{SMS}, V''_O, V''_{SMS} \rangle,$$

$$S'_{OSMS} = S_{OSMS}[S'_{SMS}/o_1][S''_{SMS}/o_2], \quad V'_{OS} = V_{OS}[V'_O/o_1][V''_O/o_2],$$

$$V'_{OSMS} = V_{OSMS}[V'_{SMS}/o_1][V''_{SMS}/o_2]$$

$$\langle obj^*, class^*, chan^*, S_{OSMS}, V_{OS}, V_{OSMS}, B \rangle \xrightarrow{sgn(ce^*)}_{OBJS} \langle S'_{OSMS}, V'_{OS}, V'_{OSMS}, B \rangle$$

$$cl_2 \in class^*, \quad o_2 \in obj^*, \quad o_2 \equiv on_2 : cl_2,$$

$$chn(type^*) \text{ async lossless from } on_1.pn_1 \text{ to } on_2.pn_2 \in chan^*,$$

$$\langle cl_2, S_{OSMS}(on_2), V_{OS}(o_2), V_{OSMS}(o_2) \rangle \xrightarrow{\text{receive } sgn(ce^*) \text{ from } pn_2}_{CL} \langle S'_{SMS}, V'_O, V'_{SMS} \rangle,$$

$$S'_{OSMS} = S_{OSMS}[S'_{SMS}/o_2], \quad V'_{OS} = V_{OS}[V'_O/o_2], \quad V'_{OSMS} = V_{OSMS}[V'_{SMS}/o_2],$$

$$B(\langle chn, o_1, o_2 \rangle) = \langle sgn, ce^* \rangle, \quad B' = B[nil/\langle chn, o_1, o_2 \rangle]$$

$$\langle obj^*, class^*, chan^*, S_{OSMS}, V_{OS}, V_{OSMS}, B \rangle \xrightarrow{\text{receiving } sgn(ce^*)}_{OBJS} \langle S'_{OSMS}, V'_{OS}, V'_{OSMS}, B' \rangle$$

Fig. 11. A subset of all deduction rules for compositions of objects

$$m = mn \ obj^* \ class^* \ chan^*,$$

$$\langle obj^*, class^*, chan^*, S_{OSMS}, V_{OS}, V_{OSMS}, B \rangle \xrightarrow{l}_{OBJS} \langle S'_{OSMS}, V'_{OS}, V'_{OSMS}, B' \rangle$$

$$\langle m, S_{OSMS}, V_{OS}, V_{OSMS}, B \rangle \xrightarrow{l}_M \langle m, S'_{OSMS}, V'_{OS}, V_{OSMS}, B' \rangle$$

Fig. 12. Deduction rule for models

A wide range of equivalence relations on LTSs have been proposed [10]. Some of them, e.g. strong bisimulation, are appropriate for concrete behavior, when every action of the system is observable. For some of the defined SLCO transformations, this is indeed sufficient. However, these relations are often too fine when part of the behavior is preferred to be abstracted away and considered unobservable. Some of the SLCO model transformations, as explained in the previous sections, add more detail to the behavior and therefore, some parts of the behavior introduced by the transformation need to be abstracted away to mimic the behavior before the transformation. In view thereof, we choose branching bisimulation [11] as the equivalence relation we use for the correctness criterion. Branching bisimulation is a relation between configurations of LTSs for which some transitions are considered *internal* (*unobservable*), represented by labeling them with τ ($\tau \notin \Lambda$). Intuitively, two configurations are branching bisimilar if every transition step that can be executed in one configuration can be mimicked in the other, possibly after a finite number of internal steps. Branching bisimulation possesses many useful properties, one of which is that related models possess the same properties that can be expressed in the temporal logic CTL* without the next state modality [6], including safety and liveness. In our case, this means that if a certain property has been established for the source model, which is usually much smaller then its implementations and thus easier to analyze, and if we apply a (well-composed) chain of model transformations for which our correctness criterion hold, then the generated model inherits the property by construction, as well as all intermediate models in the chain. Thus, our model transformation correctness criterion provides effective and efficient generation of correct-by-construction code.

Definition 1. *Let T be an SLCO model transformation such that for any SLCO model m to which T applies and for a given initialization of m, there is a renaming ρ of the labels of the $LTS(T(m))$ such that $LTS(m)$ and $\rho(LTS(T(m)))$ are branching bisimilar. Then T is a* correct model transformation.[1]

The renaming function in the definition above is needed to rename some labels into τ, but also to unify, if needed, the labels of transitions that are supposed to mimic each other. Therefore, to establish the correctness result for an SLCO transformation, we may need to find an appropriate renaming of the labels first. This is rather obvious when comparing the transitions that are to be transformed in Figure 5 with the transitions in Figure 6 created by the transformation, which are required to capture the same observable behavior.

In this section, we discuss the main lines of the correctness proof for the simple variant of the *Synchronous to Asynchronous* transformation, T^S_{s2a}. We chose this one because the proof for this transformation has all the important aspects that need to be taken into account, yet the established relation between configurations is simple enough to be given completely. Besides transformation T^S_{s2a}, there are four more transformations that require a substantial amount of cases to be considered for their correctness proof. The fourth column of Table 1 in Section 3 lists the number of proof obligations for each transformation. The last row lists two numbers, one for each version of the *Synchronous to Asynchronous* transformation. Fortunately, the correctness proofs of the majority of transformations involve a single proof obligation only, relating each configuration for input models to exactly one equivalent configuration for output models. This is a clear benefit of designing transformations that are as fine-grained as possible.

Referring back to the definition of T^S_{s2a} in Section 3.2, let Sgn be the set of all signal names used in the sending and receiving statements of all tr_s and tr_r-like transitions. We define a label-renaming function ρ as: for every $sgn \in Sgn$, $\rho(\textbf{sending } ssgn()) = \tau$, $\rho(\textbf{receiving } ssgn()) = sgn()$, $\rho(\textbf{sending } asgn()) = \tau$, $\rho(\textbf{receiving } asgn()) = \tau$, and $\rho(\epsilon) = \tau$. Renaming ρ is straightforwardly extended on $LTS(T^S_{s2a}(m))$. By renaming **receiving** $ssgn()$ to $sgn()$, we indicate that these two labels represent successful communication. The other labels are renamed to τ because they represent the implicit synchronization in the source model, and should result in unobservable behavior of the target model.

Theorem 1. *For a given SLCO model $m = mn$ class* obj* chan*, and a channel $ch_s = chn(t)$ **sync from** $on_1.pn_1$ **to** $on_2.pn_2 \in$ chan*, $T^S_{s2a}(m, chs)$ is a correct model transformation.*

Proof. We need to show that $LTS(m)$ and $\rho(LTS(T^S_{s2a}(m)))$ are branching bisimilar. As usual, the main difficulty of the proof lies in properly relating

---

[1] It is essential here that all transitions of $LTS(m)$ are observable.

the configurations from $LTS(m)$ and those from $\rho(LTS(T_{s2a}^S(m)))$. Before we define the relation, it is worth noticing that each configuration of $LTS(m)$ is also a configuration of $\rho(LTS(T_{s2a}^S(m)))$, but for the latter the buffer function is extended over the triples $\langle cha, on_1, on_2 \rangle$ and $\langle cha, on_2, on_1 \rangle$.

Let $cf = \langle m, S_{OSMS}, V_{OS}, V_{OSMS}, B \rangle$ be a configuration of $LTS(m)$, and $cf' = \langle T_{s2a}^S(m), S'_{OSMS}, V'_{OS}, V'_{OSMS}, B' \rangle$ a configuration of $\rho(LTS(T_{s2a}^S(m)))$. We define a relation R between the configurations as follows: $(cf, cf') \in R$ if and only if $V_{OS} = V'_{OS}$, $V_{OSMS} = V'_{OSMS}$ and

1. $S_{OSMS} = S'_{OSMS}$, $B'(\langle cha, on_1, on_2 \rangle) = \mathbf{nil}$, $B'(\langle cha, on_2, on_1 \rangle) = \mathbf{nil}$, and $B' = B$ otherwise, or

2. $S_{OSMS}(o_1)(sm_1) = ss_1$, $S'_{OSMS}(o_1)(sm_1) = ss_{nw}$, $S_{OSMS} = S'_{OSMS}$ otherwise, $B'(\langle cha, on_1, on_2 \rangle) = (sgn, \varepsilon)$, $B'(\langle cha, on_2, on_1 \rangle) = \mathbf{nil}$, and $B' = B$ otherwise, only if there is a tr_s-like transition from ss_1 with signal name $sgn()$, or

3. $S_{OSMS}(o_1)(sm_1) = ss_2$, $S_{OSMS}(o_2)(sm_2) = sr_2$, $S'_{OSMS}(o_1)(sm_1) = ss_{nw}$, $S'_{OSMS}(o_2)(sm_2) = sr_{nw}$, $S_{OSMS} = S'_{OSMS}$ otherwise, $B'(\langle cha, on_1, on_2 \rangle) = \mathbf{nil}$, $B'(\langle cha, on_2, on_1 \rangle) = \mathbf{nil}$, and $B' = B$ otherwise, only if there is a tr_s-like transition in sm_1 to ss_2, and there is a tr_r-like transition in sm_2 to sr_2, or

4. $S_{OSMS}(o_1)(sm_1) = ss_2$, $S'_{OSMS}(o_1)(sm_1) = ss_{nw}$, $S_{OSMS} = S'_{OSMS}$ otherwise, $B'(\langle cha, on_1, on_2 \rangle) = \mathbf{nil}$, $B'(\langle cha, on_2, on_1 \rangle) = (asgn, \varepsilon)$, and $B' = B$ otherwise, if there is a tr_s-like transition to ss_2 with signal $sgn()$.

Next, for each pair of configurations we have to show that they can mimic each other, using the SOS rules. For example, let us consider case **2**. If $cf \xrightarrow{l} cf_1$, for some label l and a configuration cf_1 of $LTS(m)$, then either $l \not\equiv sgn()$ or $l \equiv sgn()$. In the first case, this transition certainly does not involve state machine sm_1 of object o_1, since it can only synchronize in this configuration. In this case, the same state machine(s) of the same object(s) can induce the same transition $cf' \xrightarrow{l} cf'_1$, and since the updates of the functions do not change $S_{OSMS}(o_1)(sm_1)$ nor $S'_{OSMS}(o_1)(sm_1)$, it follows that $(cf_1, cf'_1) \in R$.

If $l \equiv sgn()$, then sm_2 has to be in a state sr_1 for some tr_r-like transition, according to the synchronization SOS rule in Figure 11. According to the SOS rule for an asynchronous signal reception, also in Figure 11, $cf' \xrightarrow{sgn()} cf'_1$ (which is renamed **receiving** $ssgn()$ transition). Furthermore, in cf_1, sm_1 is in state ss_2 and sm_2 is in state sr_2. In cf'_1, sm_1 is in state ss_{nw} and sm_2 is in state sr_{nw}. According to **3.**, $(cf_1, cf'_1) \in R$. Note here that lifting the constraint on sm_1 and allowing it to have other transitions besides the synchronizing signal sending in state ss_1, breaks bisimilarity of R.

If we assume a transition made by cf', i.e., $cf' \xrightarrow{l} cf'_1$, then in a similar way, we can conclude that this transition is either also executed by cf, or that this transition is mimicked by a $cf \xrightarrow{sgn()} cf_1$ transition, depending on label l.

By a careful inspection of transitions generated by the SOS rules for the other three cases of pairs of R-related configurations, we can prove that R is indeed a branching bisimulation which relates $LTS(m)$ and $\rho(LTS(T_{s2a}^S(m)))$. $\qquad\square$

5   Related Work

Various aspects of the correctness of model transformations have been considered, and different approaches have been proposed. A number of approaches are based on graph-based transformation techniques.

In [9], input and output models are related during the specification of a transformation, and then a theorem prover is used to show semantic equivalence between the input and output of this transformation. The source and target language discussed are relatively small, leading to a more straightforward transformation compared to the sequences of transformations we consider. However, the use of a theorem prover to automate parts of the correctness proofs has clear advantages over manual proofs. The approach in [15] also uses a theorem prover for assistance with correctness proofs. In this case, properties that are proved for the given model transformation are more of a structural nature. It is interesting that here the author advocates the advantages of having a single homogenous formalism for description of transformations, which we also see advantageous in our approach. The framework proposed in [18] allows for defining a set of graph transformation rules to describe the operational semantics of a DSML, which is used, similar to our approach, to generate an LTS representation of models in the DSML, which then can be model checked. However, the translation framework works only on a particular given model instance of the language, while we aim at general results at the level of the entire language.

Instance-based verification of model transformations is described also in [13]. The approach entails generating a certificate for each model that is transformed. These certificates are used to show that the model transformation preserves certain properties for the given input model, but cannot be used to show that properties are preserved for arbitrary input models.

The approaches in [7] and [8] are most closely related to the work presented in this paper. In [7], preservation of behavior by model transformation is considered. Besides the models in the source language, also the language semantics is transformed, and the result is compared with the semantics of the target language. The paper states conditions that input models and model transformations should fulfill to preserve the semantics. In [8], correctness of model transformations stated in terms of a bisimulation relation is considered. Here, the languages are first given operational semantics in terms of graph-transformation rules as well. Although our approach to the correctness of transformations is similar to this one, the two languages considered in [8] are much simpler than the language we used to demonstrate our approach.

6   Conclusions and Future Work

We described a formal framework for reasoning about the correctness of endogenous model transformations for a small but non-trivial DSML. Using this framework, we can asses whether sequences of transformations are well-composed, and whether individual transformations are provably correct. By designing transformations that are as fine-grained as possible, we improved the reusability of these transformations both between sequences of transformations and within such sequences. Furthermore, this design decision increased the number of transformations with straightforward correctness proofs, and reduced the number of proof obligations for the proofs of the larger transformation steps. The presented approach is independent of any specific model transformation language.

Given the large number of straightforward proofs, we consider investigating the application of automated theorem proving to be a promising direction for future research. Additionally, we want to investigate the generalization of the approach to other DSMLs.

References

1. van Amstel, M.F., van den Brand, M.G.J., Engelen, L.J.P.: An Exercise in Iterative Domain-Specific Language Design. In: Proc. of EVOL/IWPSE 2010, pp. 48–57. ACM (2010)
2. van Amstel, M.F., van den Brand, M.G.J., Engelen, L.J.P.: Using a DSL and Fine-Grained Model Transformations to Explore the Boundaries of Model Verification. In: Proc. ICSTW 2011, pp. 63–66. IEEE Computer Society (2011)
3. van Amstel, M.F., van den Brand, M.G.J., Protić, Z., Verhoeff, T.: Transforming Process Algebra Models into UML State Machines: Bridging a Semantic Gap? In: Vallecillo, A., Gray, J., Pierantonio, A. (eds.) ICMT 2008. LNCS, vol. 5063, pp. 61–75. Springer, Heidelberg (2008)
4. Andova, S., van den Brand, M.G.J., Engelen, L.: Prototyping the Semantics of a DSL using ASF+SDF: Link to Formal Verification of DSL Models. In: AMMSE, pp. 65–79 (2011)
5. Baum, D.: NQC Programmer's Guide (2003)
6. De Nicola, R., Vaandrager, F.W.: Three logics for branching bisimulation. Journal of the ACM 42, 458–487 (1995)
7. Ehrig, H., Ermel, C.: Semantical Correctness and Completeness of Model Transformations Using Graph and Rule Transformation. In: Ehrig, H., Heckel, R., Rozenberg, G., Taentzer, G. (eds.) ICGT 2008. LNCS, vol. 5214, pp. 194–210. Springer, Heidelberg (2008)
8. Hülsbusch, M., König, B., Rensink, A., Semenyak, M., Soltenborn, C., Wehrheim, H.: Showing Full Semantics Preservation in Model Transformation - A Comparison of Techniques. In: Méry, D., Merz, S. (eds.) IFM 2010. LNCS, vol. 6396, pp. 183–198. Springer, Heidelberg (2010)
9. Giese, H., Glesner, S., Leitner, J., Schäfer, W., Wagner, R.: Towards Verified Model Transformations. In: Proc. MoDeVa 2006, pp. 78–93 (2006)
10. van Glabbeek, R.J.: The Linear Time–Branching Time Spectrum II: The Semantics of Sequential Systems with Silent Moves. In: Best, E. (ed.) CONCUR 1993. LNCS, vol. 715, pp. 66–81. Springer, Heidelberg (1993)
11. van Glabbeek, R.J., Weijland, P.: Branching Time and Abstraction in Bisimulation Semantics. Journal of the ACM 43, 555–600 (1996)
12. Holzmann, G.J.: The model checker SPIN. IEEE Transactions on Software Engineering 23(5), 279–295 (1997)
13. Karsai, G., Narayanan, A.: On the Correctness of Model Transformations in the Development of Embedded Systems. In: Kordon, F., Sokolsky, O. (eds.) Monterey Workshop 2006. LNCS, vol. 4888, pp. 1–18. Springer, Heidelberg (2007)

14. Plotkin, G.D.: A Structural Approach to Operational Semantics. Technical Report DAIMI FN-19, University of Aarhus (1981)
15. Schätz, B.: Verification of Model Transformations. ECEASST 29 (2010)
16. Schmidt, D.C.: Model-Driven Engineering. Computer 39(2), 25–31 (2006)
17. Theelen, B.D., et al.: Software/Hardware Engineering with the Parallel Object-Oriented Specification Language. In: Proc. MEMOCODE 2007, pp. 139–148. IEEE (2007)
18. Varró, D.: Automated formal verification of visual modeling languages by model checking. Software and System Modeling 3(2), 85–113 (2004)

Multifocal: A Strategic Bidirectional Transformation Language for XML Schemas

Hugo Pacheco and Alcino Cunha

HASLab / INESC TEC & Universidade do Minho, Braga, Portugal
{hpacheco,alcino}@di.uminho.pt

Abstract. Lenses are one of the most popular approaches to define bidirectional transformations between data models. However, writing a lens transformation typically implies describing the concrete steps that convert values in a source schema to values in a target schema. In contrast, many XML-based languages allow writing structure-shy programs that manipulate only specific parts of XML documents without having to specify the behavior for the remaining structure. In this paper, we propose a structure-shy bidirectional two-level transformation language for XML Schemas, that describes generic type-level transformations over schema representations coupled with value-level bidirectional lenses for document migration. When applying these two-level programs to particular schemas, we employ an existing algebraic rewrite system to optimize the automatically-generated lens transformations, and compile them into Haskell bidirectional executables. We discuss particular examples involving the generic evolution of recursive XML Schemas, and compare their performance gains over non-optimized definitions.

Keywords: coupled transformations, bidirectional transformations, two-level transformations, strategic programming, XML.

1 Introduction

Data transformations are often coupled [16], encompassing software transformation scenarios that involve the modification of multiple artifacts such that changes to one of the artifacts induce the reconciliation of the remaining ones in order to maintain global consistency. A particularly interesting instance of this class are *two-level transformations* [18,5], that concern the type-level transformation of schemas coupled with the value-level transformation of documents that conform to those schemas. A typical example of two-level transformations are format evolution scenarios [18,11], such as schema changes occurring during maintenance operations or imposed by the natural evolution of the applications. These schema evolutions call for the coupled evolution of the underlying documents and related artifacts so that they remain consistent with the new schema.

Most existing XML transformation and querying languages, such as XSLT, XQuery or XPath, allow writing structure-shy programs that provide specific behavior only for the interesting bits of a (possibly huge) XML document without

Z. Hu and J. de Lara (Eds.): ICMT 2012, LNCS 7307, pp. 89–104, 2012.
© Springer-Verlag Berlin Heidelberg 2012

having to specify how to traverse the remaining structure. Due to their generic form, such programs are easier to write and can be applied to documents satisfying different schemas. Nevertheless, they are not two-level. For example, using XSLT we can separately specify a transformation between XML schemas (since these can be represented as regular XML documents) and between XML documents, but the second is not a byproduct of the first. Thus, consistency between both levels must be manually verified, while a two-level transformation provides both transformations such that they are consistent by construction.

Another prominent instance of coupling are *bidirectional transformations*, as a "mechanism for maintaining the consistency of two (or more) related sources of information" [9]. For example, after a format evolution both old and new documents may co-exist and evolve independently. In a bidirectional transformation, the coupling occurs between forward and backward value transformations such that changes made to one of the data instances can be propagated to its connected pair in order to recover consistency.

Similarly to two-level transformations, a good approach is to design intrinsic bidirectional languages in which a program can be read both as a forward or a backward transformation, so that these are correct for the respective semantic space. Following this notion, many bidirectional languages have emerged in the most diverse computing domains, including many focused on the transformation of tree-structured data and with a particular application to XML documents [15,3,10,21,20,14]. Among these, one of the most successful approaches are the so-called lenses, introduced by Foster *et al* [10] to solve the classical view-update problem: if a source model is abstracted into a view, how can updates made to the view be propagated back to the original model? They propose the *Focal* tree transformation language that allows users to build lenses with sophisticated synchronization behavior in a compositional way.

Still, the aforementioned bidirectional languages are at best typed but not two-level. On top of that, the programming style that grants them bidirectionality is usually more biased towards structure-sensitive constructs, to be able to identify precisely the concrete steps required to translate between source and target documents.

In this paper, we propose *Multifocal*, a generic structure-shy two-level transformation language for XML Schema evolution whose underlying value-level functions are bidirectional lens transformations that translate XML documents conforming to the old and new schemas. In comparison to a *Focal* lens transformation, that describes a bidirectional view between two particular tree structures, a *Multifocal* transformation describes a general type-level transformation (over XML Schemas) that provides multiple focus points, in the sense that it produces a different view schema and a corresponding bidirectional lens for each XML Schema to which it is applied successfully.

To describe such two-level transformations, we will use a generic style familiar of strategic rewriting languages [24,19,17], where the combination of a standard set of basic rules allows the design of flexible rewrite strategies in a compositional

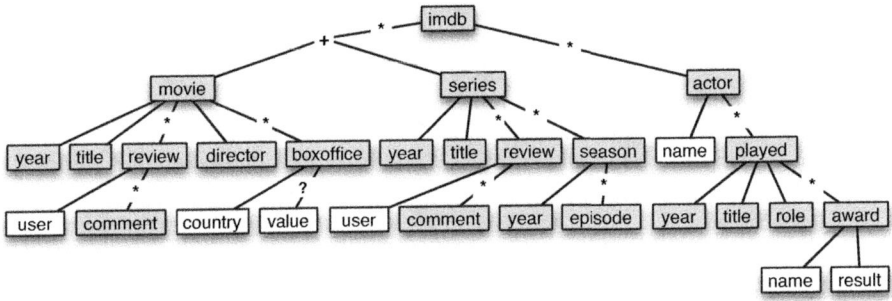

Fig. 1. Representation of a movie database schema inspired by IMDb (http://www.imdb.com). Grey boxes denote elements and white ones model attributes.

way, such as generic traversals that apply type-level transformations at arbitrary levels inside schema representations.

A known disadvantage of generic programs is their worse performance in comparison to analogous non-generic ones, since they must undergo runtime checks and blindly traverse whole input structures. In our framework, we mitigate this issue by encoding the underlying bidirectional lens transformations in a point-free[1] language with powerful algebraic laws and allowing automatic optimization by calculation [23], so that the optimized lens programs are able to efficiently propagate updates on XML documents.

In Section 2 we motivate our framework with an example. Section 3 presents the *Multifocal* language and discusses the design of our framework for the specification, optimization and execution of *Multifocal* transformations. The implementation of the framework (using the functional programming language Haskell) is shown in Section 4. Section 5 illustrates by example how our framework can tackle various application scenarios involving the generic evolution of recursive XML schemas, and compares the speedups achieved by an automatic optimization phase. In Section 6 we survey related work and Section 7 concludes the paper with a synthesis of the main contributions and directions for future work.

2 Motivating Example

Consider the XML Schema from Figure 1 representing an IMDb-like database for storing information about movies and actors. Imagine that we want to summarize this schema according to the following steps:

1. Delete all `series` elements.
2. For each `movie`, replace its `reviews` by a `popularity` attribute counting the number of `comments` and replace its `boxoffice` elements with a `profit` attribute summing the total `value` elements.
3. For each `actor`, keep its `name` and a list of `award` names renamed to `awname`.

---

[1] The point-free style is characterized by the lack of explicit "points" or variables.

The resulting schema is shown in Figure 2. However, not only do we want to transform the XML schema, but also to migrate conforming XML documents and propagate updates in both directions: if a source document is modified, then a new view document must be computed; and if a view document is modified, then those changes shall be translated back into a modified source document. Consider, for example, the source XML document from Figure 3(a) containing one movie, one se-

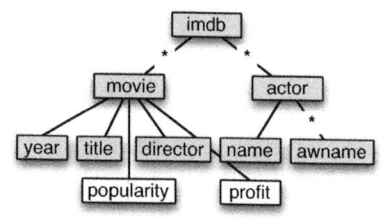

Fig. 2. A view of the movie database schema from Figure 1

ries and one actor. The forward transformation of our running example would produce the XML view shown in Figure 3(b). If we modify the view by correcting Uma Thurman's name and award information, and add an entry for the new Sherlock Holmes movie before the single `actor` element (Figure 3(c)), then the backward transformation shall correct the actress' name at the appropriate location and insert a new `movie` element with default `review` and `boxoffice` elements that are consistent with the modified view (Figure 3(d)).

In the remainder of this paper, we will propose a generic XML transformation language for XML schemas in which we can express the above transformation in a concise way close to its informal definition. Plus, the transformations for XML documents will come for free as conforming bidirectional lenses, satisfying strong round-tripping properties and supporting automatic optimization.

3   The Multifocal Framework

We now provide an overview of the *Multifocal* language and the respective framework for strategic two-level bidirectional transformation. We start with a formal definition of the bidirectional lenses at the core of our framework:

Definition 1 (Lens [10]). *A lens* $l:S \trianglerighteq V$ *comprises two total transformations* $get: S \to V$ *and* $put: V \times S \to S$, *satisfying the following properties:*

$$get\ (put\ (v,s)) = v \quad \text{PUTGET} \qquad put\ (get\ (s),s) = s \quad \text{GETPUT}$$

To give an idea of the bidirectional programs we are considering, these round-tripping properties guarantee that a lens is indeed an abstraction, i.e., the source schema S contains more information than the view schema V, and that backward propagation without modifications preserves the original documents.

Our two-level language over XML schemas is defined by instantiating a well-known suite of combinators for strategic programming [24,19], together with specific XML transformers. The full syntax of *Multifocal* is defined as follows:

$$strat ::= \texttt{nop} \mid strat >> strat \mid strat \mathbin{|\!|} strat \mid \texttt{many}\ strat \mid \texttt{try}\ strat$$
$$\mid \texttt{all}\ strat \mid \texttt{once}\ strat \mid \texttt{everywhere}\ strat \mid \texttt{outermost}\ strat$$

```
<imdb>
 <movie>
  <year>2003</year>
  <title>Kill Bill: Vol. 1</title>
  <review user="emma">
   <comment>Gorgeous!</comment></review>
  <director>Quentin Tarantino</director>
  <boxoffice country="USA" value="22089322"/>
  <boxoffice country="Japan" value="3521628"/>
 </movie>
 <series><year>2011</year>
  <title>Game of Thrones</title>
  <season><year>2011</year>
   <episode>Winter is Coming</episode>
  </season></series>
 <actor name="Umma Thurman">
  <played><year>2003</year>
   <title>Kill Bill: Vol. 1</title>
   <role>The Bride</role>
   <award name="Saturn" result="Won"/>
  </played></actor>
</imdb>
```

(a) Source XML document

```
<imdb>
 <movie> ... </movie>
 <series> ... </series>
 <movie><year>2012</year>
  <title>Sherlock Holmes: Game of Shadows</title>
  <review user="" comment=""/>
  <review user="" comment=""/>
  <director>Guy Ritchie</director>
  <boxoffice country="" value="15"/>
 </movie>
 <actor name="Uma Thurman">
  <played><year>2003</year>
   <title>Kill Bill: Vol. 1</title>
   <role>The Bride</role>
   <award name="Saturn Best Actress" result="Won"/>
  </played></actor>
</imdb>
```

(d) Modified source XML document

```
<imdb>
 <movie popularity="1" profit="25610950">
  <year>2003</year>
  <title>Kill Bill: Vol. 1</title>
  <director>Quentin Tarantino</director>
 </movie>
 <actor name="Umma Thurman">
  <awname>Saturn</awname>
 </actor>
</imdb>
```

(b) View XML document

```
<imdb>
 <movie> ... </movie>
 <movie popularity="2" profit="15">
  <year>2012</year>
  <title>Sherlock Holmes: Game of Shadows</title>
  <director>Guy Ritchie</director>
 </movie>
 <actor name="Uma Thurman">
  <awname>Saturn Best Actress</awname>
 </actor>
</imdb>
```

(c) Modified view XML document

Fig. 3. Example of a bidirectional transformation between XML documents

```
| at '"' tag '"' strat | when '"' tag '"' strat
| hoist | plunge '"' tag '"' | rename '"' tag '"'
| erase | select '"' xpath '"'
```

The set of strategic combinators allows to apply transformations sequentially (>>), alternatively (||), repetitively (many) or, more challengingly, at arbitrary depths inside schema representations. It also includes combinators for identity (nop) and optional rule application (try). Likewise [17] and other generic programming languages, instead of defining generic traversals by induction on the structure of types, we define a small set of traversal combinators. The all combinator applies a transformation to all immediate children of the current schema element (for the imdb element from Figure 1, these would be all movie, series and actor elements). The once traversal applies a given transformation exactly once somewhere inside a schema representation at an arbitrary depth, by traversing the schema in a top-down approach. Using all, we can define the everywhere combinator that traverses a schema representation in a bottom-up fashion and applies the given transformation to all its descendants. The `outermost` traversal performs top-down exhaustive rule application and can be defined at the cost of `once`.

To control the application of certain rules, it is useful to identify locations inside schemas. The `at` combinator applies a given rule if the name of the current element matches a given XML element tag[2]. On the other hand, `when` takes the name of an XML Schema element and performs the following pattern matching: if the given element name is defined as a top-level element in the source schema, it converts its structure into a type-level predicate; then, if the predicate succeeds when applied to the current top-level element in the input schema (such that its structure matches the structure of the pattern element) it applies the argument rule, otherwise rule application fails. Other local combinators inspired in *Focal* [10] are: `hoist` that untags the current element, `plunge` that names a new XML element and `rename` that renames the current element.

As a language for defining views of schemas, *Multifocal* also supports specific abstraction combinators. To delete part of a schema, we simply call `erase` at the appropriate location. So far, our language builds generic transformations that describe the explicit changes that are performed on the source schema. An alternative way to specify generic programs is to perform queries that traverse arbitrary structures to collect values of a specific type, as in for example the XPath language for selecting particular nodes from XML documents. To apply an XPath query to a schema, we invoke the `select` combinator that attempts to bidirectionalize the XPath expression by converting it into a lens transformation that abstracts the schema into the desired result type.

As an example, the evolution scenario from Section 2 can be encoded as the following *Multifocal* transformation:

```
everywhere (try (at "series" erase))
>> everywhere (try (at "movie" (
 outermost (when "reviews" (
  select "count(//comment)" >> plunge "@popularity"))
 >> outermost (when "boxoffices" (
  select "sum(//@value)" >> plunge "@profit")))))
>> everywhere (try (at "actor" (
 outermost (at "played" (select "award/@name" >> all (rename "awname")))))))
```

This transformation deletes `series` elements by applying an `erase` (constrained by `at`) everywhere in the source schema, and the `popularity` and `profit` attributes are calculated using XPath queries (constrained by `when`) and tagged with `plunge`. The list of award names of an `actor` are selected with another XPath query, and such resulting `name` elements are renamed to `awname` by applying `rename` within the `all` traversal. In this transformation, the `reviews` and `boxoffices` tags used by the `when` combinator denote top-level XML Schema elements (Figure 4) that must be defined in the source XML Schema. They match lists of elements named `review` and `boxoffice` (using the schema representations introduced in Section 4, they denote the types $[\mu_{review}F]$ and $[\mu_{boxoffice}G]$, for arbitrary functors F and G), respectively.

---

[2] As in XPath, XML node names preceded by an ampersat "@" denote attributes.

```
<xs:group name="reviews"><xs:sequence>
  <xs:element name="review" minOccurs="0" maxOccurs="unbounded"/>
</xs:sequence></xsd:group>
<xs:group name="boxoffices"><xs:sequence>
  <xs:element name="boxoffice" minOccurs="0" maxOccurs="unbounded"/>
</xs:sequence></xs:group>
```

Fig. 4. XML Schema top-level elements modeling specific type patterns

The general architecture of our framework is illustrated in Figure 5. A two-level transformation defined as a *Multifocal* expression is executed in two stages: first, it is evaluated as a type-level transformation by applying it to a source XML Schema, producing a target XML Schema and a bidirectional lens; second, the lens is compiled into an executable file that can be used to propagate updates between XML documents conforming to the source and target schemas. In our

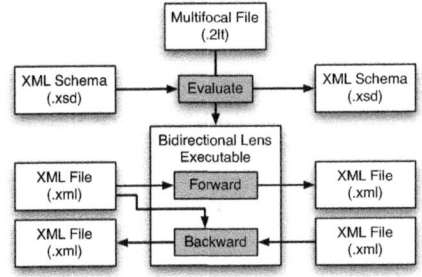

Fig. 5. Architecture of the *Multifocal* framework

scenario, optimization is done at the second stage: we optimize the value-level lenses once for each input schema and generate optimized executables that efficiently propagate updates between XML documents.

4   Implementation

This section unveils the implementation of the *Multifocal* framework in Haskell. Haskell is a general-purpose functional programming language with strong static typing, where structures are modeled by algebraic data types and programs are written as well-typed functions through pattern matching on their input values. This embedding is supported by front-ends that translate XML Schemas and XML documents into Haskell types and values, and vice-versa. A more technical description of similar XML-Haskell front-ends can be found in previous work [2].

Two-level transformations written in *Multifocal* are translated into a core library of Haskell combinators that operate on Haskell type representations. After translating the source XML Schema into an Haskell type, the framework applies the type-level transformation to produce as output a target type and a lens representation as Haskell values. From these, it generates a target XML Schema and an Haskell executable file containing the lens transformation and the data type declarations that represent all the source and target XML elements. The main function of this file parses XML documents complying to the schemas, converts them into internal Haskell values, runs the lens transformation either in the forward or backward direction to propagate source-to-target or target-to-source updates, and finishes by pretty-printing an updated XML document.

$$\circ : (B \rhd C) \to (A \rhd B) \to (A \rhd C) \qquad\qquad id : A \rhd A$$
$$\times : (A \rhd C) \to (B \rhd D) \to (A \times B) \rhd (C \times D) \qquad \pi_1 : A \times B \rhd A$$
$$+ : (A \rhd C) \to (B \rhd D) \to (A + B) \rhd (C + D) \qquad \pi_2 : A \times B \rhd A$$
$$\triangledown : (A \rhd C) \to (B \rhd C) \to (A + B) \rhd C \qquad\qquad\ ! : A \rhd 1$$
$$(\!(g)\!)_F : (F\ A \rhd A) \to (\mu F \rhd A) \qquad\qquad\qquad out_F : \mu F \rhd F\ (\mu F)$$
$$[\![g]\!]_F : (A \rhd F\ A) \to (A \rhd \mu F) \qquad\qquad\qquad in_F : F\ (\mu F) \rhd \mu F$$
$$map : (A \rhd B) \to ([A] \rhd [B]) \quad concat : [[A]] \rhd [A] \quad length : [A] \rhd Int$$
$$filter\_l : [A + B] \rhd [A] \qquad\qquad\qquad\qquad\qquad\qquad filter\_r : [A + B] \rhd [B]$$

Fig. 6. Point-free lens combinators

Encoding of Schemas and Lenses. In Haskell, sums + and products × correspond to xs:sequence and xs:choice elements in XML Schema notation. Primitives include base types, such as the unit type 1, integers *Int* or strings *String*, and lists [A] of values of type A that model XML sequences. To accommodate recursive schemas, we represent user-defined types (denoting XML elements) as fixpoints $\mu_{tag} F$ of a polynomial functor F, for a given name tag^3. A polynomial functor is either the identity *Id* (for recursive invocation), the constant $\underline{A}$, the lifting of sums $\oplus$, products $\otimes$ and lists [ ] or the composition of functors $\odot$. For example, the top-level element of the non-recursive schema from Figure 1 is represented as $\mu_{imdb}((([] \odot (\mu_{movie}M \oplus \mu_{series}S)) \otimes ([] \odot \mu_{actor}A))$, where M, S and A are the functors of the movie, series and actor elements. Application of a polynomial functor F to a type A yields an isomorphic sum-of-products type F A.

In our framework, bidirectionality is achieved by defining the value-level semantics of our two-level programs according to the point-free lens language developed in [22] and summarized in Figure 6. Each of these lens combinators possesses a *get* and a *put* function satisfying the bidirectional properties from Definition 1[4]. Fundamental lenses are identity (*id*) and composition (∘). The $!, \pi_1$ and π_2 combinators project away parts of a source type, while $\triangledown$ applies two lenses alternatively to distinct sides of a sum. The × and + combinators map two lenses to both sides of a pair or a sum, respectively. The out_F and in_F isomorphisms expose and encapsulate the top-level structure of an inductive type with functor F. The well-known fold $(\!(\cdot)\!)_F$ and unfold $[\![\cdot]\!]_F$ recursion patterns recursively consume and produce values of an inductive type. In this paper, we treat some typical operations over lists such as mapping, concatenation, length and filtering as primitive lenses. Their recursive definitions can be found in [23].

---

[3] This is actually one of many possible representations of algebraic data types for use in generic programming. For a detailed discussion see [13].

[4] In [22], some of the lens combinators admit additional parameters to control value generation. In this paper, we substitute such parameters with suitable defaults.

$nop, erase :: Rule$
$nop\ a = return\ (id, a)$
$erase\ a = return\ (!, 1)$

$at :: String \rightarrow Rule \rightarrow Rule$
$at\ name\ r\ a@(\mu_{tag}f)\ |\ name \equiv tag = r\ a$
$at\ name\ r\ a = mzero$

$all :: Rule \rightarrow Rule$
$all\ r\ Int = return\ (id, Int)$
$all\ r\ [a] = \mathbf{do}\ (l, b) \leftarrow r\ a$
$\qquad\qquad\quad return\ (map\ l, [b])$
$all\ r\ \mu_{tag}f = \mathbf{do}\ (l, g) \leftarrow allF\ r\ f$
$\qquad\qquad\quad return\ (\langle\!| in_g \circ l |\!\rangle_f, \mu_{tag}g)$
...

$allF :: Rule \rightarrow RuleF$
$allF\ r\ Id = return\ (id, Id)$
$allF\ r\ [] = return\ (id, [])$
$allF\ r\ \underline{a} = \mathbf{do}\ (l, b) \leftarrow r\ a$
$\qquad\qquad\quad return\ (l, \underline{b})$
$allF\ r\ (f \otimes g) = \mathbf{do}\ (l1, h) \leftarrow allF\ r\ f$
$\qquad\qquad\qquad\quad (l2, i) \leftarrow allF\ r\ g$
$\qquad\qquad\qquad\quad return\ (l1 \times l2, h \otimes i)$
...

$everywhere\ r = all\ (everywhere\ r) \gg r$

$once :: Rule \rightarrow Rule$
$once\ r\ Int = r\ Int$
$once\ r\ [a] = r\ [a]\ `mplus`$
$\qquad\qquad \mathbf{do}\ (l, b) \leftarrow once\ r\ a$
$\qquad\qquad\quad return\ (map\ l, [b])$
$once\ r\ a@(\mu_{tag}f) = r\ a\ `mplus`$
$\qquad\qquad \mathbf{do}\ (l, g) \leftarrow onceF\ r\ f$
$\qquad\qquad\quad return\ (\langle\!| l \circ out_f |\!\rangle_g, \mu_{tag}g)$
...

$\mathbf{type}\ RuleF = Fctr \rightarrow Maybe\ (Lens, Fctr)$

$onceF :: Rule \rightarrow RuleF$
$onceF\ r\ Id = mzero$
$onceF\ r\ [] = \mathbf{do}\ (l, g\ \bullet) \leftarrow r\ [\bullet]$
$\qquad\qquad\quad return\ (l, g)$
$onceF\ r\ (f \otimes g) =$
$\quad \mathbf{do}\ (l, h\ \bullet) \leftarrow r\ ((f \otimes g)\ \bullet)$
$\quad\quad return\ (l, h)$
$\quad `mplus`\ \mathbf{do}\ (l, h) \leftarrow onceF\ r\ f$
$\qquad\qquad\qquad return\ (l \times id, h \otimes g)$
$\quad `mplus`\ \mathbf{do}\ (l, i) \leftarrow onceF\ r\ g$
$\qquad\qquad\qquad return\ (id \times l, f \otimes i)$
...

$outermost\ r = many\ (once\ r)$

Fig. 7. Encoding of some strategic combinators as Haskell rewrite rules

Two-Level Lens Transformations. Multifocal combinators can be encoded as rewrite rules that, given a source type representation, yield a lens representation and a target type representation:[5]:

$$\mathbf{type}\ Rule = Type \rightarrow Maybe\ (Lens, Type)$$

In our implementation, types and lenses are represented as values of type *Type* and *Lens* (a grammar for lenses built using the combinators from Figure 6). The *Maybe* Haskell type models partiality of rule application: *return* denotes successful application, failure is signaled with *mzero* and *mplus* implements left-biased choice. Figure 7 presents the encoding of some combinators, namely the fundamental *all* and *once* that traverse inside the functorial structure of types.

The *all* traversal applies an argument rule to all children of the current type and has the most interesting behavior for user-defined types: it invokes the auxiliary rule *allF* that propagates a rule application down to the constants, where it applies the argument rule, and returns a lens transformation (wrapped as a rewrite rule *RuleF* on functor representations *Fctr*); then, it constructs

---

[5] For a clearer presentation, we encode types and transformations with unconstrained data representations. Our actual implementation follows a type-safe encoding inspired in [5], such that the conformity between all the artifacts (schemas, documents and transformations) is enforced by the Haskell type system.

a bottom-up lens (fold) that recursively applies the lens transformation to all values of the recursive type. The once traversal applies an argument rule exactly once at an arbitrary depth in a top-down approach, and stops as soon as the argument rule can be successfully applied. To be able to apply normal type rules inside a functor, the auxiliary rule *onceF* flattens the functor by applying it to a special type mark •. When the argument rule can be successfully applied, it infers a new functor representation using • to remember the recursive invocations[6]. For recursive types, the resulting lens performs a top-down traversal (unfold) that applies the value-level transformations of the argument rule to each recursive.

Other combinators for processing user-defined types are: hoist that unpacks an user-defined type by applying *out* at the value-level; plunge that constructs a new (non-recursive) user-defined type by applying *in* at the value-level; and rename that renames an existing user-defined type and is coupled to the *id* lens. Notice that rename n is different from hoist >> plunge n, since rename works for all data types, and plunge can only create non-recursive ones.

The erase combinator deletes the current top-level type, by replacing it with the unit type and applying ! at the value-level. In order to bundle a XPath query as a two-level transformation, the select combinator specializes it for the input type and then tries to lift the specialized expression into a lens. We specialize XPath expressions by translating them into generic point-free programs than can be optimized to non-generic point-free functions using the techniques from [8,6]. To lift the resulting functions into lenses, we check if their point-free expressions are defined using only the point-free lens combinators from Figure 6, otherwise rule application fails.

Schema Normalization. To keep a minimal suite of combinators, our language supports abstractions through the erase combinator, that deletes elements locally and thus leaves "dangling" unit types in the target schema. However, these empty unnamed types are unintended and may yield XML Schemas that are deemed ambiguous by many XML processors. For example, when applying our running *Multifocal* transformation to the IMDb schema from Figure 1, deleting series inside imdb elements will result in a list $[\mu_{movie}M + 1]$. Such dangling unit types have no representation in the XML side and must be deleted from the target schema representation. Such deletion is performed by a *normalize* procedure that removes these and other ambiguities, by exhaustively applying the rules from Figure 8[7]. Normalization is silently applied by extending the all and once traversals so that they apply *normalize* after rewriting.

Lens Optimization. Although the lens transformations generated by our framework are instantiated for particular source and target schemas, they still contain many redundant computations and traverse the whole structures, as a consequence of being a two-level transformation. To improve their efficiency, we reuse

---

[6] Unlike in the pseudo-code from Figure 7, in our implementation functor inference must be performed as a separate procedure and not simply via pattern matching.

[7] The exact lens definitions of $id \triangledown nil$ and $nil \triangledown id$ can be found in [23].

$$\pi_1 : A \times 1 \vartriangleright A \qquad \pi_2 : 1 \times A \vartriangleright A \qquad \text{-- Products}$$
$$id \bigtriangledown nil : [A] + 1 \vartriangleright [A] \qquad nil \bigtriangledown id : 1 + [A] \vartriangleright [A] \qquad \text{-- Sums}$$
$$filter\_l : [A + 1] \vartriangleright [A] \qquad filter\_r : [1 + A] \vartriangleright [A] \qquad \text{-- Lists}$$
$$id \bigtriangledown id : A + A \vartriangleright A \qquad concat : [[A]] \vartriangleright [A] \qquad \text{-- Ambiguous types}$$

Fig. 8. Rules for normalization of XML Schema representations

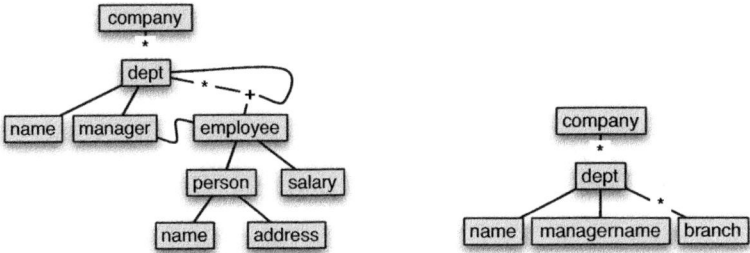

Fig. 9. A company hierarchized payroll XML schema inspired in [17]

Fig. 10. A view of the company schema

a rewrite system for the optimization of point-free lenses [23] (similar to the one for the optimization of XPath queries) that employs powerful algebraic point-free laws for fusing and cutting redundant traversals. After rewriting, the resulting transformations work directly between the source and target types and are significantly more efficient, as demonstrated in Section 5. In our framework, we provide users with the option to optimize the generated bidirectional programs at the time of generation of the Haskell bidirectional executable, if they intend to repeatedly propagate updates between XML documents conforming to the same schemas. This could be the case, for example, when the schemas represent the configuration of a live system that replies to frequent requests. In such cases, the once-a-time penalty of an additional optimization phase for a specific schema is amortized by a larger number of executions.

5   Application Scenarios

We now demonstrate two XML evolution scenarios (the IMDb example from Section 3 and another example for the evolution of a recursive XML Schema), and compare the performance of the lenses resulting from the execution of the two-level transformations with their automatically optimized definitions.

A classical schema used to demonstrate strategic programming systems is the so called "paradise benchmark" [17]. Suppose one has a recursive XML Schema to model a company with several departments, each having a name, a manager and a collection of employees or sub-departments, illustrated in Figure 9. Our second evolution example consists in creating a view of this schema according to the following transformation:

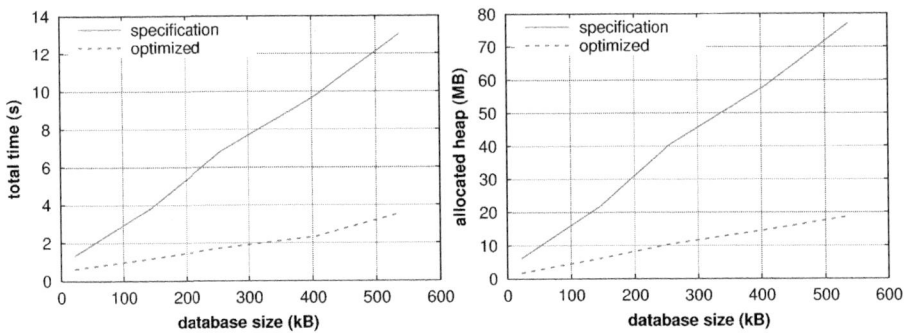

Fig. 11. Benchmark results for the *IMDb* example

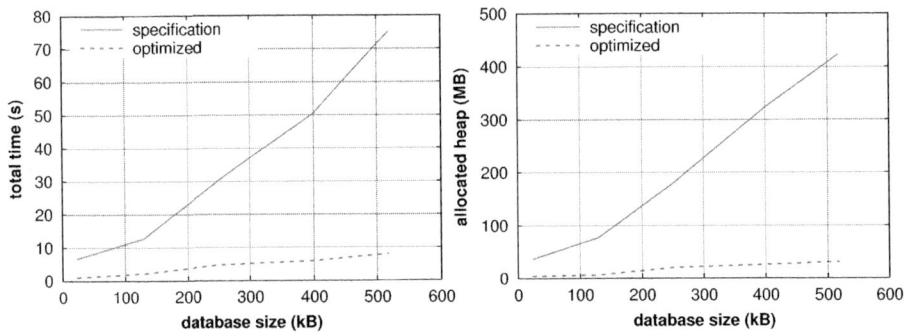

Fig. 12. Benchmark results for the *paradise* example

```
everywhere (try (at "manager" (
  all (select "(//name)[1]") >> rename "managername")))
>> everywhere (try (at "employee" erase))
>> once (at "dept" (hoist >> outermost (at "dept" (
  select "name" >> rename "branch")) >> plunge "dept"))
```

For each top-level department, this transformations keeps only the names of managers (renamed to managername), deletes all employees and collects the names of direct sub-departments renamed to branch. The resulting non-recursive schema is depicted in Figure 10. There are some details worth noticing. First, it is easier to keep only manager names using a generic query instead of a transformation that would need to specify how to drop the remaining structure. The XPath filter "[1]" guarantees a sole result if multiple names existed under managers. Second, since dept is a recursive type, we unfold its top-level recursive structure once using hoist to be able to process sub-branches, and create a new non-recursive dept element with plunge.

Performance Analysis. Unfortunately, the lenses resulting from the above transformations are not very efficient. For instance, in the IMdb example the traversals over `series`' movies and `actors` are independent and can be done in parallel. Also, the transformations of `reviews` and `boxoffices` and the extra normalizing step that filters out unit types (resulting from erased `series` elements) can be fused into a single traversal. For the paradise example, all the three steps and the extra normalization step (for erased `employee` elements) can be fused into a single traversal. Also, the first two steps, that traverse all departments recursively due to the semantics of `all` (invoked by `everywhere`) for recursive types, are deemed redundant for sub-departments by the last step.

All these optimizations can be performed by our lens optimization phase. We have measured space and time consumption of the lenses generated by our two examples, and the results are presented in Figures 11 and 12. To quantify the speedup achieved by the optimizations, we have compared the runtime behavior of their backward transformations for non-optimized (*specification*) and optimized lens definitions (*optimized*)[8]. To factor out the cost of parsing and pretty-printing XML documents, we have tested the *put* functions of the lenses with pre-compiled input databases of increasing size (measured in kBytes needed to store their Haskell definitions), randomly generated with the *QuickCheck* testing suite [4]. We compiled each function using `GHC 7.2.2` with optimization flag `02`. As expected, the original specification performs much worse than the optimized lens, and the loss factor grows with the database size. Considering the biggest sample, the loss factors are of 3.7 in time and 4.1 in space for the IMDb example and of 9.4 in time and 13.4 in space for the paradise example. The more significant results (and the worse overall performance) for the paradise example are justified by the elimination of the recursive traversals over sub-departments.

6   Related Work

In [18], Lämmel *et al* propose a systematic approach to XML schema evolution, where the XML-based formats are transformed in a step-wise fashion and the transformation of coupled XML documents can be largely induced from the schema transformations. They study the properties of such transformations and identify categories of XML schema evolution steps, taking into account many XML-specific issues, but do not propose a formalization or implementation of such a general framework for two-level transformation. The X-Evolution system [11] provides a graphical interface for the evolution of XML Schemas coupled with the adaptation of conforming XML documents. Document migration is automated for the cases when minimal document changes can be inferred from the schema evolution steps, while user intervention through query-based adaptation techniques is required to appropriately handle more complex schema changes.

Two famous bidirectional languages for XML are XSugar [3] and biXid [15], that describe XML-to-ASCII and XML-to-XML mappings, respectively. In both, bidirectional transformations are specified using pairs of intertwined grammars describing the source and target formats, from which a forward transformation

---

[8] Note that parsed XPath expressions are already optimized in the non-optimized lens, since their successful "lensification" depends on their specialization.

is obtained by parsing according to the rules in one grammar and a backward transformation by pretty printing according to the rules in the other. However, while the emphasis of XSugar is on bijective transformations, biXid admits ambiguity in the transformations and only postulates that translated documents shall be consistent up to the grammar specification.

The *Focal* lens language [10] provides a rich set of lens combinators, from general functional programming features (composition, mapping, recursion) to tree-specific operations (splitting, pruning, merging) for the transformation of tree-structured data. In [20], Liu *et al* propose Bi-X, a functional lens language closely resembling the XQuery Core language that can serve as the host language for the bidirectionalization of XQuery. The main feature of Bi-X is its support for variable binding, allowing lenses that perform implicit duplication. Using variable referencing, structure-shy combinators such as XPath's descendant axis can also be translated into equivalent Bi-X programs. Although their development is done in an untyped setting, they define a type system of regular expressions that is used to refine backward behavior.

In previous work [2], we proposed a two-level bidirectional transformation framework (2LT) implemented in Haskell for the strategic refinement of XML Schemas into SQL databases. Later [7], we showed how point-free program calculation can be used to optimize such bidirectional programs and how these can be combined to provide structure-shy query migration. In this paper, we tackle the dual problem of XML schema abstraction, supporting the execution and optimization of coupled bidirectional lenses. While refinement scenarios are inherent to strategic rewriting techniques, like the automatic mapping of abstract schemas to more concrete ones [2], view definition scenarios typically involve more surgical steps that abstract or preserve specific pieces of information and motivate a different language of primitive evolution steps. Contrarily to the 2LT framework, where two-level transformations are written within Haskell using a combinator library, we propose the *Multifocal* XML transformation language, mixing strategic and specific XML transformers, to write "out of the box" views of XML Schemas. Another new feature of our approach is the specification and optimization of generic two-level transformations over recursive XML Schemas.

7   Conclusion

In this paper we have proposed *Multifocal*, a generic two-level bidirectional transformation language for XML Schema evolution with document-level migrations based on the bidirectional framework of lenses. By using strategic programming techniques, these coupled transformations can be specified in a concise and generic way, mimicking the typical coding pattern of XML transformation languages such as XSLT, that allow to easily specify how to modify only selected nodes via specific templates. When applied to input schemas, our schema-level transformations produce new schemas, as well as bidirectional lens transformations that propagate updates between old and new documents. In our framework, we release such bidirectional transformations as independent programs that can be used to translate updates for particular source and target schemas. We also provide users with an optional optimization phase that improves the efficiency of the generated lens programs for intensive usage scenarios.

Our framework has been fully implemented in Haskell, and is available through the Hackage package repository (`http://hackage.haskell.org`) under the name `multifocal`. It can be used both as a stand-alone tool for XML Schema evolution and as a combinator library for the two-level bidirectional evolution of arbitrary inductive data type representations.

Although our language already supports combinators in the style of XSLT transformations and XPath queries, the expressiveness of the underlying bidirectional transformations is naturally limited by the language of point-free lenses in use. That said, the translation of some XPath features that are not perfect abstractions, such as value-level filtering, is not considered in our approach. In future work, we plan to extend this language to support more XPath features. This would require, however, to loosen either the round-tripping laws or the requirement that lens functions must be totally defined for documents conforming to the schemas. Other directions for future work that we are investigating are: (i) the processing of annotations in the input XML Schemas such as `xs:key` to identify reorderable chunks in the source document and provide extra alignment information to guide the translation of view updates in the style of [1]; (ii) the leveraging of the underlying bidirectional framework from asymmetric lenses to other symmetric formulations such as [12] that guarantee weaker properties but do not impose a particular abstract-or-refine data flow.

In this work, we propose a way of replacing three unidirectional XML transformations (a schema-level transformation and two transformations between XML documents) with a single two-level bidirectional *Multifocal* transformation. In order to bring *Multifocal* closer to standard XML transformation tools, we plan to develop translations from XSLT-like idioms to *Multifocal*. For a successful integration, a comparative study on the usefulness, expressiveness and efficiency of *Multifocal* transformations would be needed.

Acknowledgments. This work is funded by the ERDF through the programme COMPETE and by the Portuguese Government through FCT (Foundation for Science and Technology), project reference FCOMP-01-0124-FEDER-020532.

References

1. Barbosa, D.M.J., Cretin, J., Foster, J.N., Greenberg, M., Pierce, B.C.: Matching lenses: alignment and view update. In: ICFP 2010, pp. 193–204. ACM (2010)
2. Berdaguer, P., Cunha, A., Pacheco, H., Visser, J.: Coupled Schema Transformation and Data Conversion for XML and SQL. In: Hanus, M. (ed.) PADL 2007. LNCS, vol. 4354, pp. 290–304. Springer, Heidelberg (2006)
3. Brabrand, C., Møller, A., Schwartzbach, M.I.: Dual syntax for xml languages. Information Systems 33, 385–406 (2008)
4. Claessen, K., Hughes, J.: QuickCheck: a lightweight tool for random testing of Haskell programs. In: ICFP 2000, pp. 268–279. ACM (2000)

5. Cunha, A., Oliveira, J.N., Visser, J.: Type-Safe Two-Level Data Transformation. In: Misra, J., Nipkow, T., Karakostas, G. (eds.) FM 2006. LNCS, vol. 4085, pp. 284–299. Springer, Heidelberg (2006)
6. Cunha, A., Pacheco, H.: Algebraic specialization of generic functions for recursive types. ENTCS 229(5), 57–74 (2011)
7. Cunha, A., Visser, J.: Strongly typed rewriting for coupled software transformation. ENTCS 174(1), 17–34 (2007)
8. Cunha, A., Visser, J.: Transformation of structure-shy programs with application to xpath queries and strategic functions. Science of Computer Programming 76(6), 512–539 (2011)
9. Czarnecki, K., Foster, J.N., Hu, Z., Lämmel, R., Schürr, A., Terwilliger, J.F.: Bidirectional Transformations: A Cross-Discipline Perspective. In: Paige, R.F. (ed.) ICMT 2009. LNCS, vol. 5563, pp. 260–283. Springer, Heidelberg (2009)
10. Foster, J.N., Greenwald, M.B., Moore, J.T., Pierce, B.C., Schmitt, A.: Combinators for bidirectional tree transformations: A linguistic approach to the view-update problem. TOPLAS 29(3), 17 (2007)
11. Guerrini, G., Mesiti, M.: X-evolution: A comprehensive approach for xml schema evolution. In: DEXA 2008, pp. 251–255. IEEE (2008)
12. Hofmann, M., Pierce, B.C., Wagner, D.: Symmetric lenses. In: POPL 2011, pp. 371–384. ACM (2011)
13. Holdermans, S., Jeuring, J., Löh, A., Rodriguez, A.: Generic Views on Data Types. In: Yu, H.-J. (ed.) MPC 2006. LNCS, vol. 4014, pp. 209–234. Springer, Heidelberg (2006)
14. Hu, Z., Mu, S.-C., Takeichi, M.: A programmable editor for developing structured documents based on bidirectional transformations. Higher Order and Symbolic Computation 21(1-2), 89–118 (2008)
15. Kawanaka, S., Hosoya, H.: Bixid: a bidirectional transformation language for xml. In: ICFP 2006, pp. 201–214. ACM (2006)
16. Lämmel, R.: Coupled Software Transformations (Extended Abstract). In: 1st International Workshop on Software Evolution Transformations (2004)
17. Lämmel, R., Jones, S.P.: Scrap your boilerplate: A practical design pattern for generic programming. In: TLDI 2003, pp. 26–37. ACM (2003)
18. Lämmel, R., Lohmann, W.: Format Evolution. In: RETIS 2001, vol. 155, pp. 113–134. OCG (2001)
19. Lämmel, R., Visser, J.: A Strafunski Application Letter. In: Dahl, V. (ed.) PADL 2003. LNCS, vol. 2562, pp. 357–375. Springer, Heidelberg (2002)
20. Liu, D., Hu, Z., Takeichi, M.: Bidirectional interpretation of xquery. In: PEPM 2007, pp. 21–30. ACM (2007)
21. Mu, S.-C., Hu, Z., Takeichi, M.: An Algebraic Approach to Bi-directional Updating. In: Chin, W.-N. (ed.) APLAS 2004. LNCS, vol. 3302, pp. 2–20. Springer, Heidelberg (2004)
22. Pacheco, H., Cunha, A.: Generic Point-free Lenses. In: Bolduc, C., Desharnais, J., Ktari, B. (eds.) MPC 2010. LNCS, vol. 6120, pp. 331–352. Springer, Heidelberg (2010)
23. Pacheco, H., Cunha, A.: Calculating with lenses: optimising bidirectional transformations. In: PEPM 2011, pp. 91–100. ACM (2011)
24. Visser, E.: Stratego: A Language for Program Transformation Based on Rewriting Strategies System Description of Stratego 0.5. In: Middeldorp, A. (ed.) RTA 2001. LNCS, vol. 2051, pp. 357–361. Springer, Heidelberg (2001)

Bidirectional Transformation
of Model-Driven Spreadsheets*

Jácome Cunha[1], João P. Fernandes[1,2], Jorge Mendes[1],
Hugo Pacheco[1], and João Saraiva[1]

[1] HASLab / INESC TEC, Universidade do Minho, Portugal
[2] Universidade do Porto, Portugal
{jacome,jpaulo,jorgemendes,hpacheco,jas}@di.uminho.pt

Abstract. Spreadsheets play an important role in software organizations. Indeed, in large software organizations, spreadsheets are not only used to define sheets containing data and formulas, but also to collect information from different systems, to adapt data coming from one system to the format required by another, to perform operations to enrich or simplify data, etc. In fact, over time many spreadsheets turn out to be used for storing and processing increasing amounts of data and supporting increasing numbers of users. Unfortunately, spreadsheet systems provide poor support for modularity, abstraction, and transformation, thus, making the maintenance, update and evolution of spreadsheets a very complex and error-prone task.

We present techniques for model-driven spreadsheet engineering where we employ bidirectional transformations to maintain spreadsheet models and instances synchronized. In our setting, the business logic of spreadsheets is defined by *ClassSheet* models to which the spreadsheet data conforms, and spreadsheet users may evolve both the model and the data instances. Our techniques are implemented as part of the MDSheet framework: an extension for a traditional spreadsheet system.

Keywords: Software Evolution, Data Evolution, Bidirectional Transformation, Model Synchronization, Spreadsheets.

1 Introduction

Spreadsheets are widely used in the development of business applications. Spreadsheet systems offer end users a high level of flexibility, making initiation easier for new users. This freedom, however, comes at a price: spreadsheets are notoriously error prone as shown by numerous studies reporting that up to 90% of real-world spreadsheets contain errors [21].

* This work is funded by the ERDF through the Programme COMPETE and by the Portuguese Government through FCT - Foundation for Science and Technology, projects ref. PTDC/EIA-CCO/108613/2008 and PTDC/EIA-CCO/120838/2010. The three first authors were also supported by FCT grants SFRH/BPD/73358/2010, SFRH/BPD/46987/2008 and BI4-2011PTDC/EIA-CCO/108613/2008, respectively.

Z. Hu and J. de Lara (Eds.): ICMT 2012, LNCS 7307, pp. 105–120, 2012.
© Springer-Verlag Berlin Heidelberg 2012

In recent years, the spreadsheet research community has recognized the need to support end-user *model-driven spreadsheet development* (MDSD), and to provide spreadsheet developers and end users with methodologies, techniques and the necessary tool support to improve their productivity. Along these lines, several techniques have been proposed, namely the use of templates [1], *ClassSheet* models [10] and class diagrams [14]. These proposals guarantee that end users can safely edit their spreadsheets and introduce a form of model-driven software development: they allow to define a spreadsheet business model from which a customized spreadsheet application holding the actual data is generated. The consistency of the spreadsheet data with the overlying model is guaranteed, often by limiting the editing options on the data side to ensure that the structure of the spreadsheet remains unchanged.

A significant drawback of such approaches lies in the fact that the evolution of both spreadsheet models and the instances generated from them is considered in isolation. That is to say that, after obtaining a spreadsheet instance from a particular model, a simple evolution step on the model side may break the conformity with its instance, and vice versa. A first attempt to overcome this limitation was proposed in [7], where it was shown how to co-evolve spreadsheet instances upon a model evolution defined according to a well-behaved set of possible transformations. The approach presented in [7], however, has two important drawbacks: *i)* the evolutions that are permitted at the model level can only be refinement steps: it is not possible to perform model evolutions that are frequent in spreadsheets such as removing a column, for example; and *ii)* it does not allow users to directly evolve spreadsheet instances having the corresponding model automatically co-evolved.

The goal of this paper is to study a more general setting where editing operations on models can be translated into conforming editing operations on spreadsheets and editing operations on spreadsheets can be translated into respective editing operations on models. For this purpose, we develop independent editing languages for both models and spreadsheets and bind them together using a symmetric bidirectional framework [9, 16] that handles the edit propagation. Among other properties, the fundamental laws governing the behavior of such bidirectional transformations guarantee that the conformity of spreadsheet instances and models can always be restored after a modification. Both the model and instance evolution steps are available as an extension of OpenOffice.

2   *ClassSheets* as Spreadsheet Models

Erwig *et al.* [10] introduced the language of *ClassSheets* to model spreadsheets at a higher abstraction level, thus allowing for spreadsheet reasoning to be performed at the conceptual level. *ClassSheets* have a visual representation very similar to spreadsheets themselves: in Figure 1, we present a possible model for a **Budget** spreadsheet, which we adapted from [10].[1]

---

[1] We assume colors are visible in the digital version of this paper.

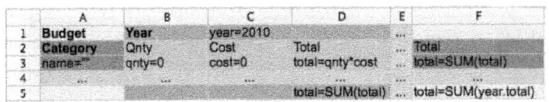

Fig. 1. Budget spreadsheet model

This model holds two classes where data is to be inserted by end users: *i)* **Year**, with a default value of 2010, for the budget to accommodate multi-year information and *ii)* **Category**, for assigning a label to each expense. The actual spreadsheet may hold several repetitions of any of these elements, as indicated by the ellipsis. For each expense we record its quantity and its cost (with 0 as default value), and we calculate the total amount associated with it. Finally, (simple) summation formulas are used to calculate the global amount spent per year (cell D5), the amount spent per expense type in all years (cell F3) and the total amount spent in all years (cell F5) are also calculated.

Erwig *et al.* not only introduced *ClassSheets*, but they also developed a tool - the Gencel tool [12] - that given a *ClassSheet* model generates an instance (*i.e.* a concrete spreadsheet) that conforms to the model. Figure 2 presents a possible spreadsheet as generated by Gencel given the *ClassSheet* shown in Figure 1 (and after the end user manually introduced soma data). In this particular case, the spreadsheet is used to record the annual budget for travel and accommodation expenses of an institution.

	A	B	C	D	E	F	G	H	I
1	Budget	Year	2010		Year	2011		...	
2	Category	Qnty	Cost	Total	Qnty	Cost	Total		Total
3	Travel	2	320	640	7	420	2940	...	3580
4	Accommodation	5	140	700	8	185	1480	...	2180
5	...	...	...		...	...	...		
6				1340			4420	...	5760

Fig. 2. Budget spreadsheet instance

Since the spreadsheet is generated using all the information in the model, it is able of providing some correctness guarantees: formulas are kept consistent while new years are added, for example. Note also that, throughout the years, cost and quantity are registered for two types of expenses: travel and accommodation, and that formulas are used to calculate the total expense amounts.

Spreadsheet Evolution. At the end of 2011, the spreadsheet of Figure 2 needs to be modified to accommodate 2012 data. Most spreadsheet users would typically take four steps to perform this task: *i)* insert three new columns; *ii)* copy all the labels ("Year", "Qnty", "Cost" and "Total"); *iii)* copy all the formulas (to compute the total amount spent per expense type in 2012, and the total expense for that same year) and *iv)* update all the necessary formulas in the last column to account for the new year information. More experienced users would possibly shortcut these steps by copy-inserting, for example, the 3-column block of 2011 and changing the label "2011" to "2012" in the copied block. Still, the range of the multi-year totals must be manually extended to include the new year information. In any (combination) of these situations, a conceptually unitary edition, *add year*, needs to be executed via an error-prone combination of steps.

This is precisely the main advantage of model-driven spreadsheet development: it is possible to provide unitary transformations such as the addition of class instances (e.g., a year or a category) as one-step procedures, while all the structural impacts of such transformations are handled automatically (e.g., the involved formulas being automatically updated). This advantage is exploited to its maximum when the model and the instance are part of the same spreadsheet development environment, as it was proposed for OpenOffice in [5].[2] Besides automation, it is also guaranteed that this type of instance level operations does not affect the model-instance conformity binding.

There are, however, several situations in which the user prefers to change a spreadsheet instance (or a particular model) in such a way that, after the edit, it will no longer conform to the previously defined model (or the respective instance). For example, if the user wants to add a column containing a possible expense discount for a particular year only, this is a trivial operation to perform at the data level which is actually not simple to perform at the model level. Therefore choosing to evolve the original spreadsheet, we may obtain the one given in Figure 3, which no longer conforms to the model of Figure 1 (the discount was added in column K).

	A	B	C	D	E	F	G	H	I	J	K	L	M	N
1	Budget	Year	2010		Year	2011		...	Year	2012				
2	Category	Qnty	Cost	Total	Qnty	Cost	Total	...	Qnty	Cost	Discount	Total	...	Total
3	Travel		2	320	640	7	420	2940	...	1	420	0.1	378	...	3958
4	Accommodation		5	140	700	8	185	1480	...	8	185	0	1480	...	3660
5	...		...	...	...	...	...	...	...	...			...	...
6				1340			4420	...				1858	...	7618

Fig. 3. Budget spreadsheet instance with an extra column

One possible model evolution that regains conformity is shown in Figure 4: a class for years with 3 columns is kept, while a new class for years with an extra discount column is introduced.

	A	B	C	D	E	F	G	H	I	J	K
1	Budget	Year	year=2010		...	Year	year=2010				...
2	Category	Qnty	Cost	Total	...	Qnty	Cost	Discount	Total		Total
3	name=""	qnty=0	cost=0	total=qnty*cost	...	qnty=0	cost=0	disc=0	total=qnty*cost*(1-disc)	...	total=SUM(total)
4	...		...		...	...				...	...
5				total=SUM(total)	...				total=SUM(total)	...	total=SUM(year.total)

Fig. 4. Budget spreadsheet model with an extra column

In the remainder of this paper, we study the evolution of spreadsheet models and instances in a systematic way. As a result of our work, we present a bidirectional framework that maintains the consistency between a model and its instance. By being bidirectional, it supports either manually evolving the spreadsheet instances, as we have described in this section, or editing the model instead. In any case, the correlated artifact is automatically co-evolved, so that their conformity relationship is always preserved.

---

[2] Actually, Figure 1 and Figure 2 present a *ClassSheet* model and a spreadsheet instance as defined in the embedding of *ClassSheets* in spreadsheets [5].

3   Spreadsheet Evolution Environment

This section presents a bidirectional spreadsheet evolution environment. This environment combines the following techniques:

- firstly, we embed *ClassSheet* models in a spreadsheet system. Since the visual representation of *ClassSheets* very much resembles spreadsheets themselves, we have followed the traditional embedding of a domain specific language (*ClassSheets*) in a general purpose programming language (spreadsheet system). In this way, we can interact with both the models and the instances in the same environment as described in [4, 7].
- secondly, we construct a framework of bidirectional transformations for *Class-Sheets* and spreadsheet instances, so that a change in an artifact is automatically reflected to its correlated one. This framework provides the usual end-user operations on spreadsheets like adding a column/row or deleting a column/row, for example. These operations can be realized in either a model or an instance, and the framework guarantees the automatic synchronization of the two. This bidirectional engine, that we describe in detail in the next section, is defined in the functional programming language HASKELL [18].
- finally, we extend the widely used spreadsheet system Calc, which is part of OpenOffice, in order to provide a bidirectional model-driven environment to end-users. Evolution steps at the model and the instance level are available as new buttons that extend the functionalities originally built-in the system. A script in OpenOffice Basic was developed to interpret the evolution steps and to make the bridge with the HASKELL framework. An OpenOffice extension is available at the SSaaPP project web page: http://ssaapp.di.uminho.pt/.

In Figure 5, we present an overview of the bidirectional spreadsheet environment that we propose. On the left, the embedded *ClassSheet* is presented in a Model worksheet while the data instance that conforms to it is given on the right, in a Data worksheet. Both worksheets contain buttons that perform the evolution steps at the model and instance levels.

Every time a (model or instance) spreadsheet evolution button is pressed, the system responds automatically. Indeed, it was built to propagate one update at a time and to be highly interactive by immediately giving feedback to users. A global overview of the system's architecture is given in Figure 6.

Also, our bidirectional spreadsheet engine makes some natural assumptions on the models it is able to manipulate and restricts the number of operations that are allowed on the instance side.

Model Evolution Steps. On the model side, we assume the *ClassSheet* on which an editing operation is going to be performed is well-formed. By being well-formed we mean a model respecting the original definition of *ClassSheets* [10], where all references made in all formulas are correctly bound. Also, a concrete model evolution operation is only actually synchronized if it can be applied to the initial model and the evolved model remains well-formed; otherwise, the

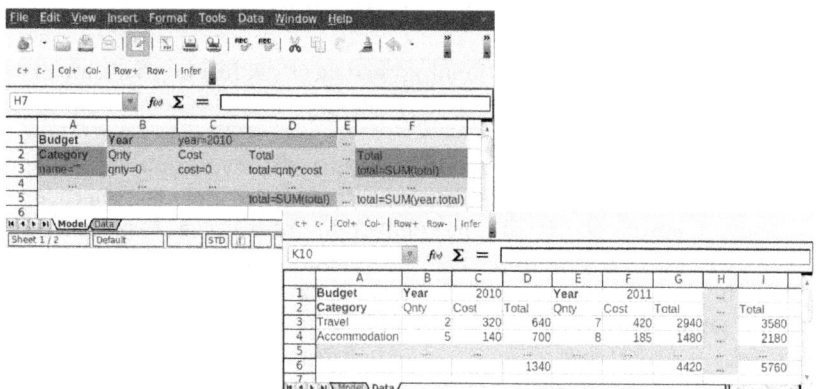

Fig. 5. A bidirectional model-driven environment for the budget spreadsheet

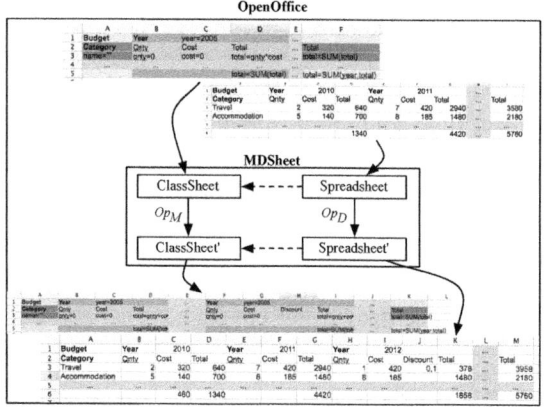

Fig. 6. Architecture of the MDSheet addon for OpenOffice

operation is rejected and the original model remains unchanged. The intuition behind this is that operations which cannot be applied to one side should not be put to the other side, but rejected by the environment. A similar postulate is made in [16].

For example, an operation that would produce an ill-formed *ClassSheet* model occurs when removing the cost column in Figure 1. In this case, the formula that computes the total expense amount per year would point to a non-existing cost value. This is precisely one of the cases that our system rejects.

Data Evolution Steps. The success of spreadsheets comes from the simplicity and freedom that they provide to end-users. This freedom, however, is also one of the main causes of errors in spreadsheets. In our evolution environment, we need to restrict the number of operations that a user may request. The reason for this is the following: for any supported operation, we guarantee that there exists a concrete model that the evolved instance will conform to; and for this we need to reduce the huge number of operations that is available in a spreadsheet system such as Calc, so that we can ensure model-instance conformity at all times.

As an example of an operation on an instance that we are not able to propagate to its conforming model is the random addition of data rows and columns. Indeed, if such an edit disrespects the structure of the original spreadsheet, we will often be unable to infer a new model to which the data conforms. Therefore, the operations (such as addColumn) that may affect the structure of a spreadsheet instance need to be performed explicitly using the corresponding button of our framework. The remaining editing operations are performed as usually.

4   The MDSheet Framework

In this section, we present our framework for bidirectional transformations of spreadsheets. The implementation is done using the functional programming language HASKELL and we will use some of its notation to introduce the main components of our framework. After defining the data types that encode models and instances, we present two distinct sets of operations over models and instances. We then encode a bidirectional system providing two unidirectional transformations *to* and *from* that map operations on models into operations on instances, and operations on instances to operations on models, respectively. Figure 7 illustrates our bidirectional system:

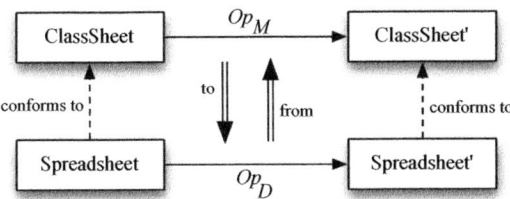

Fig. 7. Diagram of our spreadsheet bidirectional transformational system

Given a spreadsheet that conforms to a *ClassSheet* model, the user can evolve the model through an operation of the set Op_M, or the instance through an operation of the set Op_D. The performed operation on the model (data) is then transformed into the corresponding operation on the data (model) using the *to* and *from* transformations, respectively. A new model and data are obtained with the new data conforming to the new model.

4.1   Specification of Spreadsheets and Models

The operations defined in the next sections operate on two distinct, but similar, data types. The first data type, named *Model*, is used to store information about a model. It includes the list of classes that form the model and a grid (a matrix)

that contains the definition of the *ClassSheet* cells. The second type, *Data*, is used to store information about model instances, *i.e.*, the spreadsheet data. It also stores a list of classes and a matrix with the cell contents. The definition of these data types is as follows:

data *Model = Model* { *classes* :: [*ModelClass*], *grid* :: *Grid* }
data *Data = Data* { *classes* :: [*DataClass*], *grid* :: *Grid* }

The difference between the two data types lies in the kind of classes used. Models define classes with repetitions, but do not have repetitions *per se*. However, the data can have several instances, and that information is stored within the class. For a class, we store its name, the position of its top-left and bottom-right points (respectively *tl* and *br* in the data structure below) and the kind of expansion of the class.

data *ModelClass = ModelClass* {
    classname :: *String*
    , *tl*   :: (*Int*, *Int*)
    , *br*   :: (*Int*, *Int*)
    , *expansion* :: *Expansion* () }

data *DataClass = DataClass* {
    classname :: *String*
    , *tl*   :: (*Int*, *Int*)
    , *br*   :: (*Int*, *Int*)
    , *expansion* :: *Expansion Int* }

In *DataClass*, the number of instances is stored in the *expansion* field. In *ModelClass*, the *expansion* is used to indicate the kind of expansion of a class. It is possible to store the number of instances for horizontal and for vertical expansions, and for classes that expand both horizontally and vertically. It is also possible to represent static classes (*i.e.*, that do not expand).

Having introduced data types for *ClassSheet* models and spreadsheet instances, we may now define operations on them. The next two sections present such operations.

4.2 Operations on Spreadsheet Instances

The first step in the design of our transformational system is to define the operations available to evolve the spreadsheets. The grammar shown next defines the operations the MDSheet framework offers.

data Op_D : *Data* → *Data* =
    $addColumn_D$ *Where Index*   -- add a column
    | $delColumn_D$ *Index*   -- delete a column
    | $addRow_D$ *Where Index*   -- add a row
    | $delRow_D$ *Index*   -- delete a row
    | $AddColumn_D$ *Where Index*   -- add a column to all instances
    | $DelColumn_D$ *Index*   -- delete a column from all instances
    | $AddRow_D$ *Where Index*   -- add a row to all instances
    | $DelRow_D$ *Index*   -- delete a row from all instances
    | $replicate_D$ *ClassName Direction Int Int*   -- replicate a class
    | $addInstance_D$ *ClassName Direction Model*   -- add a class instance
    | $setLabel_D$ (*Index*, *Index*) *Label*   -- set a label
    | $setValue_D$ (*Index*, *Index*) *Value*   -- set a cell value

| $SetLabel_D$ | $(Index, Index)$ $Label$ | -- set a label in all instances |
| $SetValue_D$ | $(Index, Index)$ $Value$ | -- set a cell value in all instances |

To each entry in the grammar corresponds a particular function with the same arguments. The application of an update $op_D : Op_D$ to a data instance $d : Data$ is denoted by op_D $d : Data$.

The first operation, $addColumn_D$, adds a column in a particular place in the spreadsheet. The *Where* argument specifies the relative location (*Before* or *After*) and the given *Index* defines the position where to insert the new column. This solves ambiguous situations, like for example when inserting a column between two columns from distinct classes. The behavior of $addColumn_D$ is illustrated in Figure 8.

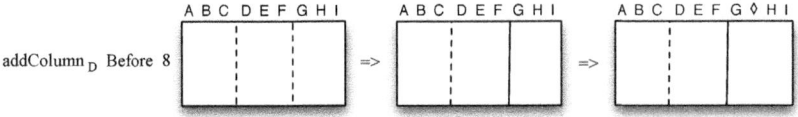

Fig. 8. Application of the data operation $addColumn_D$

In an analogous way, the second operation, $delColumn_D$, deletes a given column of the spreadsheet. The operations $addRow_D$ and $delRow_D$ behave as $addColumn_D$ and $delColumn_D$, but work on rows instead of on columns. An operation in all similar to $addColumn_D$ ($delColumn_D$, $addRow_D$ and $delRow_D$) is $AddColumn_D$ ($DelColumn_D$, $AddRow_D$ and $DelRow_D$). This operation is in fact a mapping of $addColumn_D$ over all instances of a class: it adds a column to each instance of an expandable class. The operation $replicate_D$ allows to replicate (or duplicate) a class, with the two last integer arguments being the number of instances of the class provided as first argument and the number of the instance to replicate, respectively. This operation will be useful for our bidirectional transformation functions, and will be explained in more detail later. The operation $addInstance_D$ performs a more complex evolution step: it adds a new instance of a given class to the spreadsheet. For the example illustrated in Figure 2, it could be used to add a new instance of the year class. The operations described so far work on sets of cells (*e.g.* rows and columns), whereas the last two Op_D operations, $setLabel_D$ and $setValue_D$, work on a single cell. The former allows to set a new label to a given cell while the latter allows to update the value of a cell. Versions of these last two operations that operate on all instances are also available: $SetLabel_D$ and $SetValue_D$, respectively.

When adding a single column to a particular instance with several columns, the chosen instance becomes different than the others. Therefore, this operation is based on two steps: firstly, the chosen instance is separated from the others (note that the second dashed line becomes a continuous line); secondly, a new column, indicated by ◊, is inserted in the specified index. This operation can be used to evolve the data of the budget example as suggested at the end of Section 2 (and illustrated in Figure 3).

4.3   Operations on Models

In this section we present the operations that allow transformations on the model side. The grammar shown next declares the existing operations:

data $Op_M : Model \rightarrow Model =$
$addColumn_M$	*Where Index*	-- add a new column
$\mid delColumn_M$	*Index*	-- delete a column
$\mid addRow_M$	*Where Index*	-- add a new row
$\mid delRow_M$	*Index*	-- delete a row
$\mid setLabel_M$	*(Index, Index) Label*	-- set a label
$\mid setFormula_M$	*(Index, Index) Formula*	-- set a formula
$\mid replicate_M$	*ClassName Direction Int Int*	-- replicate a class
$\mid addClass_M$	*ClassName (Index, Index) (Index, Index)*	-- add a static class
$\mid addClassExp_M$	*ClassName Direction (Index, Index) (Index, Index)*	
		-- add an expandable class

As it occurred with Op_D for instances, the Op_M grammar represents the functions operating on models. The application of an update $op_M : Op_M$ to a model $m : Model$ is denoted by $op_M\ m : Model$.

The first five operations are analogous to the data operations with the same name. New operations include $setFormula_M$ which allows to define a formula on a particular cell. On the model side, a formula may be represented by an empty cell, by a default plain value (*e.g.*, an integer or a date) or by a function application (*e.g.*, cell F5 in Figure 2 is defined as the result of SUM(Year.total)). The operation $replicate_M$ allows to replicate (or duplicate) a class. This will be useful for our bidirectional transformation functions. The last two operations allow the addition of a new class to a model: $addClass_M$ adds a new static (non-expandable) class and $addClassExp_M$ creates a new expandable class. The *Direction* parameter specifies if it expands horizontally or vertically.

To explain the operations on models, we present in Figure 9 an illustration of the execution of the composition of two operations: firstly, we execute an addition of a row, $addRow_M$ (where the new row is denoted by $\Diamond$); secondly, we add a new expandable class ($addClassExp_M$) constituted by columns B and C (denoted by the blue rectangle and the grey column labeled with the ellipsis).

addRow$_M$ Before 3; addClassExp$_M$ "BlueClass" Horizontal (2,1) (3,4)

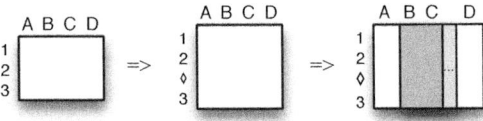

Fig. 9. Application of the model operations $addRow_M$ and $addClassExp_M$

The first operation, $addRow_M$, adds a new row to the model between rows 2 and 3. The second operation, $addClassExp_M$, adds a new class to the model. As a first argument, this operation receives the name of the class, which is "Blue-Class" in this case. Its second argument specifies if the class expands vertically or horizontally (the latter in this case). The next two arguments represent the upper-left and the right-bottom indexes limiting the class, respectively. The last argument is the model itself.

These operations allow to evolve models like the one presented in Figure 1. However, the problem suggested in Figures 3 and 4, where the data evolves as a result of an end-user operation and the model automatically co-evolves, is not yet handled by transformations we presented so far. In the next section, we give a bidirectional setting where co-evolution is automatic.

4.4 Bidirectional Transformation Functions

In this section, we present the bidirectional core of our spreadsheet transformation framework. In our context, a bidirectional transformation is defined as a pair of unidirectional transformations with the following signature:

$$to \quad : Model \times Op_M \rightarrow Op_D^\star$$
$$from : Data \quad \times Op_D \rightarrow Op_M^\star$$

Since these transformations are used in an online setting, in which the system reacts immediately to each user modification, the general scheme of our transformations is to transform a single update into a sequence of updates. That said, the forward transformation to propagates an operation on models to a sequence of operations on underlying spreadsheets, and the backward transformation $from$ propagates an operation on spreadsheets to a sequence of operations on overlying models. We denote a sequence of operations on models $op^\star$ as being either the empty sequence $\emptyset$, an unary operation op, or a sequence of operations $op_1^\star; op_2^\star$. Our transformations take an additional model or instance to which the original modifications are applied. This is necessary because most of the operations calculate the indexes of the cells to be updated based on the operation itself, but also based on the previous model or spreadsheets, depending on the kind of operation.

We now instantiate the to and $from$ transformations for the operations on models and instances defined in the previous two sections. We start by presenting the to transformation: it receives an operation on models and returns a sequence of operations on data. Although, at least currently, the translation of model operations returns an empty or singleton sequence of data operations, we keep its signature consistent with $from$ to facilitate its understanding and evidence the symmetry of the framework.

$$
\begin{aligned}
&to : Op_M \rightarrow Op_D^\star \\
&to\ (addColumn_M \quad w\ i \qquad\qquad) = AddColumn_D\ w\ (columnIndex_D\ i) \\
&to\ (delColumn_M \quad w\ i \qquad\qquad) = DelColumn_D \qquad (columnIndex_D\ i) \\
&to\ (addRow_M \qquad w\ i \qquad\qquad) = AddRow_D \qquad w\ (rowIndex_D\ i) \\
&to\ (delRow_M \qquad w\ i \qquad\qquad) = DelRow_D \qquad (rowIndex_D\ i) \\
&to\ (setLabel_M \qquad (i,j)\ l \qquad\ ) = SetLabel_D\ (position_D\ (i,j))\ l \\
&to\ (setFormula_M \quad (i,j)\ f \qquad\ ) = SetValue_D\ (position_D\ (i,j))\ f \\
&to\ (replicate_M \qquad cn\ dir\ n\ inst) = replicate_D\ dir\ cn\ n\ inst \\
&to\ (addClass_M \qquad cn \qquad p_1\ p_2 \quad) = \emptyset \\
&to\ (addClassExp_M\ cn\ dir\ p_1\ p_2 \quad) = \emptyset
\end{aligned}
$$

The first five model operations have a direct data transformation, that is, a transformation with the same name which does the analogous operation. An interesting transformation is perfomed by $replicate_M$: it duplicates a given class, but not the data. Instead, a new empty instance is added to the newly created class. Another interesting case is the transformation of the model operation $addClass_M$. This model transformation does not have any impact on data instances. Thus, to returns an empty set of data transformations. In fact, the same happens with the $addClassExp_M$ operation.

We now present the *from* transformation, which maps data operations into model operations:

$$from : Op_D \to Op_M^\star$$
$$from\ (addColumn_D\ w\ i) =$$
$$\quad replicate_M\ className\ Horizontal\ classInstances\ instanceIndex_M$$
$$\quad ; addColumn_M\ w\ columnOffsetIndex_M$$
$$from\ (delColumn_D\ i) =$$
$$\quad replicate_M\ className\ Horizontal\ classInstances\ instanceIndex_M$$
$$\quad ; delColumn_M\ columnOffsetIndex_M$$
$$from\ (addRow_D\ w\ i) =$$
$$\quad replicate_M\ className\ Vertical\ classInstances\ rowIndex_M$$
$$\quad ; addRow_M\ w\ rowOffsetIndex_M$$
$$from\ (delRow_D\ i) =$$
$$\quad replicate_M\ className\ Vertical\ classInstances\ rowIndex_M$$
$$\quad ; delRow_M\ rowOffsetIndex_M$$
$$from\ (setLabel_D\ (i,j)\ l) =$$
$$\quad replicate_M\ className\ Horizontal\ classInstances\ columnIndex_M$$
$$\quad ; replicate_M\ className\ Vertical\quad classInstances\ rowIndex_M$$
$$\quad ; setLabel_M\ positionOffset_M\ l$$
$$from\ (setValue_D\ (i,j)\ l\qquad) = \emptyset$$
$$from\ (addInstance_D\ cn\ dir\ m) = \emptyset$$

The transformations in this case are more complex than the model-to-data ones. In fact, most of them produce a sequence of model operations. For instance, the first transformation ($addColumn_D$) results in the replication of a class followed by the addition of a new column. The argument *classInstances* is actually a function that calculates the number of data class instances based on the data to be evolved. On the other hand, the operation to set a value of a particular cell, $setValue_D$, does not have any impact on the model. The same happens to the operation $addInstance_D$ which adds a new instance of an expandable class. The definition of *from* for the (non-empty) data operations in the range of *to* (*e.g.*, $AddColumn_D, DelRow_D$) is simply the inverse of *to*.

4.5 Bidirectional Transformation Properties

Since the aim of our bidirectional transformations is to restore the conformity between instances and models, a basic requirement is that they satisfy *correctness* [22] properties entailing that propagating edits on spreadsheets or on models leads to consistent states:

$$\frac{d :: m}{(to\ m\ op_M)\ d :: op_M\ m}$$
to-CORRECT

$$\frac{d :: m}{op_D\ d :: (from\ d\ op_D)\ m}$$
from-CORRECT

Here, we say that a data instance d conforms to a model m if $d :: m$, for a binary consistency relation $(::) \subseteq Data \times Model$.

As most interesting bidirectional transformation scenarios, spreadsheet instances and models are not in bijective correspondence, since multiple spreadsheet instances may conform to the same model, or vice versa. Another bidirectional property, *hippocraticness* [22], postulates that transformations are not allowed to modify already consistent pairs, as defined for *from*:

$$\frac{d :: m \quad op_D\ d :: m}{from\ d\ op_D = \emptyset}$$
from-HIPPOCRATIC

Reading the above law, if an operation on data op_D preserves conformity with the existing model, then *from* produces an empty sequence of operations on models $\emptyset$. Operationally, such a property is desired in our framework. For example, if a user adds a new instance to an expandable class, preserving conformity, the original model is expected to be preserved because it still reflects the structure of the data. However, hippocracticness for *to* is deemed too strong because even if the updated model is still consistent with the old data, we still want to update the data to reflect a change in the structure, making a better match with the model [8]. For example, if the user adds a column to the model, the intuition is to insert also a new column to the data, even if the old data remains consistent.

Usually, bidirectional transformations are required to satisfy "round-tripping" laws that ensure a certain degree of *invertibility* [9, 13]. In our application, spreadsheet instances refine spreadsheet models, such that we can undo the translation of an operation on models with an application of *from* (except when the operation on models only concerns layout, such as $addClass_M$, and is not reflexible on the data):

$$\frac{to\ m\ op_M = op_D^\star \quad op_D^\star \neq \emptyset \quad d :: m}{from^\star\ d\ op_D^\star = op_M}$$
to-INVERTIBLE

However, the reverse implication is not true. For example, the transformation steps *from* $addColumn_D = replicate_M; addColumn_M$ and $to^\star$ ($replicate_M$; $addColumn_M$) = $replicate_D; AddColumn_D$ do not give equal data operation.

Since an operation on data may issue a sequence of operations on models, we introduce the transformation $to^\star$ which applies *to* to a sequence of operations:

$$\overline{to^\star\ \emptyset = \emptyset} \qquad \overline{to^\star\ op_M = to\ op_M} \qquad \frac{to^\star\ op_{M1}^\star = op_{M1}^\star \quad to^\star\ op_{M2}^\star = op_{M2}^\star}{to^\star\ (op_{M1}^\star; op_{M2}^\star) = to^\star\ op_{M1}^\star; to^\star\ op_{M2}^\star}$$

A dual definition can be given for $from^\star$. As common for incremental transformations (that translate edits rather than whole states), our sequential transformations naturally satisfy an *history ignorance* [8] property meaning that the translation of consecutive updates does not depend on the update history.

5 Related Work

Our bidirectional approach is inspired in the state-based bidirectional framework of (constraint) maintainers [20, 22], where (correct and hippocratic) forward and backward transformations propagate modifications from source to target models, and vice-versa, while preserving a consistency relation that establishes a relationship between them. However, our formulation is closer to operation-based symmetric bidirectional frameworks [9, 16]. The framework of symmetric delta lenses from [9] generalizes maintainers to transformations that operate over deltas as high-level representations of updates. Like [16], our transformations carry a more operational feeling as they transform the actual operations on data and models. Our bidirectional transformations also satisfy a similar totality law guaranteeing that if an operation does not fail on the initiating side, then the transformed sequence of operations also succeeds. For example, given a consistent state $d :: m$, propagating a data operation over d always generates operations on *ClassSheet* models that can be applied to the original model m.

A group of researchers from Tokyo developed a series of bidirectional approaches for the interactive development of XML documents [17, 23]. Similarly to our OpenOffice assisted environment, they assume an online setting where the editor reacts immediately to one operation at a time. In their setting, instead of preserving an explicit consistency relation, transformations obey one-and-a-half round-tripping laws (in the style of *to* ∘ *from* ∘ *to* = *to*) to ensure that after each modification the editor converges into a consistent state, *i.e.*, a further transformation does not alter the related documents.

The coupled evolution of metamodels (let it be grammars, schemas, formats, etc) and conforming models is a typical problem in MDE. Works such as [15, 19] assess the degree of automation of metamodel-model evolution scenarios by studying categories of metamodel modifications that are model-independent or support the co-evolution of underlying models which need to be transformed in order to become conforming to an updated version of their original metamodel. Existing tools for automated coupled metamodel-model evolution may either require users to specify sequences of simple that describe how to evolve a source metamodel into a new version [24], or assume that a new metamodel is provided externally so that the system must user model difference approaches to identify the concrete metamodel changes [3]. Both [24] and [3] support typical metamodel operations such as renaming, addition and deletion and many others to manipulate particular object-oriented features. The *to* transformation proposed in this paper tackles an instance of this problem (concerned with translating a single modification at a time), with metamodels as classsheets and models as spreadsheets. In our bidirectional setting, the *from* transformation tackles another dual but less common coupled model-metamodel evolution problem.

In [4–7], the authors introduced tools to transform spreadsheets into relational databases, and more generically, to evolve a model and automatically co-evolve the underlying data. This work, however, has some limitations: first, it does not allow users to perform non-refinement evolutions, i.e., it is not possible to remove data from spreadsheets. In our work we created a more general setting where all kinds of evolutions are possible, including the deletion of data. Second, it is not possible to evolve the structure of the spreadsheet through changes to the data, i.e., it is only possible to edit the data in such a way that it always conforms to the model. We have solved this problem by allowing users to change the data and infer a new model whenever necessary. Third, the previous work propagates modified states into new states. This work propagates editing operations themselves, and thus allows for more efficient incremental transformations.

The first approach to deliver model-driven engineering to spreadsheet users, *Gencel* [11], generates a new spreadsheet respecting a previously defined model. In this approach, however, there is no connection between the stand alone model development environment and the spreadsheet system. As a result, it is not possible to (automatically) synchronize the model and the spreadsheet data, that is, the co-evolution of the model (instance) and its instance (model) is not possible, unless it is done by hand, which is very error prone and thus not desirable. In our work we present a solution for this problem.

6   Conclusions and Future Work

In this paper we have presented a bidirectional model-driven spreadsheet environment. We constructed a bidirectional framework defining usual end-user operations on spreadsheet data and on *ClassSheet* models that always guarantee synchronization after an evolution step at the data or model level. We have created an extension to the OpenOffice Calc spreadsheet system so that it offers a model-driven software development environment. The developed spreadsheet evolution environment allows: the generation of a spreadsheet instance from a *ClassSheet* model; the evolution of the model and the automatic co-evolution of the data; the evolution of the data and the automatic co-evolution of the model.

The techniques we present propose the first bidirectional setting for the evolution of spreadsheet models and instances. Our research efforts, however, have thus far considered standalone, non-concurrent spreadsheet development only. In a computing world that is growingly distributed, developing spreadsheets is often performed in a collaborative way, by many actors. As part of our plans for future research, we have already engaged in trying to extend our work in this paper to a distributed environment.

Although there exists some empirical evidence that an approach to spreadsheet development based on models can sometimes be effective in practice [2], the global environment envisioned in this paper still lacks a concrete empirical analysis. In this line, a study with real spreadsheet users is under preparation.

References

1. Abraham, R., Erwig, M., Kollmansberger, S., Seifert, E.: Visual specifications of correct spreadsheets. In: VL/HCC, pp. 189–196. IEEE Computer Society (2005)
2. Beckwith, L., Cunha, J., Fernandes, J.P., Saraiva, J.: End-users productivity in model-based spreadsheets: An empirical study. In: IS-EUD, pp. 282–288 (2011)
3. Cicchetti, A., Ruscio, D.D., Eramo, R., Pierantonio, A.: Automating co-evolution in model-driven engineering. In: EDOC, pp. 222–231. IEEE CS (2008)

4. Cunha, J., Fernandes, J.P., Mendes, J., Saraiva, J.: MDSheet: A Framework for Model-driven Spreadsheet Engineering. In: ICSE, pp. 1412–1415. ACM (2012)
5. Cunha, J., Mendes, J., Fernandes, J.P., Saraiva, J.: Embedding and evolution of spreadsheet models in spreadsheet systems. In: VL/HCC 2011, pp. 179–186. IEEE (2011)
6. Cunha, J., Saraiva, J., Visser, J.: From spreadsheets to relational databases and back. In: PEPM, pp. 179–188. ACM, New York (2009)
7. Cunha, J., Visser, J., Alves, T., Saraiva, J.: Type-Safe Evolution of Spreadsheets. In: Giannakopoulou, D., Orejas, F. (eds.) FASE 2011. LNCS, vol. 6603, pp. 186–201. Springer, Heidelberg (2011)
8. Diskin, Z.: Algebraic Models for Bidirectional Model Synchronization. In: Czarnecki, K., Ober, I., Bruel, J.-M., Uhl, A., Völter, M. (eds.) MODELS 2008. LNCS, vol. 5301, pp. 21–36. Springer, Heidelberg (2008)
9. Diskin, Z., Xiong, Y., Czarnecki, K., Ehrig, H., Hermann, F., Orejas, F.: From State- to Delta-Based Bidirectional Model Transformations: The Symmetric Case. In: Whittle, J., Clark, T., Kühne, T. (eds.) MODELS 2011. LNCS, vol. 6981, pp. 304–318. Springer, Heidelberg (2011)
10. Engels, G., Erwig, M.: ClassSheets: automatic generation of spreadsheet applications from object-oriented specifications. In: ASE, pp. 124–133. ACM (2005)
11. Erwig, M., Abraham, R., Cooperstein, I., Kollmansberger, S.: Automatic generation and maintenance of correct spreadsheets. In: ICSE, pp. 136–145. ACM (2005)
12. Erwig, M., Abraham, R., Kollmansberger, S., Cooperstein, I.: Gencel: a program generator for correct spreadsheets. J. Funct. Program 16(3), 293–325 (2006)
13. Foster, J.N., Greenwald, M.B., Moore, J.T., Pierce, B.C., Schmitt, A.: Combinators for bi-directional tree transformations: a linguistic approach to the view update problem. In: POPL, pp. 233–246. ACM (2005)
14. Hermans, F., Pinzger, M., van Deursen, A.: Automatically Extracting Class Diagrams from Spreadsheets. In: D'Hondt, T. (ed.) ECOOP 2010. LNCS, vol. 6183, pp. 52–75. Springer, Heidelberg (2010)
15. Herrmannsdoerfer, M., Benz, S., Juergens, E.: Automatability of Coupled Evolution of Metamodels and Models in Practice. In: Czarnecki, K., Ober, I., Bruel, J.-M., Uhl, A., Völter, M. (eds.) MODELS 2008. LNCS, vol. 5301, pp. 645–659. Springer, Heidelberg (2008)
16. Hofmann, M., Pierce, B.C., Wagner, D.: Edit lenses. In: POPL (to appear, 2012)
17. Hu, Z., Mu, S.-C., Takeichi, M.: A programmable editor for developing structured documents based on bidirectional transformations. HOSC 21(1-2), 89–118 (2008)
18. Jones, S.P., Hughes, J., Augustsson, L., et al.: Report on the programming language haskell 98. Tech. rep. (February 1999)
19. Lämmel, R., Lohmann, W.: Format Evolution. In: RETIS 2001. vol. 155, pp. 113–134. OCG (2001)
20. Meertens, L.: Designing constraint maintainers for user interaction (1998), manuscript available at http://www.kestrel.edu/home/people/meertens
21. Panko, R.: Spreadsheet errors: What we know. what we think we can do. EuSpRIG (2000)
22. Stevens, P.: Bidirectional Model Transformations in QVT: Semantic Issues and Open Questions. In: Engels, G., Opdyke, B., Schmidt, D.C., Weil, F. (eds.) MODELS 2007. LNCS, vol. 4735, pp. 1–15. Springer, Heidelberg (2007)
23. Takeichi, M.: Configuring bidirectional programs with functions. In: IFL (2009)
24. Vermolen, S., Visser, E.: Heterogeneous Coupled Evolution of Software Languages. In: Czarnecki, K., Ober, I., Bruel, J.-M., Uhl, A., Völter, M. (eds.) MODELS 2008. LNCS, vol. 5301, pp. 630–644. Springer, Heidelberg (2008)

Domain-Specific Optimization in Digital Forensics

Jeroen van den Bos[1,2] and Tijs van der Storm[1]

[1] Centrum Wiskunde & Informatica, Amsterdam, The Netherlands
[2] Netherlands Forensic Institute, Den Haag, The Netherlands
jeroen@infuse.org, storm@cwi.nl

Abstract. File carvers are forensic software tools used to recover data from storage devices in order to find evidence. Every legal case requires different trade-offs between precision and runtime performance. The resulting required changes to the software tools are performed manually and under the strictest deadlines.

In this paper we present a model-driven approach to file carver development that enables these trade-offs to be automated. By transforming high-level file format specifications into approximations that are more permissive, forensic investigators can trade precision for performance, without having to change source.

Our study shows that performance gains up to a factor of three can be achieved, at the expense of up to 8% in precision and 5% in recall.

1 Introduction

Digital forensics is a branch of forensic science that attempts to answer legal questions based on the analysis of information recovered from digital devices. These digital devices are typically computers or mobile phones confiscated from a suspect, found near a crime scene or otherwise expected to have information stored that is relevant to an investigation. In the context of this paper we are interested in *file carvers*: tools that recover data from storage devices without the help of (file system) storage metadata [16].

The current growth in size of storage devices requires that file carvers scale to analyze data in the terabyte range. Moreover, forensic investigations are often performed under very strict deadlines, making the runtime performance of such tools critical. Additionally, the large diversity in (variants of) file formats encountered on devices requires these tools to be easy to modify and extend.

Because each case may require different trade-offs with respect to precision and runtime performance, file carvers often need to be modified on a case-by-case basis. Currently, this kind of just-in-time "carver hacking" is performed by hand, which is error prone and time consuming; it is also inherently incompatible with very strict deadlines.

In previous work we have developed a model-driven approach to digital forensics tool construction [5]. In this work the file formats of interest, e.g., JPEG, GIF etc., are declaratively modeled using a domain-specific language (DSL) called DERRIC. These descriptions are then input to a code generator that produces

Z. Hu and J. de Lara (Eds.): ICMT 2012, LNCS 7307, pp. 121–136, 2012.

highly efficient and accurate format validators that form an essential part of our file carver EXCAVATOR.

EXCAVATOR competes with file carvers widely used in practice, and is much easier to maintain due to the high-level DERRIC language. Nevertheless, the generated components encode a particular trade-off between precision and runtime performance. In this work we apply model transformations on DERRIC descriptions in order to make this trade-off configurable. We present three model transformations that successively obtain format validators that are more permissive (i.e., produce more false positives) but exhibit better runtime performance. As a result forensic investigators can choose between precision and runtime performance without having to change any code.

We have evaluated EXCAVATOR using the different format validators at each permissiveness configuration for the file formats JPEG, GIF and PNG on a representative test image of 1TB. Our results show that performance gains up to a factor of three can be achieved, at the expense of up to 8% in precision and 5% in recall.

This paper makes the following contributions:

- We present three model transformations to automatically derive format validators that trade precision for better runtime performance.
- We evaluate our approach on a representative test image in the terabyte range showing that substantial performance gains can be achieved.

Organization of this Paper. The rest of this paper is organized as follows. Section 2 discusses file carving and analyzes the development, performance and scalability challenges in the engineering of digital forensics software. We introduce our model-driven approach to building file carvers and discuss how it addresses the challenges. This includes an overview of DERRIC, our domain-specific language (DSL) for file format description. Section 3 defines three model transformations on DERRIC descriptions. Section 4 evaluates the effect of the model transformations on the runtime performance and precision of the generated carvers. In Section 5 we discuss our results. Related work is discussed in Section 6. We summarize our research and results in Section 7.

2   Background

2.1   File Carving

When recovering data from a storage device, all available metadata such as file system records and application logs are used to identify locations where data is stored. After this initial step, there is usually a significant amount of *unallocated space* left on the storage device. This space may contain only zeros (or some other factory default value), but may also contain deleted files, operating system caches or data that has been hidden on purpose. To recover this data, a content-based technique called *file carving* can be used.

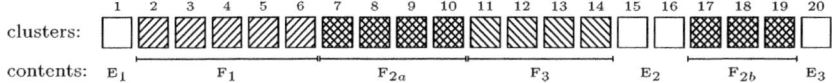

Fig. 1. An example set of contiguous clusters on a storage device

A typical modern file carver consists of a set of format validators used by one or more file reconstruction algorithms. In its most basic form the format validators consist of checking for format-specific constants at the start and end of a stream (called *header/footer matching*) and the file reconstruction algorithm simply moves through the input stream in a single pass, invoking all format validators at each offset to determine whether a file is located there. On each hit, the identified file is saved for further analysis.

Apart from generating a large amount of false positives, this approach has another drawback: it is unable to recover files that are split into multiple parts and stored in non-contiguous locations. This so-called file fragmentation is common, usually as a result of performance optimization by the operating system and implementation details of the file system.

To recover fragmented files but avoid a combinatorial explosion, file carvers implement file reconstruction algorithms, such as bifragment gap carving [10]. However, to improve precision and reduce the amount of required iterations to reconstruct a single file, they also use more advanced format validators that validate (part of) the format's structure and content.

Common optimizations include running multiple format validators on the same block of data concurrently and applying data classification techniques to reduce the search space (e.g., removing blocks of zeros). These techniques are not discussed further in this paper.

File Carving Example. An example set of contiguous clusters commonly found on storage devices is shown in Figure 1. Clusters 1, 15, 16 and 20 contain only zeros. The remaining clusters contain three files: F_1 (clusters 2–6), F_2 (fragmented, clusters 7–10 and 17–19) and F_3 (clusters 11–14).

A traditional file carver that performs a single pass over the data checking for headers and footers only will probably recover F_1, since it will find a header in cluster 2 and a correct following footer in cluster 6. Fragmented file F_2 is problematic, as the first footer following the header in cluster 7 is F_3's footer in cluster 14. As a result, both F_2 and F_3 are not recovered.

A more sophisticated format validator may detect a problem around cluster 11 or 12 and report this to the file carver. The file carver can then decide to look for suitable footers within a certain range, possibly finding both F_3's footer in cluster 14 as well as F_2's footer in cluster 19. Some shuffling of the clusters between the original error location in cluster 11 and the potential footers may lead the file carver to consider clusters 7–10 and 17–19, which the format validator will accept. From the remaining clusters, F_3 will then be easy to recover as well.

2.2 Software Engineering Challenges

From a software engineering perspective, the challenges in file carver construction can be classified into three areas, described in the following subsections.

Modifiability. Digital forensics tools must be continually adapted to new versions and variants of storage formats encountered during investigations. For instance, even when using a standardized format such as the JPEG image file format, different vendors of, for instance, digital cameras may store the actual files in different ways, often deviating from the standard. When forensic investigators encounter traces on some device that they want to recover or analyze, they often need to adapt their tools to these new, modified or different storage formats in order to maximize recoverable evidence.

Runtime Performance. Strict time constraints means that analyses must be completed as quickly as possible, even when the amount of data to analyse grows very fast. Brute force algorithms are intractable when it comes to reconstructing a file by finding its parts in a set of millions of fragments. Hence, the challenge is to use as much domain-specific knowledge as possible for optimization. This includes knowledge about hardware, operating systems, file system implementation, file formats and typical fragmentation patterns [10].

Scalability. Digital forensics tools must be scalable to deal with relatively large data sizes. Common hard drive sizes in desktop computers are already in the terabyte range. Support for these data sizes imposes additional constraints on the design and implementation of tools. Recovering evidence from a set of data of which 1% barely fits into working memory requires custom approaches. Most analyses must use a streaming architecture to collect information while reading through the data from beginning to end in a single pass.

2.3 Model-Driven Digital Forensics

To address the challenges described in the previous section, we have developed a model-driven approach to file carver construction, called EXCAVATOR. The architecture of EXCAVATOR consists of three parts and is shown in Figure 2.

The first part is a domain-specific language called DERRIC that allows file formats to be specified in a declarative way. A simplified example of a DERRIC specification of the PNG image file format is shown in Figure 3, which will be discussed in more detail below. A DERRIC file format description captures the information to be used by a file carver to recognize (fragments of) files in a data stream. DERRIC file format descriptions are tailored to digital forensics applications; they may leave out details of a file format that would be relevant for implementing a file viewer, for instance, but are not important for file carving.

The DERRIC file format descriptions are input to the second component, a code generator to obtain format validators. A format validator is used to check that a certain sequence of bytes indeed can be recognized as part of a file format.

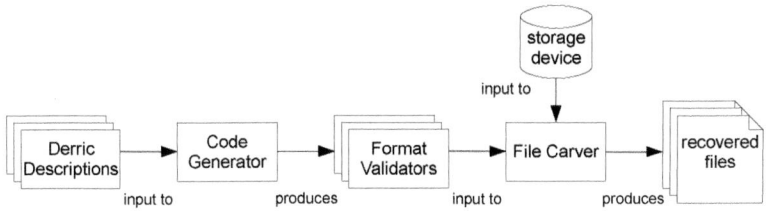

Fig. 2. Overview of the EXCAVATOR architecture

The code generator performs domain-specific optimizations to make the resulting code as efficient as possible, such as skipping over blocks of data that will not be interpreted and only generating variables for values read from the input data that will actually be referenced. Both the DERRIC DSL[1] and the EXCAVATOR code generator have been developed using RASCAL[2], a DSL for source code analysis and transformation [13]. The code generator produces Java source code.

The third part is the file carver itself, which employs dedicated algorithms and heuristics for locating candidate files in the data stream. This component uses the generated format validators to verify if a candidate file is an instance of a file format. This component can be considered the runtime system of EXCAVATOR. The runtime system is implemented in Java using the latest IO libraries for maximum throughput.

EXCAVATOR can be configured to run with or without file reconstruction capabilities. The algorithm it implements is bifragment gap carving with a configurable maximum gap size, with a default value of 2MB. It supports a variable cluster size with a default value of 4096 bytes. It does not support parallelism or filtering through data classification.

Our model-driven approach to digital forensics tool construction addresses the aforementioned challenges in the following way:

- **Modifiability.** Using high-level file format descriptions separates the "what" from the "how": if a new variant or version of a file format has to be accommodated, only the file format description has to be changed; the code generator and runtime system remain unchanged.
- **Runtime Performance.** The code generator can apply sophisticated optimizations to obtain fast code. Because this concern is now isolated in the code generator, it does not affect the description of file formats. Traditionally, optimizations in digital forensics tools are tangled with the matching logic of file format structure.
- **Scalability.** The runtime system effectively captures the way data is processed, independently from the generated validators. This means that a file carver can be made to run in streaming fashion by changing the runtime

---

[1] http://www.derric-lang.org/
[2] http://www.rascal-mpl.org/

system. Additionally, state-of-the-art file carving algorithms (e.g., [8]) can be plugged into the system without affecting the other components.

Still, there is room for improvement. Digital forensics tools are often adapted to a certain situation in order to trade quality and completeness of the results for increased performance. On the one hand, if a recovery tool produces many false positives, this may be problematic, because they all have to be inspected manually. On the other hand, this may be preferable to not having any results at all before the deadline. In order to make this trade-off configurable we can apply model transformations to DERRIC file format descriptions to obtain a faster file carver at the cost of some precision. These transformations are described in Section 3.

2.4   Example: PNG Image File Format

As an illustration of DERRIC, we present a description of a simplified version of the PNG image file format in Figure 3. It omits the details of optional data structures but is complete enough to be transformed into a validator that properly recognizes PNG files.

At the beginning of the format description, the name of the format is specified (line 1) along with a set of storage-related defaults, such as string encoding (line 2) and default numerical type (lines 3–6), in this case single-byte unsigned integers.

Next is the definition of the format's *sequence* (lines 8–11), which defines the ordering of data structures in a valid file. In this example only a single operator appears (asterisk), which specifies that the structure must appear zero or more times. Additional constructs exist such as selection (parentheses), subsequencing (square brackets), optionality (question mark) and exclusion (exclamation mark).

The final part is the *structures* block (lines 13–54), defining the structures mentioned in the *sequence*. Each structure has a name and a list of field descriptions between curly braces. For example, the *Chunk* structure on lines 18–27 has four fields: *length* (line 19), *chunktype* (line 20), *chunkdata* (line 21) and *crc* (lines 22–26).

The *Chunk* structure's fields demonstrate some of DERRIC's specification constructs. The *length* field has the length of the *chunkdata* field as value, and its type is a 32-bit unsigned integer. The *chunktype* field is four bytes in size and may contain any value except the ASCII string "IDAT". The *chunkdata* field does not specify its value but constrains that its size must correspond to the value of the *length* field. Circular references like this are common in format descriptions and are useful in situations where only part of a data structure has been recovered; each value can be used to validate the other.

Finally, the *crc* field has a fixed size of four bytes and defines a value that must be calculated using the "crc32-ieee" algorithm (line 22) using the values of the *chunktype* and *chunkdata* fields (line 25).

Additionally, DERRIC supports structure inheritance. This is shown on line 28 where the *IHDR* structure inherits the fields of the *Chunk* structure and then

```
 1 format PNG                          28 IHDR = Chunk {
 2   strings ascii                     29   chunktype: "IHDR";
 3   sign false                        30   chunkdata: {
 4   unit byte                         31     width: !0 size 4;
 5   size 1                            32     height: !0 size 4;
 6   type integer                      33     bitdepth: 1|2|4|8|16;
 7                                      34     colourtype: 0|2|3|4|6;
 8 sequence                            35     compression: 0;
 9   Signature IHDR                    36     filter: 0;
10   Chunk* IDAT IDAT* Chunk*          37     interlace: 0|1;
11   IEND                              38   }
12                                     39 }
13 structures                          40
14 Signature {                         41 IDAT = Chunk {
15   marker: 137,80,78,71,13,10,26,10; 42   chunktype: "IDAT";
16 }                                   43   chunkdata: compressed(
17                                      44              algorithm="deflate",
18 Chunk {                             45              layout="zlib",
19   length: lengthOf(chunkdata) size 4; 46            fields=chunkdata)
20   chunktype: !"IDAT" size 4;        47          size length;
21   chunkdata: size length;           48 }
22   crc: checksum(algorithm="crc32-ieee", 49
23          init="allone",start="lsb", 50 IEND {
24          end="invert",store="msbfirst", 51   length: 0 size 4;
25          fields=chunktype+chunkdata) 52   chunktype: "IEND";
26        size 4;                      53   crc: 0xAE, 0x42, 0x60, 0x82;
27 }                                   54 }
```

Fig. 3. Structure of the simplified PNG image file format

overrides the *chunktype* and *chunkdata* fields (lines 29–38). Its *length* and *crc* fields remain the same as in *Chunk*.

3   Transforming Derric Models

In order to make the trade-off between precision and runtime performance configurable we have implemented three model-transformations on DERRIC descriptions, based on an analysis of validation techniques in file carving [3]. Each transformation removes constraints so that more permissive specifications are obtained. The transformations consist of replacing computationally expensive operations with cheaper versions that resemble the original technique, or skip over data entirely instead of processing it. They can be applied successively so that in the end four format validators can be derived from a DERRIC specification. The transformations are source-to-source transformations; as a result, the generic code generator of EXCAVATOR can be reused to obtain a working format validator from each transformed description.

Using the transformations, we can distinguish four configurations of format validator precision:

- **Base:** base validator (the most precise validator, based on the complete file format description).
- **NoCA:** removal of all content analysis (e.g., removal of CRC checks, data decompression, etc.).
- **NoDD:** removal of all data dependencies (e.g., a field's value becomes undefined if it used to be equal to the contents of some other field's value).
- **Header:** removal of all matching except header and footer patterns.

Although each transformation could be applied independently, for the purpose of this paper we only consider the consecutive application of each transformation. The effect of other combinations of transformations is left as future work. The transformations are described in more detail below.

Remove Content Analysis. The most computationally expensive technique is content analysis, which is the interpretation and validation of a file's content, as opposed to matching structural metadata. For instance on lines 22–26 of Figure 3 a CRC32 over each *Chunk* of PNG data is defined using the `checksum` keyword. Additionally, lines 43–46 describe the compression scheme used by the *IDAT* structure using the `compressed` keyword. Removing these expensive analyses will reduce running time significantly at the cost of missing some fragmented files due to lower precision.

Removing content analysis consists of one of two rewrites, based on the field the content analysis is defined on:

- If the field has an externally defined size, i.e., if it has a fixed value (such as the CRC32's four bytes) or references an outside value (such as the *IDAT*'s reference to its *length* field), the field's value specification is removed. As a result, the data will be skipped over instead of processed.
- When the end of a field is specified by an end marker as part of the content analysis itself, the end marker is lifted out of the content analysis specification to be used to specify the end of the field.

More precisely, the transformation is defined by the following two rules:

$$f\colon CA(\overline{x}) \ \texttt{size} \ n; \ \Rightarrow \ f\colon \texttt{size} \ n;$$
$$f\colon CA(\overline{x}, \ \texttt{terminator=}c); \ \Rightarrow \ f\colon \texttt{terminatedBy} \ c;$$

The first rule replaces a fixed-length field f which requires content-analysis CA with a field of unknown data but of the same length. If the field f has no fixed length, but a terminator constant c is specified in the content-analysis, the content-analysis is removed, and field f is now `terminatedBy` c.

Remove Data Dependencies. The second transformation removes data dependencies. All references to values or sizes defined elsewhere in the description are removed. An example of this is the *chunkdata* field as shown on line 21 in

Figure 3 where *size* depends on the value of *length* on line 19. There are two types of data dependencies that are dealt with differently. First, if the contents of a field are defined by reference to another field, the reference is removed by clearing the content specification. The field's value becomes "undefined". The transformation rule implementing this transformation is as follows:

$$f: E[f'] \text{ size } n \ \Rightarrow \ f: \text{ size } n;$$

If the value of a fixed-length field f is defined by some expression E referencing field f', the value specification is simply removed.

Second, if the size specification of a field depends on another field, the transformation is more involved. It is not possible to clear the size specification of a field just like with value dependencies, since then the position of a following field or structure becomes undefined. Instead, we remove the entire field from its containing structure. To ensure that the generated validator still works, we locate the first field f' that defines a constant value c that is required to follow the removed field f; if s does not define such a field itself, we find the first following structure that does, using the format's sequence. We replace the definition of f' with f': `terminatedBy` c;. To prevent backtracking in the generated validator, we remove any non-mandatory structures (indicated by *, ?, and ()) inbetween f and f'. To find the first mandatory field that defines a constant, we use a simple algorithm, similar to the computation of first-sets of context-free grammars [1].

Figure 4 shows the effect of a single transformation step to remove the size dependency of the *chunktype* field of PNG's *IDAT* structure[3]. In this example the content-analysis and value dependencies have already been removed. In this step, the *chunkdata* field has been removed from *IDAT*. Additionally, the *length* field of *IEND* has been changed to include the `terminatedBy` modifier, because it is the first mandatory constant field following the removed *chunktype* field.

```
IDAT {
  length: lengthOf(chunkdata) size 4;
  chunktype: "IDAT";
  chunkdata: size length;
  crc: size 4;
}
IEND {
  length: 0 size 4
  ...
}
```

⇒

```
IDAT {
  length: size 4;
  chunktype: "IDAT";
  crc: size 4;
}
IEND {
  length: terminatedBy 0 size 4
  ...
}
```

Fig. 4. Example of *Remove Data Dependencies*

---

[3] Note that the *IDAT* structure no longer inherits from the *Chunk* structure; the inheritance hierarchy has been flattened during normalization.

```
sequence
  s e

structures
  s { header: 137, 80, 78, 71, 13, 10, 26, 10; }
  e { footer: terminatedBy 0, 0, 0, 0, "IEND", 0xAE, 0x42, 0x60, 0x82; }
```

Fig. 5. Example of *Reduce to Header/Footer*

Reduce to Header-Footer Matching. The third and last model transformation reduces a format description to two patterns: one for the beginning and one for the end of the file. This is the same strategy that is employed by the SCALPEL carver [17]. It requires file formats to have a clearly defined header and footer, using only constants. As a result, a validator based on this description will hardly ever reject data since for every header some footer is very likely to be found (assuming a large amount of files or fragments in the input data). Fragmentation in the input data will lead almost certainly to false positives. However, all recovered files are collected in a single linear pass over the input data.

The transformation operates as follows. Let S be the largest sequence of non-optional consecutive structures starting from the beginning of the sequence definition of the file format. Let E be a similar list of structures, but now starting backwards, from the end of the sequence definition. Now collapse both S and E into single structures s and e by taking the largest sequence of constant fields starting from the beginning and the end respectively, and concatenating consecutive field constants into single constants a and b. Then define the structures s and e as s { header: a; } and e { footer: **terminatedBy** b; }. Finally, construct a new file format with sequence s e. The resulting file format searches for the constant header pattern a, and (if found) subsequently searches for the constant footer pattern b.

Figure 5 shows the result of applying this transformation to the full PNG description of Figure 3. Note that all consecutive constant fields in the *IEND* structure have been merged into the single field *footer* to construct the largest possible constant.

4 Evaluation

To evaluate the effect of the transformations we have applied them on three DERRIC file format specifications, namely for JPEG, GIF and PNG. We have run the resulting $3 \times 4 = 12$ carver configurations on a representative disk image of 1TB, containing over a million recoverable files. We have then compared the difference in runtime performance, precision and recall between the configurations.

4.1   Development of Benchmark Disk Image

The largest publicly available disk image for exercising file carvers is 40GB in size[4]. This, however, is not large enough to properly assess how an application deals with scalability issues in practice. We have therefore developed our own 1TB test set based on data downloaded from Wikipedia. The size of Wikipedia means we could get enough files to fill at least a significant part of the 1TB data set we wanted to create. We used the latest available static dump of all images on Wikipedia, which dates from 2008[5]. Attempting to download all files from that list resulted in around 50% errors due to missing files. The end result was a usable set of over 1.2 million files with a total size of 357GB. An overview of how the files are distributed over each type (JPEG, GIF and PNG) and their total sizes is shown in the first column of Table 1.

These files were written into the test image file, spread out across the entire 1TB. Space between files (or fragments) was filled using 543GB of random data and 100GB of only zeros. Although there is little known about the amount and size of zero data blocks on hard drives, we believe 10% is a low estimate, which means the test image is more challenging for file carvers (since zeros are relatively easy to disqualify).

93% of the files have been written into the test image in contiguous blocks and are therefore not fragmented. 3% has been split into two parts and the remaining 4% has been divided into four equal size groups of 3, 4, 5–10 and 11–20 fragments, corresponding to observations of fragmentation in the wild [10]. Splitting was done at random locations in the files, but always on a cluster boundary of 4096 bytes, corresponding to the smallest common cluster size.

Table 1. Results per configuration for all three file formats

Format	Configuration	Running time	True positives	False positives	Precision	Recall
JPEG	Base	742m	882,511	0	100.0%	94.9%
input data:	NoCA	295m	860,022	22,007	97.5%	92.4%
total files: 930,424	NoDD	231m	837,382	46,561	94.7%	90.0%
total size: 327GB	Header	231m	837,382	46,561	94.7%	90.0%
GIF	Base	320m	34,078	0	100.0%	93.2%
input data:	NoCA	267m	33,210	702	97.9%	90.8%
total files: 36,576	NoDD	231m	32,912	2,780	92.2%	90.0%
total size: 3GB	Header	231m	32,912	2,780	92.2%	90.0%
PNG	Base	691m	222,660	0	100.0%	94.2%
input data:	NoCA	280m	219,001	8,073	96.4%	92.6%
total files: 236,457	NoDD	231m	212,911	13,905	93.9%	90.0%
total size: 27GB	Header	231m	211,790	14,577	93.6%	89.6%

---

[4] http://digitalcorpora.org/corpora/disk-images
[5] http://static.wikipedia.org/downloads/2008-06/en/images.lst

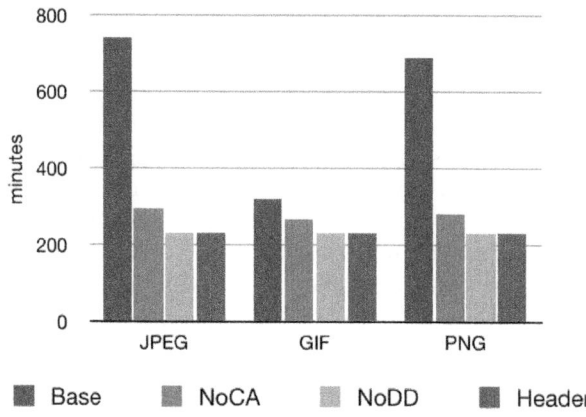

Fig. 6. Effect of each carver configuration on runtime performance

4.2   Execution of the Benchmark

The 12 carver configurations have been run on a 3.4GHz Intel Core i7-2600 with 8GB of RAM and an attached 2TB 10.000RPM SATA harddrive. The operating system used was Ubuntu Linux 11.04, with Oracle's JDK 1.6.0 update 13. The results of each run are shown in Table 1. For each file type and configuration it shows the wall clock running time in minutes in the third column. The fourth and fifth column of each table display the number of true and false positives respectively. True positive means a file has been recovered that was actually present in the disk image. False positive means that the file carver recovered a file erroneously, for instance, by combining a file header with the wrong footer. The last two columns give precision and recall percentages. An overview of the effect on runtime performance is shown graphically in Figure 6.

4.3   Analysis of Results

The fastest two configurations, NoDD and Header, require the same amount of time to complete for each format. The 231m corresponds to the time required to read through a terabyte of data on the hardware used, indicating that when using the NoDD and Header configurations, the application is bound by the read performance of the underlying platform. In other words, reading all data in a single linear pass would take the same amount of time.

Additionally, on JPEG and GIF, both the NoDD and Header configurations return exactly the same results, indicating that the final transformation does not impact the quality of the results or runtime performance. However, on PNG the situation is different: the NoDD configuration returns a little more true positives and fewer false positives.

This difference can be traced to the fact that the descriptions for JPEG and GIF both have a large variable block in the middle that is effectively eliminated by the *remove data dependencies* transformation, while the PNG description does have a fixed structure at a variable location between the first and final structure (the *IDAT* structure). This causes the PNG NoDD configuration to be more discriminating than the Header configuration. The result is slightly higher precision and recall.

For all three formats, the Base configuration returns no false positives, reaching 100% precision. The Base descriptions are complete, which leads to validation of all the contents of a candidate match. Since all three formats are compressed, even a single missing or misplaced fragment will lead to errors during validation and be rejected by the validator.

Another point of interest is the running time of the Base configuration. For JPEG and PNG, this is both at least twice the time required to run the NoCA configuration and at least three times the amount of time required to run the NoDD and Header configurations. Two factors contribute to this. The first factor is the relatively expensive operations by the validators. An example of this is CRC calculation. Although an optimized implementation is used, due to fragmentation, the CRC is sometimes calculated over large blocks that end up not being matches.

The second factor is the effect of fragment reordering in EXCAVATOR. Whenever a validator rejects a candidate match, an additional check is performed to determine whether a possible footer of the same file format is relatively close to the error location. If this is the case, the clusters between the error location and the matching footer are partially reordered and removed, running the validator on each combination to determine possible hits. To prevent a combinatorial explosion, reordering is only enabled when the distance between error location and footer is smaller than 2MB. Consequently, it is triggered by the most precise validators. In the more permissive validators the gap size is either too large or it is entirely undetected (and leads to a false positive in the results).

5 Discussion

Effects on Analysis Time. It can be argued that, although more permissive validators will run faster, in practice, they may end up requiring more of the investigator's time, because there are more false positives to inspect. This time could also be spent running the analysis using a higher precision validator. Depending on the legal case, however, it might be more valuable to have results more quickly: even with more false positives, a crucial piece of evidence could be found earlier.

With our current results we believe the transformed validators are a useful alternative to the most precise validators, since the loss of precision and recall (8% and 5% respectively) is relatively small compared to the gain in performance (between 40% and 320%). For example, for PNG, the fastest carver returns 211,790 true positives and 14,577 false positives but it requires only 1/3rd of the running time of the most precise carver.

At the same time, the fastest validators do not make the original validators obsolete, considering that, after the fastest validator has finished, the most precise JPEG validator is able to recover 45,129 true positives in the extra 510 minutes.

An alternative approach is to use the more precise validators for only a short period of time and use their intermediate results when time runs out. While this is possible, there is a chance that the more precise validator will spend a lot of time near the beginning of the disk image recovering a fragmented file, while the fastest validator (which does not reject anything) will skip over it and return all the relatively simple matches directly.

Another alternative approach is to use one of the fastest validators and run the most precise validator on the results to remove false positives. This may help all carver configurations achieve 100% precision.

Other File Formats. Our experiment takes three popular image file formats and shows how the described model transformations affect runtime performance and precision of the generated validators from their descriptions. A question is whether this approach works as well on other file formats. There is a strong indication that they will perform similarly, considering that most forensically interesting file formats tend to either be multimedia, document or container files. All three of these types of files often have features comparable to the image file types we used: extensive metadata, compressed contents and well-defined headers and footers. Examples of forensically interesting file types that are structured similarly are AVI and MPEG for multimedia, XLS and PDF for documents, and ZIP and RAR for containers. In future work we will apply EXCAVATOR and the model transformations on DERRIC descriptions of these file formats.

6   Related Work

Transformation for optimization is as old as the theory of compiler construction [2]. Moreover, transformation is considered to be one of the cornerstones of model-driven engineering [18,4] and generative programming [9]. In both areas the objective is to specify the essential variability of an application domain at high levels of abstraction, and then generating the low-level code automatically. The commonality of an application domain is captured by such transformations. We have applied this well-known pattern in the context of digital forensics.

Domain-specific analysis, verification, optimization, parallelization and transformation (AVOPT) are well-known reasons for DSL development [14]. In particular, for optimization, the explicit representation of high-level domain concepts can be used by a compiler in order to generate code that is more efficient. Such optimizations are very hard to obtain in the context of ordinary, hand-written programs, since the high-level domain concepts are lost in low-level code. In this paper we have shown how to use domain concepts of DERRIC (content analysis, data dependencies and header/footer) in order to obtain faster file carvers.

In [7] the authors present a model and strategy for transforming source code in order to reduce the energy consumption of a program. It includes an explicit cost model of both the transformations and the object program. Our transformations themselves are very inexpensive, and the cost model for file carving is based solely on the most expensive operations at runtime. Another instance of applying model transformation for optimization is presented in [6]. The authors apply a number of successive transformations on BIP (Behavior, Interaction, Priorities) models to obtain a single monolithic, efficient program. The DERRIC model transformations operate in the same way in that they remove overhead elements from the input model. What makes our transformations different from such approaches, however, is that the transformations are not (strictly) semantics preserving, as they discard information. As such the transformations can be considered approximations, in a similar way that context-free grammars can be approximated by regular expressions [15].

Our software tool EXCAVATOR represents the state-of-the-art in digital forensics data recovery, implementing fragmented file recovery [10,8] and a stream-based processing model [11]. Furthermore, our model-driven approach distinguishes itself by allowing high-level specification of elaborate data structures not implemented in popular file carvers. By comparison, PHOTOREC [12] requires handwritten format validators and SCALPEL [17] employs regular expressions for format validation.

7 Conclusion

Modifiability, runtime performance and scalability are the major challenges in digital forensics software construction. Moreover, forensic investigations are often constrained by very strict deadlines. As a result digital forensics software is often modified on a case-by-case basis. This just-in-time "carver hacking" is error prone and time consuming.

In previous work we have introduced a model-driven approach to digital forensics software development, DERRIC, which improves performance and modifiability by generating efficient code from high-level file format descriptions. In this paper we introduced three source-to-source model transformations on DERRIC descriptions in order to make the trade-off between precision and runtime performance configurable. This allows investigators to choose performance over precision if time constraints should require so, or the other way around,—without having to change any code.

The effect of the model transformations is evaluated on a 1TB disk image containing over a million recoverable files, specifically constructed to resemble a realistic file carving scenario. Our results show that performance gains up to a factor of three can be achieved. This comes at a loss of up to 8% in precision and 5% in recall.

References

1. Aho, A.V., Lam, M.S., Sethi, R., Ullman, J.: Compilers: Principles, Techniques, and Tools, 2nd edn. Prentice Hall (2006)
2. Allen, F., Cocke, J.: A Catalogue of Optimizing Transformations. In: Design and Optimization of Compilers, pp. 1–30. Prentice-Hall (1972)

3. Aronson, L., van den Bos, J.: Towards an Engineering Approach to File Carver Construction. In: 2011 IEEE 35th Annual Computer Software and Applications Conference Workshops (COMPSACW), pp. 368–373. IEEE (2011)
4. Bézivin, J.: Model Driven Engineering: An Emerging Technical Space. In: Lämmel, R., Saraiva, J., Visser, J. (eds.) GTTSE 2005. LNCS, vol. 4143, pp. 36–64. Springer, Heidelberg (2006)
5. van den Bos, J., van der Storm, T.: Bringing Domain-Specific Languages to Digital Forensics. In: Proceedings of the 33rd International Conference on Software Engineering (ICSE 2011), pp. 671–680. ACM (2011)
6. Bozga, M., Jaber, M., Sifakis, J.: Source-to-Source Architecture Transformation for Performance Optimization in BIP. IEEE Trans. Industrial Informatics 6(4), 708–718 (2010)
7. Chung, E.Y., Benini, L., De Micheli, G.: Source Code Transformation based on Software Cost Analysis. In: Proceedings of the 14th International Symposium on Systems Synthesis (ISSS 2001), pp. 153–158. ACM (2001)
8. Cohen, M.I.: Advanced Carving Techniques. Digital Investigation 4(3-4), 119–128 (2007)
9. Czarnecki, K., Eisenecker, U.: Generative Programming: Methods, Tools, and Applications. Addison Wesley (2000)
10. Garfinkel, S.L.: Carving Contiguous and Fragmented Files with Fast Object Validation. Digital Investigation 4(S1), 2–12 (2007)
11. Garfinkel, S.L.: Digital Forensics Research: The Next 10 Years. Digital Investigation 7(S1), S64–S73 (2010)
12. Grenier, C.: PhotoRec, http://www.cgsecurity.org/
13. Klint, P., van der Storm, T., Vinju, J.: Rascal: A Domain Specific Language for Source Code Analysis and Manipulation. In: Proceedings of the Ninth IEEE International Working Conference on Source Code Analysis and Manipulation (SCAM 2009), pp. 168–177. IEEE (2009)
14. Mernik, M., Heering, J., Sloane, A.M.: When and how to develop domain-specific languages. ACM Comput. Surv. 37, 316–344 (2005)
15. Mohri, M., Nederhof, M.J.: Regular approximation of context-free grammars through transformation. In: Robustness in Language and Speech Technology, ch. 9, pp. 251–261. Kluwer (2000)
16. Pal, A., Memon, N.: The Evolution of File Carving. IEEE Signal Processing Magazine 26(2), 59–71 (2009)
17. Richard III, G.G., Roussev, V.: Scalpel: A Frugal, High Performance File Carver. In: Proceedings of the Fifth Annual DFRWS Conference (2005)
18. Schmidt, D.C.: Model-Driven Engineering. Computer 39, 25–31 (2006)

Empirical Assessment of Business Model Transformations Based on Model Simulation

María Fernández-Ropero[1], Ricardo Pérez-Castillo[1],
Barbara Weber[2], and Mario Piattini[1]

[1] Instituto de Tecnologías y Sistemas de la Información, University of Castilla-La Mancha
Paseo de la Universidad 4 13071, Ciudad Real, Spain
{marias.fernandez,ricardo.pdelcastillo,mario.piattini}@uclm.es
[2] University of Innsbruck
Technikerstraße 21a, 6020, Innsbruck, Austria
barbara.weber@uibk.ac.at

Abstract. Business processes are recognized by organizations as one of the most important intangible assets, since they let organizations improve their competitiveness. Business processes are supported by enterprise information systems, which can evolve over time and embed particular business rules that are not present anywhere else. Thus, there are many organizations with inaccurate business processes, which prevent the modernization of enterprise information systems in line with the business processes that they support. Therefore, business process mining techniques are often used to retrieve reliable business processes from the event logs recorded during the execution of enterprise systems. Unfortunately, such event logs are represented with purpose-specific notations such as Mining XML and still don't apply the recent software modernization standard: ISO 19506 (KDM, Knowledge Discovery Metamodel). This paper presents an exogenous model transformation between these two notations. The main advantage is that process mining techniques can be effectively reused within software modernization projects according to the standard notation. This paper is particularly focused on the empirical evaluation of this transformation by simulating different kinds of business process models and several event logs with different sizes and configurations from such models. After analyzing all the model transformation executions, the study demonstrates that the transformation can provide suitable KDM models in a linear time in accordance with the size of the input models.

Keywords: Business Processes, Event Logs, Knowledge Discovery Metamodel, Model Simulation.

1 Introduction

Most companies recognize business processes as a valuable asset to carry out their daily operation with the aim of achieving their business goals [1]. Business processes management helps companies to continuously adapt their operation in order to maintain their degree of competitiveness.

Z. Hu and J. de Lara (Eds.): ICMT 2012, LNCS 7307, pp. 137–151, 2012.
© Springer-Verlag Berlin Heidelberg 2012

Most parts of business processes are automatically supported by means of enterprise information systems [2]. These information systems unfortunately undergo software erosion overtime as a result of uncontrolled maintenance, and they become Legacy Information Systems (LIS). LIS embed much business knowledge that is not present anywhere else, which may imply that the business process representations of a company are misaligned with the actual business processes.

Software modernization is a suitable solution to address software erosion problems. Software modernization is the concept of evolving LIS with a focus on all aspects of the current system's architecture and the ability to transform current architectures into target architectures [3]. Software modernization improves the Return on Investment (ROI) by extending the lifecycle of systems, since it advocates preserving the embedded business knowledge. Business process mining techniques facilitate the preservation of business knowledge, since such techniques retrieve the actual, embedded business processes [4].

Business process mining techniques work with event logs recorded from the system execution, which represent the sequence of business activities executed by an enterprise system. Event logs models are often represented according to the Mining XML (MXML) metamodel [5]. These event logs represented according to MXML are suitable for most of the process mining techniques but they are not to be used in whole software modernization projects. For example, the discovered business processes model cannot have additional information about relationships between source code elements and the respective discovered business activities. This kind of information is necessary to understand and modernize LIS in line with the actual business processes supported by them [6].

Moreover, software modernization advocates the usage of the Knowledge Discovery Metamodel (KDM), which was recognized as the standard ISO 19506 [7], to represent different legacy software artifacts. KDM is organized into various orthogonal concerns (metamodel packages) that are in turn organized in different abstraction layers. The KDM event package allows representing event models alternatively to MXML.

This paper presents a declarative model transformation implemented using QVTr (Query/View/Transformation Relations) for transforming MXML event models into KDM event models [8]. The main advantage is that event logs transformed into KDM models can be integrated into software modernization processes so that synergies between event models and the remaining kinds of models (e.g., code model, database model, etc.) can be exploited together and in a homogeneous and standardized way.

This paper provides a formal experiment to empirically validate the model transformation. The experiment systematically simulates several event log models following different configurations. The study analyzes the effects of simulating factors (e.g., size of logs, complexity, etc.) on the efficiency of the model transformation. The result of this study demonstrates the scalability and suitability of the model transformation to be applied for obtaining KDM event models from MXML models in a linear time in accordance with the size and the complexity of the input model.

The remainder of this paper is organized as follows: Section 2 summarizes related work. Section 3 presents the model transformation under study. Section 4 describes the experiment based on model simulation. Section 5 provides the analysis and interpretation of results. Finally, Section 6 discusses conclusions and future work.

2   Related Work

Business process mining describes a family of a posteriori analysis techniques exploiting the information recorded in an event log [9]. Event logs sequentially record the business activities executed in process aware information systems. There are several works that use process mining dealing with the construction of business processes when there is no a priori business process model. For example, *Van der Aalst et al.* [10] propose the α-algorithm to discover the control flow of business processes from event logs. Similarly, *Madeiros et al.* [11] suggest a genetic algorithm for business process discovery.

Other proposals deal with the registration of event logs, e.g., *Ingvaldsen et al.* [12] focus on ERP (*Enterprise Resource Planning*) systems to obtain event logs from the SAP's transaction data logs. *Günther et al.* [13] provide a generic import framework for obtaining event logs from different kinds of systems. Other authors such as *Pérez-Castillo et al.* [14] propose an approach to obtain event logs by means of the injection of traces in legacy source code to enable the collection of event logs in non-process-aware systems.

All these proposals focus on the development and application of business process mining techniques. However, the mentioned approaches do not address the effective use of business processes to modernize legacy information systems being aligned with the actual business process. *Zou et al* [15] developed a framework that statically analyzes the legacy source code and applies a set of heuristic rules to recover the underlying business processes. Other works focus on recovering business processes by dynamically tracing the system execution driven by use cases (e.g., *Cai et al.* [16]), or driven by the users' navigation in graphical user interfaces (e.g., *di Francescomarino et al* [17]). The goal of these works is to obtain the actual, embedded business processes to be used during software modernization.

Unfortunately, all these works [15-17] propose *ad hoc* techniques that do not follow the KDM standard. As a consequence, the reuse as well as the scalability of these techniques to be applied to large and complex LIS is limited. In this sense, *Pérez-Castillo et al.* [8] present a preliminary method to integrate MXML event logs into KDM repositories, which is the starting point of this research. Nevertheless, this method has not been empirically validated, for example, through model simulation.

Model simulation is often applied in other research fields such as aerospace, healthcare, etc. Literature contains some proposals that use model simulation for empirically assessing model transformations. For instance, *Wong et al.* [18] use model simulation to empirically validate the translation of business process diagrams into executable BPEL (*Business Process Execution Language*) processes. *Syriani et al.* [19] use simulation to validate models of reactive systems such as modern computer games. *Biermann et al.* [20] propose simulation environments based on a model's concrete syntax definition for visual languages. The validation of the proposed model transformation follows a model simulation approach similar to such studies.

3   MXML to KDM Transformation

The proposed model transformation takes an MXML model and obtains an equivalent
KDM model at the same abstraction level. MXML is the notation commonly used to
represent event logs to be exploited in business process mining techniques [5] (see
Fig. 1). An MXML model represents an individual log (*WorkflowLog*). The log
consists of a set of business processes (*Process*) that collect, in turn, several instances
of such processes (*ProcessInstance*). Each process instance represents a certain
execution of a business process using particular data. For example, in a bank
company, process could be different execution instances for different customers. Each
process instance has a sequence of events (*AuditTrailEntry*). Each event consists of
four elements: (i) the business activity executed (*WorkflowModelElement*); (ii) the
type of the event (i.e., start or complete) (*EventType*); (iii) the user who started or
completed the business activity (*Originator*); and finally (iv) the time when the event
was recorded (*Timestamp*). All these elements can contain additional information
through *Data* and *Attribute* elements.

On the other hand, The KDM Event metamodel (see Fig. 2) defines the
EventModel metaclass to depict an event model in KDM. Each event model
aggregates a set of event resources of a LIS (*EventResource*). Particularly, event
resources can be states, transitions or events themselves (*Event*). Each event has two
features: the *name* of the event and the *kind* (i.e., start or complete). Event resources
and other elements can be related by means of event relationships
(AbstractEventRelationship) which can depict next states, transitions, consumed
events, etc. Moreover, the KDM event metamodel extends the KDM *action* package
metamodel by defining a set of event actions that can be associated with event
resources (see Fig. 2). For example, events that are produced by particular code
elements can be represented with *ProducesEvent* elements. These elements contain
references to pieces of source code (*CodeElement*) by means of the feature
implementation. It enables the integration of KDM event models with the remaining
of KDM models ensuring its appropriate usage in modernizations projects.

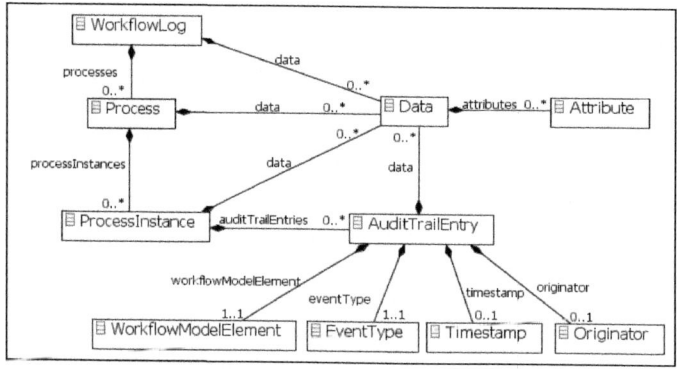

Fig. 1. The MXML Metamodel

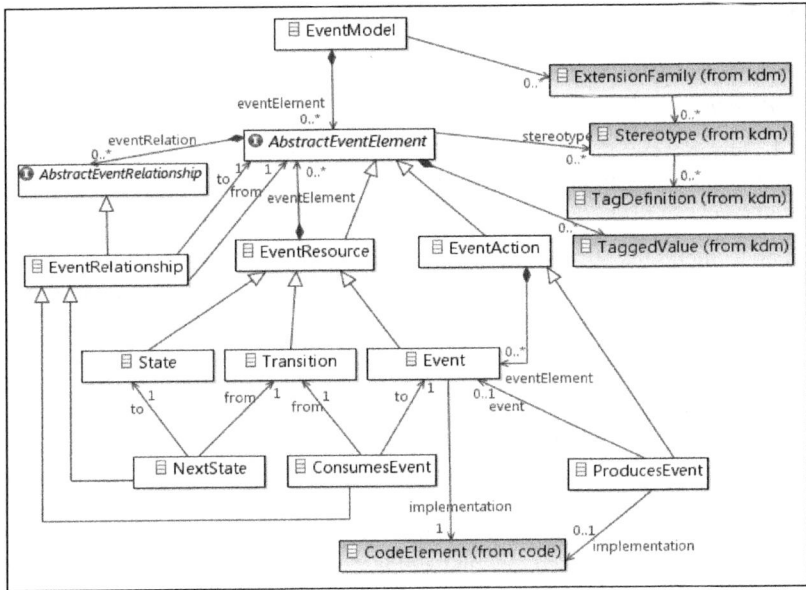

Fig. 2. The KDM event metamodel and extensions

3.1   Transformation Rules

The MXML to KDM model transformation consists of a set of eight declarative transformation rules. First of all, a KDM event model must be created from the MXML event log model (Rule 1). Events entail the key element of MXML event log models, thus, events must be transformed into the KDM event model (Rule 2). Furthermore, the information concerning the four components of an event (i.e., business activity name, type, originator and timestamp) must be represented in the KDM event model. Besides, the name and type of the events in the MXML models are represented in the KDM event models by using the features of the *Event* metaclass respectively (Rule 3). The information concerning the originator and timestamp cannot be directly represented in the KDM model according to the KDM event package. For this reason, the KDM event model must be extended with additional metaclasses so that it can support this information. Events (which represent executed business activities) are mapped to the pieces of source code that support those activities (Rule 4). This is possible because the KDM event model can be linked with other KDM models (e.g., the KDM code model) by means of the features *implementation* that link code elements (see Fig. 2).

Rule 1. *Each instance of the WorkflowLog metaclass is transformed into an instance of the EventModel metaclass in the output model.*

Rule 2. *Each instance of the AuditTrailEntry metaclass is transformed into an instance of the Event metaclass in the output model.*

Rule 3. *Instances of the WorkflowModelElement and EventType metaclass, belonging to an instance of the AuditTrailEntry metaclass, are respectively incorporated into the features 'name' and 'kind' of the respective instance of the Event metaclass (see Rule 2).*

Rule 4. *Instances of the Attribute metaclass with the name feature 'implementation' are transformed into instances of the CodeElement metaclass within the respective instance of the Event metaclass in the output model (see Rule 2).*

In order to represent all the information registered in a MXML model in the KDM model, the KDM event metamodel is extended by means of the *ExtensionFamily* metaclass, the standard extension mechanism of KDM (see highlighted metaclasses in Fig. 2). The extension family defines a set of stereotypes containing a set of tag definitions. Stereotypes define a wide concern while tag definitions specify the new elements that will be used in normal elements of the KDM event metamodel through tagged values. Tagged values allow changing or adjusting the meaning of those elements by associating a value with a previously defined tag. According to the extension mechanism, Rule 5 refines R1 by adding the extension family within the event model. The extension family has four stereotypes: *<process>*, *<processInstance>*, *<originator>* and *<timestamp>*. Event resources are tagged with *<process>* (Rule 6) and *<processInstance>* (Rule 7) to respectively collect business processes and their instances from event logs. Finally, both originator and timestamp are represented by incorporating tagged values to the respective event (Rule 8).

Rule 5. *An instance of the ExtensionFamily metaclass is created for each instance of the EventModel metaclass in the output model (see Rule 1). This instance contains four instances of the Stereotype metaclass. In turn, each Stereotype instance contains an instance of the TagDefiniton metaclass. The values of these four stereotypes are: <process>, <processInstance>, <originator> and <timestamp>.*

Rule 6. *Each instance of the Process metaclass is transformed in the output model into an instance of the EventResource metaclass with an instance of the TaggedValue metaclass. The tag feature of this instance links to the <Process> stereotype, and the value feature represents process name.*

Rule 7. *Each instance of the ProcessInstance metaclass is transformed in the output model into an instance of the EventResource metaclass with an instance of the TaggedValue metaclass. The tag feature of this instance links to the <ProcessInstance> stereotype, and the value feature represents the name of the business process instance.*

Rule 8. *Instances of the Originator and Timestamp metaclass are transformed into two instances of the TaggedValue metaclass which are added to the respective instance of the Event metaclass (see Rule 2). The instances of the TaggedValue metaclass respectively define their tag features as <Originator> and <Timestamp> stereotype, and their value feature with the name of the originator and timestamp registered in the input model.*

An executable version of the model transformation has been implemented using QVTr (Query/View/Transformation relations) [21], which provides a declarative and rule-based specification. Due to the space limitation this paper shows the *'auditTrailEntry2Event'* relation as an example (see Fig. 3). The full transformation is available online [22]. The *checkonly* domain of the relation is defined on instances of the *AuditTrailEntry* metaclass. This input domain checks the existence of the four elements in an event (i.e., the business activity, type, originator and timestamp). The input domain also evaluates the existence of the process instance, process and the event log where the *AuditTrailEntry* element belongs. The *enforce* domain creates an instance of the *Event* metaclass according to Rule 2. This event is created within the respective log, process and process instance. The originator and timestamp are added with the appropriate stereotype according to Rule 8. Finally, the *when* clause invokes the *'processInstance2eventResource'* to check, as a pre-condition, that the respective process instance was previously created by means of the invoked relation (see Fig. 3).

```
top relation auditTrailEntry2Event {           enforce domain event eventModel:event::EventModel{
  xEventName : String;                           name = xModelName,
  xEventType : String;                           eventElement = eRes:event::EventResource {
  xOriginatorName : String;                        name = xProcessName,
  xDate : String;                                  eventElement = eRes2:event::EventResource {
  xProcessInstanceName : String;                     name = xProcessInstanceName,
  xProcessName : String;                             eventElement = event : event::Event {
  xModelName : String;                                 name = xEventName,
  checkonly domain mxml ate : mxml::AuditTrailEntry {   kind = xEventType,
    workflowModelElement = wme : mxml::WorkflowModelElement {   taggedValue = originatorTag : kdm::TaggedValue {
      name = xEventName                                  tag = ot : kdm::TagDefinition {
    },                                                     tag = 'Originator'
    eventType = type : mxml::EventType {                 },
      type = xEventType                                  value = xOriginatorName
    },                                                 },
    originator = originator : mxml::Originator {       taggedValue = timestampTag : kdm::TaggedValue {
      name = xOriginatorName                             tag = dt : kdm::TagDefinition {
    },                                                     tag = 'Timestamp'
    timestamp = timestamp : mxml::Timestamp {          },
      date = xDate                                     value = xDate
    },                                               },
    processInstance = pi : mxml::ProcessInstance {   implementation = codeElement : code::CodeElement {
      name = xProcessInstanceName,                     name = xEventName
      process = p : mxml::Process {                  }
        name = xProcessName,                       }
        workflowLog = wl : mxml::WorkflowLog {     }
          name = xModelName                      }
        }                                      };
      }                                        when {
    }                                            processInstance2eventResource (pi, eventModel);
  }                                            }
};                                           }
```

Fig. 3. The 'auditTrailEntry2Event' QVT relation

4   Experiment Description

This section presents one experiment to validate the proposed model transformation. The experiment is based on the formal protocol proposed by *Jedlitschka et al.* [23] for conducting and reporting empirical research in software engineering. According to this protocol, the following sections describe the research goal and questions, research hypothesis, variables, design and execution procedure, as well as the analysis procedure.

4.1   Research Goal and Questions

The main research goal of this experiment is the efficiency assessment of the transformation. In order to evaluate this property the experiment attempts to answer two research questions:

- RQ1: *Is the model transformation scalable to large MXML models?*
- RQ2: *Does the input model's complexity affect to the transformation performance?*

Firstly, scalability assessment (RQ1) is important to ensure the applicability of this transformation with large and complex event logs. Secondly, the study of side effects of the event log's complexity (RQ2) in the transformation performance is valuable to prove its feasibility with any kind of event log.

The study randomly simulates a set of event logs for assessing the research questions. Event logs are simulated through *Process Log Generator* (PLG) [24], a tool for the generation of business process models and simulation of different MXML logs (cf. Section 4.4).

4.2   Variables

A set of variables is defined for the assessment of the model transformation efficiency. There are two independent variables: (i) **Size**, which represents the number of events in the simulated MXML log; and (ii) **ECyM**, which represents the Extended Cyclomatic Metric (ECyM) of the MXML model [25]. ECyM determines how complicated the behavior of the model is, i.e., its complexity. The ECyM of a graph G with V vertices, E edges, and p connected components is: $ECyM = |E| - |V| + p$

The dependent variables of the study are two: (i) **Transformation Time**, which is the time spent on transforming a MXML model into a KDM model through the proposed model transformation; and (ii) **Performance,** which is the ratio between the size of the input model and the transformation time. This variable is normalized in the range [0, 1].

4.3   Research Hypothesis

In the case of RQ1, it is necessary to check if there is a linear relation between the size of the MXML model and the time of transformation through the proposed model transformation. To do this, the hypotheses are:

- $H_{RQ1,0}$: The size of the MXML model has a linear relation with the time of transformation.
- $H_{RQ1,1}$: The size of the MXML model has not a linear relation with the time of transformation

To answer RQ2 it is necessary to check if the input model's complexity affects the performance. To do this, the hypotheses are:

— $H_{RQ2,0}$: The ECyM of the log does not influence the performance.

— $H_{RQ2,1}$: The ECyM of the log influences the performance.

The goal of the statistical analysis is being able to accept these null hypotheses with an acceptable confidence level.

4.4   Design and Execution Procedure

The experiment evaluates the transformation model in several simulated event logs. The experiment's execution consists of the following steps:

1. The set of MXML logs are simulated using PLG. PLG allows users to obtain business processes with different sizes by defining the maximum number of nested branches. The study uses three sizes: 2, 3 and 4 maximum nested branches, which are respectively labeled as low, medium and high. Four business process models are created for each size. In turn, four event logs are simulated for each business process with different numbers of business process instances: 50, 100, 150, and 200. In total, 48 logs conform the sample to perform the experiment. For each log steps 2 to 3 are repeated.

2. The MXML log is analyzed for collecting relevant variables (i.e., number of events, ECyM, etc.).

3. The MXML is transformed into a KDM event model. The transformation is executed through *Medini QVT* [26], a model transformation engine supporting QVTr. The transformation is executed in a computer with a dual processor of 2.1 GHz and 4 GB of RAM memory. After the execution, transformation information is also recorded. The whole collected information is shown in Table 1.

4. After the whole execution of the sample, the collected information is statistically analyzed to answer the research questions.

4.5   Analysis Procedure

The data analysis was carried out according to the following steps:

1. The hypotheses established for RQ1 are evaluated by means of a regression line model using the Pearson linear correlation test, which quantifies the intensity of the linear relation between variables size and transformation time. Under the hypothesis that the transformation time is theoretically linear (i.e., $O(n)$ with n=number of events), a linear regression model is established to check it and find out whether the proposal is therefore scalable. The linear regression model considers the transformation time as a dependent variable and the size of the business processes as the independent variable. The obtained Pearson's correlation coefficient R^2 (between -1 and 1) indicates the degree to which the real values of the dependent variable are close to the predicted values.

Table 1. Data collected in the experiment execution

ID	Size (#events)	Complexity (ECyM)	Transf. Time (s)	Performance
1	788	27	10.03	0.50
2	1564	27	41.15	0.27
3	2300	27	94.15	0.24
4	3080	27	141.62	0.13
5	1000	21	10.55	0.69
6	2000	21	41.31	0.28
7	3000	21	91.62	0.15
8	4000	21	138.11	0.11
9	628	10	76.00	1.00
10	1294	10	349.26	0.45
11	1926	10	751.34	0.26
12	2560	10	1259.48	0.22
13	646	12	93.79	0.97
14	1282	12	382.96	0.44
15	1910	12	811.55	0.27
16	2552	12	1373.64	0.23
17	972	43	22.04	0.69
18	2046	43	91.40	0.24
19	3150	43	212.45	0.16
20	3980	43	350.19	0.12
21	1200	14	33.41	0.53
22	2400	14	134.28	0.22
23	3600	14	287.99	0.13
24	4800	14	663.58	0.04
25	1842	0	23.27	0.33
26	4066	0	74.56	0.11
27	5962	0	118.06	0.05
28	7870	0	241.19	0.02
29	1094	66	21.38	0.60
30	2188	66	107.58	0.24
31	3176	66	215.36	0.16
32	4454	66	318.04	0.09
33	2408	246	27.36	0.22
34	5110	246	114.27	0.07
35	7548	246	218.29	0.01
36	9986	246	421.56	0.00
37	1904	1037	131.08	0.31
38	3800	1037	562.49	0.11
39	5934	1037	1329.71	0.04
40	7980	1037	1959.02	0.00
41	2032	0	82.63	0.28
42	4076	0	323.87	0.09
43	6116	0	763.84	0.04
44	8216	0	1512.17	0.01
45	1906	54	87.45	0.28
46	3948	54	363.11	0.10
47	6500	54	889.43	0.04
48	7668	54	1175.69	0.02
Mean	3510	127	386.32	0.24
Std. Dev.	2352	285	468.27	0.23

2. Hypotheses of RQ2 are assessed by using the ANOVA test (*Analysis Of Variance between groups*). It is a parametric test to compare how a particular factor affects the mean of a quantitative variable. If the means of variable for each factor are equal, then the factor does not affect the variable. The factor under study is the event log complexity (ECyM) labeled as low, medium and high. Each event log is categorized in these three groups according to the percentiles $Q_{1/3}$ and $Q_{2/3}$, which divide the distribution in three sub-samples. As a result, the hypotheses of RQ2 are equivalent to the following according to the ANOVA test:

— $H_{RQ2,0}$: $\mu_{\text{performance; ECyM="low"}} = \mu_{\text{performance; ECyM="medium"}} = \mu_{\text{performance; ECyM="high"}}$. All expected means are equal.

— $H_{RQ2,1}$: $\mu_{\text{performance; ECyM="low"}} \neq \mu_{\text{performance; ECyM="medium"}} \neq \mu_{\text{performance; ECyM="high"}}$.

5   Results

The following sections show the results after analyzing the data obtained in the experiment using *R*, an open source statistical tool [27].

5.1   Scalability Testing (RQ1)

To calculate the equation of the regression line the variables were represented by a scatter plot (see Fig. 4). If the regression line is very close to most points in the scatter chart, both variables are strongly correlated. The regression line equation estimated is $y = 193.23x - 291849$.

After applying the Pearson correlation test, the value of linear correlation coefficient of Pearson was $R^2=0.94$, which is very close to 1. This value makes it possible to ensure that there is a strong positive correlation between both variables. In terms of significance, the correlation value means that 5% of the transformation time variation cannot be explained by size through the line of fit.

The result shows that the null hypothesis ($H_{RQ1,0}$) cannot be rejected since there is a linear relationship between the size of input model and transformation time.

5.2   Suitability Testing (RQ2)

The $H_{RQ2,0}$ hypothesis proposes that the means of performance for each factor of ECyM are equal. After applying the ANOVA test (see Table 2) the p-value is 0.105. Since the p-value is greater than 0.05 the null hypothesis cannot be rejected; it is accepted therefore that the means are equal at a 95% confidence level. This can also be checked graphically in Fig. 5 which shows the box chart for each distribution with low, medium and high complexity. This result proves that the complexity (ECyM) does not have an influence in the performance.

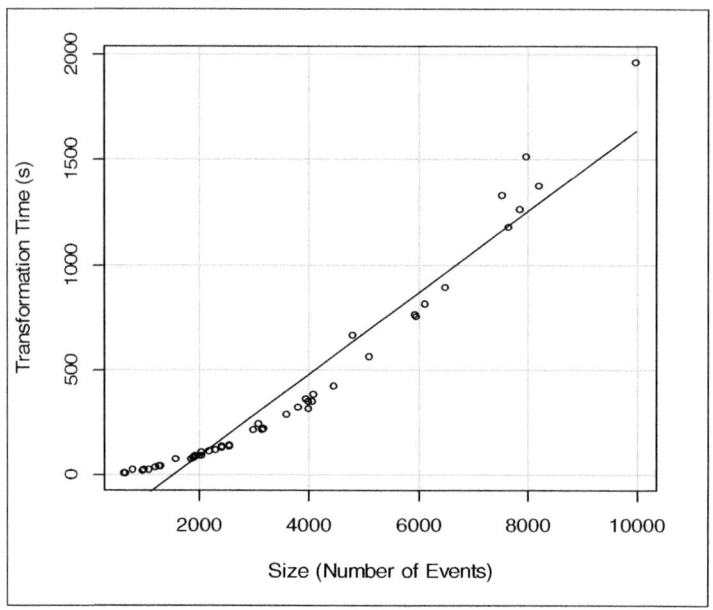

Fig. 4. Scatter plot of the size and the transformation time

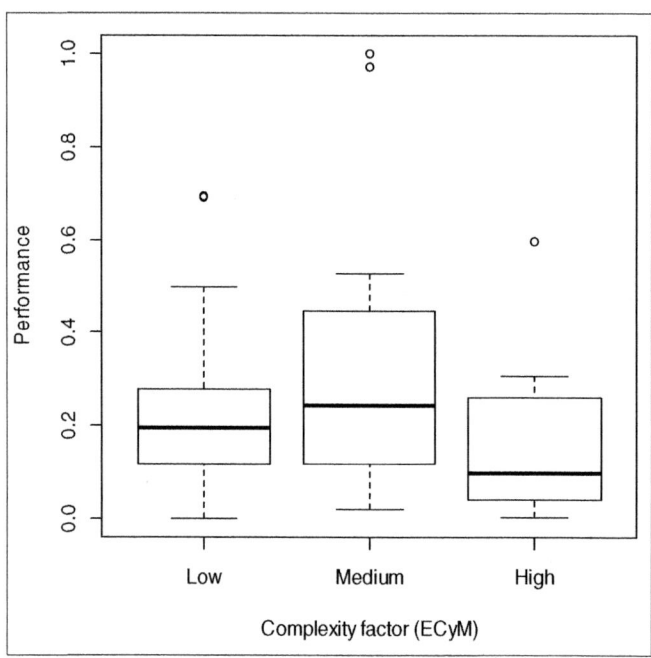

Fig. 5. Box plot of performance

Table 2. ANOVA results

	Df	Sum Sq	Mean Sq	F value	Pr (>F)
Complexity Factor	2	0.249	0.12449	2.365	**0.105**
Residuals	45	2.368	0.05263		

5.3   Validity Evaluation

This section discusses the threats to the validity of the experiment.

- **Internal Validity:** The simulation was carried out with 48 event logs simulated from randomly generated business processes. Hence, the results may differ slightly in case of generation of different business processes. In addition, the supporting tool used to obtain the business process could be a factor that may affect the values of the experiment. To mitigate this threat the experiment should be replicated by using larger samples, different tools, and then, by comparing the obtained results.

- **Construct Validity:** The selected variables were adequate to answer the research questions in an appropriate manner. However, the way in which such variables are assessed could be a threat. To mitigate this threat, other mechanisms can be considered for the evaluation of the proposed variables (e.g., complexity can be calculated using other metrics available in the literature).

- **External Validity:** The experiment considers simulated event logs, thus the obtained results could not be strictly generalized to real-life event logs. This threat may be mitigated by replicating the experiment using industrial event logs.

6   Conclusions

This paper proposes a model transformation to integrate MXML event logs into the KDM event model repository. Nowadays, KDM makes it possible to build reverse engineering tools in a KDM ecosystem where reverse engineering tools recover knowledge regarding different artifacts, and the outgoing knowledge is represented and managed in an integrated and standardized way through a KDM repository. As a result, the KDM event models can be used in combination with other embedded knowledge recovered through reverse engineering to modernize legacy information systems. This transformation therefore facilitates the applicability of business process mining techniques and algorithms within software modernization projects.

This work provides an implementation of the model transformation using QVTr as well as a supporting tool in order to facilitate its validation and adoption by the industry. In fact, the transformation is validated through an experiment based on the automatic simulation of event logs. The experiment shows that the model transformation is able to obtain KDM event models from MXML logs in a scalable and suitable way. This means that the transformation can be executed in a linear time regarding the number of events. The performance of the transformation is also independent of the complexity of the input log.

The future work will address the repeatability of the experiment using additional and different event log models in order to deal with the detected threats and to obtain strengthened conclusions.

Acknowledgments. This work was supported by the FPU Spanish Program and the R&D projects ALTAMIRA (PII2I09-0106-2463), PEGASO/MAGO (TIN2009-13718-C02-01) and MOTERO (JCCM and FEDER, PEII11-0366-9449).

References

1. Weske, M.: Business Process Management: Concepts, Languages, Architectures, Leipzig, Alemania, p. 368. Springer, Heidelberg (2007)
2. Jeston, J., Nelis, J., Davenport, T.: Business Process Management: Practical Guidelines to Successful Implementations, 2nd edn., p. 469. Butterworth-Heinemann, Elsevier Ltd., NV, USA (2008)
3. Newcomb, P.: Architecture-Driven Modernization (ADM). In: Proceedings of the 12th Working Conference on Reverse Engineering. IEEE Computer Society (2005)
4. van der Aalst, W., Weijters, A.J.M.M.: Process-aware information systems: bridging people and software through process technology. In: Dumas, M., van der Aalst, W., Ter Hofstede, A. (eds.) Process Mining, pp. 235–255. John Wiley & Sons, Inc. (2005)
5. Van der Aalst, W.M.P., et al.: ProM: the process mining toolkit. In: 7th International Conference on Business Process Management (BPM 2009) - Demonstration Track, pp. 1–4. Springer, Ulm (2009)
6. van den Heuvel, W.-J.: Aligning Modern Business Processes and Legacy Systems: A Component-Based Perspective (Cooperative Information Systems). The MIT Press (2006)
7. Pérez-Castillo, R., de Guzmán, I.G.R., Piattini, M.: Knowledge Discovery Metamodel - ISO/IEC 19506: a Standard to Modernize Legacy Systems. Computer Standards & Interfaces Journal, 519–532 (2011)
8. Pérez-Castillo, R., et al.: Integrating Event Logs into KDM Repositories. In: 27th Annual ACM Symposium on Applied Computing (SAC 2012). ACM, Riva del Garda (in Press, 2012)
9. van der Aalst, W.M.P.: Process-Aware Information Systems: Lessons to Be Learned from Process Mining. In: Jensen, K., van der Aalst, W.M.P. (eds.) ToPNoC II. LNCS, vol. 5460, pp. 1–26. Springer, Heidelberg (2009)
10. van der Aalst, W., Weijters, T., Maruster, L.: Workflow mining: discovering process models from event logs. IEEE Transactions on Knowledge and Data Engineering 16(9), 1128–1142 (2004)
11. Medeiros, A.K., Weijters, A.J., Aalst, W.M.: Genetic process mining: an experimental evaluation. Data Min. Knowl. Discov. 14(2), 245–304 (2007)
12. Ingvaldsen, J.E., Gulla, J.A.: Preprocessing Support for Large Scale Process Mining of SAP Transactions. In: ter Hofstede, A.H.M., Benatallah, B., Paik, H.-Y. (eds.) BPM Workshops 2007. LNCS, vol. 4928, pp. 30–41. Springer, Heidelberg (2008)
13. Günther, C.W., van der Aalst, W.M.P.: A Generic Import Framework for Process Event Logs. In: Eder, J., Dustdar, S. (eds.) BPM Workshops 2006. LNCS, vol. 4103, pp. 81–92. Springer, Heidelberg (2006)
14. Pérez-Castillo, R., Weber, B., García-Rodríguez de Guzmán, I., Piattini, M.: Toward Obtaining Event Logs from Legacy Code. In: Muehlen, M.z., Su, J. (eds.) BPM 2010, Part II. LNBIP, vol. 66, pp. 201–207. Springer, Heidelberg (2011)

15. Zou, Y., Hung, M.: An Approach for Extracting Workflows from E-Commerce Applications. In: Proceedings of the Fourteenth International Conference on Program Comprehension, pp. 127–136. IEEE Computer Society (2006)
16. Cai, Z., Yang, X., Wang, W.: Business Process Recovery for System Maintenance - An Empirical Approach. In: 25th International Conference on Software Maintenance (ICSM 2009), pp. 399–402. IEEE Computer Society, Edmonton (2009)
17. Di Francescomarino, C., Marchetto, A., Tonella, P.: Reverse Engineering of Business Processes exposed as Web Applications. In: 13th European Conference on Software Maintenance and Reengineering (CSMR 2009), pp. 139–148. IEEE Computer Society, Fraunhofer IESE (2009)
18. Wong, P., Gibbons, J.: On Specifying and Visualising Long-Running Empirical Studies. In: Vallecillo, A., Gray, J., Pierantonio, A. (eds.) ICMT 2008. LNCS, vol. 5063, pp. 76–90. Springer, Heidelberg (2008)
19. Syriani, E., Vangheluwe, H.: Programmed Graph Rewriting with Time for Simulation-Based Design. In: Vallecillo, A., Gray, J., Pierantonio, A. (eds.) ICMT 2008. LNCS, vol. 5063, pp. 91–106. Springer, Heidelberg (2008)
20. Biermann, E., et al.: Flexible visualization of automatic simulation based on structured graph transformation. IEEE (2008)
21. OMG, QVT. Meta Object Facility (MOF) 2.0 Query/View/Transformation Specification, OMG (2008), http://www.omg.org/spec/QVT/1.0/PDF
22. Pérez-Castillo, R.: MXML to KDM Transformation implemented in QVT Relations (2011), http://alarcos.esi.uclm.es/per/rpdelcastillo/modeltransformations/MXML2KDM.htm (cited March 29, 2011)
23. Jedlitschka, A., Ciolkowski, M., Pfahl, D.: Reporting experiments in software engineering. In: Guide to Advanced Empirical Software Engineering, pp. 201–228 (2008)
24. Burattin, A., Sperduti, A.: PLG: a Framework for the Generation of Business Process Models and their Execution Logs (2011)
25. Lassen, K.B., van der Aalst, W.M.P.: Complexity metrics for Workflow nets. Information and Software Technology 51(3), 610–626 (2009)
26. ikv++, Medini QVT (2008), ikv++ technologies ag
27. R. The R Project for Statistical Computing (2011), http://cran.r-project.org/

Traceability Visualization in Model Transformations with TraceVis[*]

Marcel F. van Amstel, Mark G.J. van den Brand, and Alexander Serebrenik

Department of Mathematics and Computer Science
Eindhoven University of Technology, Eindhoven, The Netherlands
{M.F.v.Amstel,M.G.J.v.d.Brand,A.Serebrenik}@tue.nl

Abstract. Model transformations are commonly used to transform models suited for one purpose (e.g., describing a solution in a particular domain) to models suited for a related but different purpose (e.g., simulation or execution). The disadvantage of a transformational approach, however, is that feedback acquired from analyzing transformed models is not reported on the level of the problem domain but on the level of the transformed model. Expressing the feedback on the level of the problem domain requires improving traceability in model transformations.

We propose to visualize traceability links in (chains of) model transformations, thus making traceability amenable for analysis.

1 Introduction

Domain-specific languages (DSLs) play a pivotal role in the model-driven engineering (MDE) paradigm. DSLs describe the domain knowledge [7] and offer expressive power focused on that domain through appropriate notations and abstractions [9]. A typical application of a DSL such as [1,6], however, goes beyond *describing* the domain knowledge, and includes, for example, execution or simulation (cf. the discussion of executability of DSLs in [15]). Hence, application of a DSL requires adequate tool support for execution and simulation. In MDE, typically the transformational approach is adopted to this end [18].

The main advantage of the transformational approach is the flexibility it provides by reusing existing formalisms. Adapting the DSL for a different purpose, such as simulation or execution, solely requires the implementation of another model transformation. However, the disadvantage of a transformational approach, is that analyses are not performed on the domain level, but on the level of the target language of a model transformation [14]. This raises the issue of *traceability*, i.e., results acquired from analyzing the target models of a model transformations have to be related to the source models.

Traceability plays an essential role in a number of typical model development scenarios such as debugging and change impact analysis. When debugging models, it is important to understand whether the erroneous part of the target model

---

[*] This work has been carried out as part of the FALCON project under the responsibility of the Embedded Systems Institute with Vanderlande Industries as the industrial partner. This project is partially supported by the Netherlands Ministry of Economic Affairs under the Embedded Systems Institute (BSIK03021) program.

Z. Hu and J. de Lara (Eds.): ICMT 2012, LNCS 7307, pp. 152–159, 2012.

results from the source model or from the transformation itself, and to pinpoint the corresponding parts of the source model and/or transformation functions. Moreover, when source models are about to change, one should determine the effect of proposed changes on the target model.

In this paper, we propose to apply a visualization technique to facilitate the analysis of the relation between source models, model transformations, and target models. The proposed approach is also applicable to chains of model transformations in which the target model(s) of a preceding model transformation serve as source model(s) for the subsequent one. In this way, the approach enables traceability of model transformation compositions.

On the model level, our visualization makes explicit what source model elements are the origin of a target model element and what transformation elements are involved in creating a target model element. The model developer can therefore identify source model element(s) and transformation element(s) responsible for producing (erroneous) target model element(s), as well as generated target model element(s) based on a particular source model element (cf. Section 3.4).

The remainder of this paper is structured as follows. We start by stating the requirements has to satisfy in Section 2. Next we discuss the visualization in Section 3. In Section 4, we review the related work. Section 5 concludes the paper and provides directions for further research.

2   Requirements

We start by identifying requirements the visualization approach should satisfy. Our first requirements follow from the definition of traceability: the visualization should be able to represent *structure of* (**Req1**) and *relations between* (**Req2**) the source (meta)model, the target (meta)model and the model transformation. The visualization should further be able to present model transformations with *multiple models* serving as input and as output (**Req3**).

The next group of requirements is related to our intention to apply the visualization to chains of model transformation. The visualization, hence, should allow the user to inspect *traceability across multiple transformation steps* (**Req4**), e.g., to identify all source model elements that indirectly are responsible for producing an (erroneous) target element. Moreover, the user should be able to *ignore (some of) the intermediate transformation steps* (**Req5**), if desired.

Scrutinizing these requirements we observe similarity with visualizing traceability in traditional software development: models correspond to software development artifacts such as requirements specifications, design documents, and source code. Relating test cases to requirements is essentially the same as relating source models to target models through multiple transformation steps. This realization led to the decision to use a traceability visualization tool originally developed for traditional software development process. The tool, called TraceVis [17], turned out to be perfectly suitable for our purpose.

The basic outline of the TraceVis visualization is shown in Fig. 1. It represents three hierarchies (left, middle, right) connected by means of traceability links.

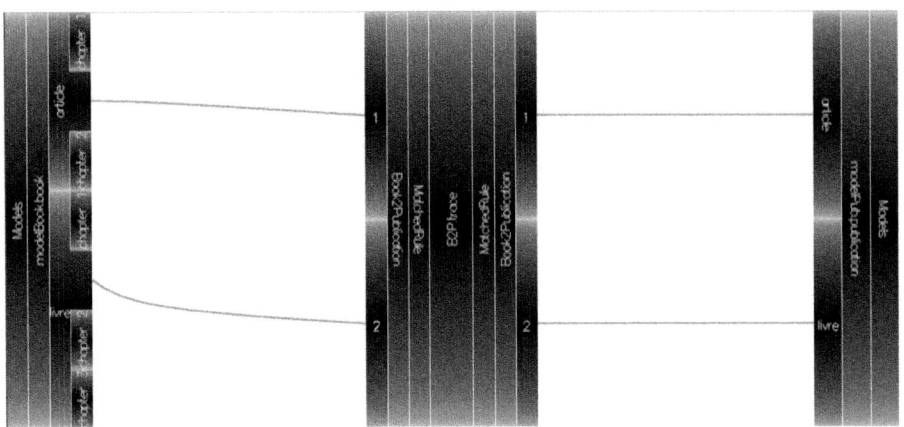

Fig. 1. TraceVis shows the relations between different hierarchies

The middle hierarchy is mirrored to allow the relations with both the left and right hierarchies to be represented. When the user selects elements from one of the hierarchies, connected elements from other hierarchies and the connecting traceability links become highlighted (cf. **Req4**). Moreover, TraceVis allows the user to hide a hierarchy (cf. **Req5**). Finally, it is known that while visualization of relations as straight lines is well-suited for smaller number of relations, it does not scale up well [12]. Therefore, TraceVis supports hierarchical edge bundling [12] specially designed for scalability.

3   Traceability on the Model Level

3.1   Basic Visualization

Fig. 1 shows part of a screen shot from the TraceVis tool applied to a "Book 2 Publication" model transformation [10]. The source model, the target model and the model transformation are represented as hierarchies: the left one, the right one and the middle one, respectively.

In hierarchies representing source and target models, the outermost columns represent the roots of the hierarchies. These are artificial nodes created for grouping (multiple) model(s) that serve as input and output of a model transformation (cf. **Req3**). The next level of the hierarchies, represented by the second leftmost and the second rightmost columns, shows the filenames of the input, and output models, respectively. The remaining levels of the hierarchy correspond to the elements of the input and output models. Model elements in these columns are shown *on top* of the elements they contain. In this way, the containment hierarchy in the models is visualized as a hierarchy.

The middle hierarchy represents a transformation. This hierarchy is mirrored to allow both the relations with the input and output models to be represented. The middle column serves for grouping all the modules of the transformation, it

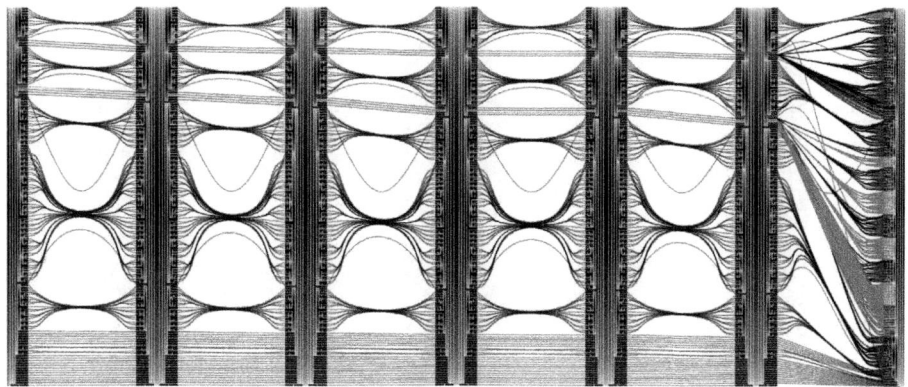

Fig. 2. TraceVis visualizes a chain of six model transformations

is labeled with the filename of the trace model that is visualized (cf. Section 3.3). The columns adjacent to the middle column are used for grouping the different kinds of transformation elements available in the model transformation language: e.g., in ATL we group helpers, matched rules, lazy matched rules, unique lazy matched rules, and called rules. Proceeding from the middle column outwards we see the column showing the actual transformation elements, e.g., ATL transformation rules. Finally, the outermost columns visualize the instances of the transformation rules, i.e., run-time applications of a rule to model elements.

3.2   Visualizing Transformation Chains

Fig. 2 shows a chain of six endogenous model transformations. The DSL on which the transformations are defined is aimed at modeling systems consisting of objects that operate in parallel and communicate with each other [1]. The transformations perform a stepwise refinement of the source model and bring it closer to the implementation by replacing, e.g., synchronous communication with asynchronous one, and lossless channels with lossy ones.

In Fig. 2 the transformation hierarchies are hidden, i.e., only the input and output models of the transformations are visualized. In all transformations, the model is copied and "slightly" modified. This is why there are many "straight" lines in the visualization. Since the model transformations refine the model, changes to the models are all local. For the first transformation, a selection (orange) is made that shows such a local change. While this change might appear as being hard to detect, it becomes apparent when zooming in on a part of the visualization (cf. Fig. 3). Only the last model transformation changes the model drastically. In the visualization, a single model element (pink) is selected in the penultimate model. This single model element gives rise to many model elements in the target model. This particular model transformation extends the model with a protocol implementation consisting of many model elements.

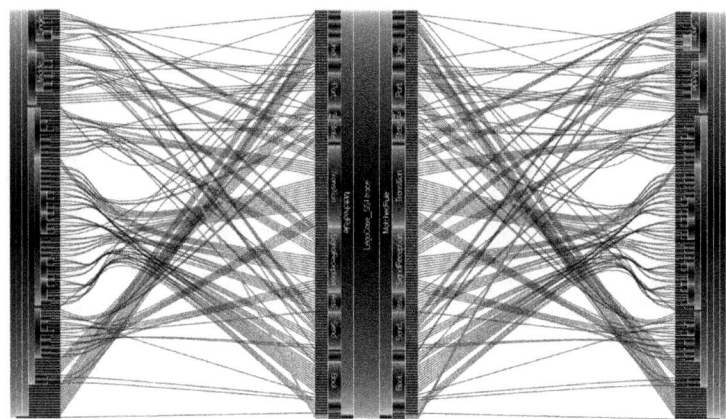

Fig. 3. Zooming in reveals that two lazy matched rules modify the model

3.3   From Model Transformations to TraceVis

Since TraceVis has originally not been designed for model transformations, we
had to implement a complementary tool for automatically generating TraceVis
input from model transformations. To make the distinction between different
applications of the same transformation element, we have to analyze the trace
model. To obtain the trace model the transformation should be executed, since
without executing the transformation there is no target model and, hence, no
trace model. Various approaches can be chosen to acquire a trace model from a
model transformation execution. One can adapt the transformation engine such
that it generates a trace model. In spirit of MDE we have opted, however, for a
model transformation rather than transformation engine adaptation. The entire
tool chain for generating trace models is presented in Fig. 4.

First, a higher-order model transformation, called *tracer adder* and implemented in ATL, takes as an input the model transformation being visualized,
say T, and augments T such that the trace model is generated as an additional output model of T. This transformation is based on the one described by
Jouault [13]. Next, the augmented transformation is applied to input models,
resulting in output models and a trace model that contains links both to the
input models and to the output models. Finally, the input models, the output
models and the trace model are transformed to an XML file that serves as input
for the TraceVis tool. This transformation has been implemented in Java and
can be reused if other model transformation languages are considered.

3.4   Applications

Debugging. In the introduction, we observed that domain-specific models are
typically analyzed by transforming them to a formalism suitable for analysis.
The disadvantage of this approach is that feedback from the analysis model is

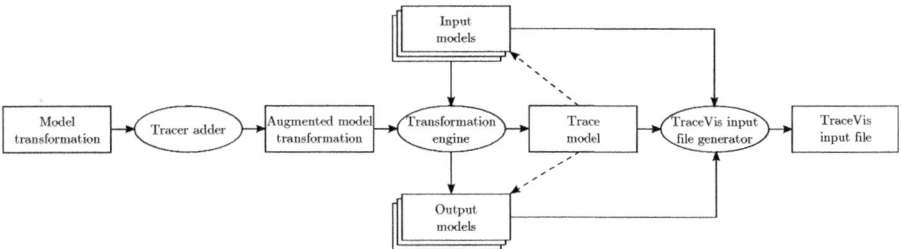

Fig. 4. Tool architecture

not reported in terms of the DSL, but in terms of the generated model. Suppose an error manifests in an executable target model. To fix this error, it has to be related to the elements in the domain-specific source model it is generated from. Using TraceVis, the visualized trace links can be followed from the model elements that caused the error to the model elements in the source model to establish the origin of the error. This is referred to as origin tracking [8].

It may be the case that the manifested error is not caused by an erroneous source model, but by an erroneous model transformation. Since the visualized trace links also show the relations between target model elements and transformation elements, errors in model transformations can easily be found. Also, when obsolete target model elements are generated, the visualization can be consulted to identify the responsible transformation element.

Impact Analysis. In addition to debugging, the visualization can also be used to facilitate change impact analysis. Change impact analysis is the process of determining the effect of proposed changes [3]. Evaluating this effect is considered to be one of the most expensive activities in the software maintenance process [4]. Most maintenance efforts include nowadays means of identifying impacts prior to making extensive software changes [5]. Using our visualization one can determine which target model elements are affected by changed source model elements.

By means of impact analysis, one can determine what part of a target model will change based on a change in the source model it was generated from. For instance, for the transformation chain described in [1], one can determine which part of the target Promela model will be affected by the change in a source SLCO model. Subsequently, this information can be used to perform incremental model checking of the Promela model (cf. [19]). Hence, conducting change impact analysis as a preliminary step reduces the verification effort. A similar argument can be made for, e.g., reducing the simulation effort.

4   Related Work

A traceability framework for model transformations was implemented in the model-oriented language Kermeta [11]. Similarly to [11], we support designers in gathering information on the transformation behavior, and make use of a transformation trace model. However, as opposed to [11], our approach focuses on visualization. Moreover, while the approach of [11] focuses on model transformations written in Kermeta, our approach is, in principle, language independent although the current implementation is restricted to ATL.

To visualize chains of model transformations, Von Pilgrim et al. place model representations on two-dimensional planes in a three-dimensional space [16]. Lines between these planes connect source model elements to target model elements. As opposed to our work, the model transformation itself is not visualized. Moreover, the choice of two-dimensional planes in a three-dimensional space leads to scalability issues when long transformation chains are considered.

In our previous work [2] we have applied TraceVis to visualize model transformation on the *metamodel* level. While traceability visualization at the metamodel level is similar to traceability visualization at the model level, the information extraction is different from the process described in Section 3.3. The reason for this is that the relation between a model transformation and the elements of its source and target metamodel can be derived from its source code directly without running the transformation. Moreover, the intended applications of the current work differ from those of [2].

5   Conclusions and Future Work

In this paper, we have presented a novel approach to visualization of traceability information in model transformations. Our approach explicates relations on a model level, and is applicable not only to single transformations but also to transformation chains. Applications of the proposed approach range from debugging and coverage analysis to change impact analysis.

The approach consists of two phases. First, the hierarchical structure of models and model transformation(s) as well as relations between these structures are extracted from the model and model transformation files. Second, the hierarchies are extracted and the relations between them are visualized using TraceVis. By visual inspection of the TraceVis visualization, model designers can answer such questions as what source model element(s) and transformation element(s) are responsible for producing what (erroneous) target model element(s), or what generated target model element(s) are based on a particular source model element? Finally, model designers can use the TraceVis visualization as a starting point of a change impact analysis that aims at understanding how changes to a source model affect the target model and the corresponding transformation.

The current version of the extraction tool supports ATL transformations only. Therefore, we consider application of the approach to other model transformation languages to be an important part of the *future work*.

Acknowledgements. We would like to thank Wiljan van Ravensteijn for providing us with TraceVis, Ivo van der Linden for his help on the implementation of data extraction, as well as Joost Gabriels for his comments on the earlier versions of this manuscript.

References

1. van Amstel, M.F., van den Brand, M.G.J., Engelen, L.J.P.: An Exercise in Iterative Domain-Specific Language Design. In: IWPSE-EVOL, pp. 48–57. ACM (2010)
2. van Amstel, M.F., Serebrenik, A., van den Brand, M.G.J.: Visualizing Traceability in Model Transformation Compositions. In: Pre-proceedings of the First Workshop on Composition and Evolution of Model Transformations (2011)
3. Arnold, R.S., Bohner, S.A.: Impact Analysis – Towards A Framework for Comparison. In: Card, D.N. (ed.) ICSM, pp. 292–301. IEEE CS (September 1993)
4. Barros, S., Bodhuin, T., Escudie, A., Queille, J.P., Voidrot, J.F.: Supporting Impact Analysis: A Semi-Automated Technique and Associated Tool. In: ICSM, pp. 42–51. IEEE CS (1995)
5. Bohner, S.A.: Extending Software Change Impact Analysis into COTS Components. In: Proceedings of the 27th Annual NASA Goddard/IEEE Software Engineering Workshop (SEW-27 2002), pp. 175–182. IEEE CS (2002)
6. van den Brand, M.G.J., van der Meer, A.P., Serebrenik, A., Hofkamp, A.T.: Formally specified type checkers for domain specific languages: experience report. In: LDTA, pp. 12:1–12:7. ACM, New York (2010)
7. Brandic, I., Dustdar, S., Anstett, T., Schumm, D., Leymann, F., Konrad, R.: Compliant Cloud Computing (C3): Architecture and Language Support for User-Driven Compliance Management in Clouds. In: CLOUD, pp. 244–251. IEEE CS (2010)
8. van Deursen, A., Klint, P., Tip, F.: Origin Tracking. Journal of Symbolic Computation 15(5-6), 523–545 (1993)
9. van Deursen, A., Klint, P., Visser, J.: Domain-Specific Languages: An Annotated Bibliography. SIGPLAN Notices 35(6), 26–36 (2000)
10. Eclipse Foundation: ATL Transformations, http://www.eclipse.org/m2m/atl/atlTransformations/
11. Falleri, J.R., Huchard, M., Nebut, C.: Towards a traceability framework for model transformations in kermeta. In: ECMDA-TW Workshop, pp. 31–40 (2006)
12. Holten, D.: Hierarchical edge bundles: Visualization of adjacency relations in hierarchical data. IEEE Trans. Vis. Comput. Graph. 12(5), 741–748 (2006)
13. Jouault, F.: Loosely Coupled Traceability for ATL. In: ECMDA (2005)
14. Mannadiar, R., Vangheluwe, H.: Debugging in Domain-Specific Modelling. In: Malloy, B., Staab, S., van den Brand, M. (eds.) SLE 2010. LNCS, vol. 6563, pp. 276–285. Springer, Heidelberg (2011)
15. Mernik, M., Heering, J., Sloane, A.M.: When and how to develop domain-specific languages. ACM Computing Surveys 37, 316–344 (2005)
16. von Pilgrim, J., Vanhooff, B., Schulz-Gerlach, I., Berbers, Y.: Constructing and Visualizing Transformation Chains. In: Schieferdecker, I., Hartman, A. (eds.) ECMDA-FA 2008. LNCS, vol. 5095, pp. 17–32. Springer, Heidelberg (2008)
17. van Ravensteijn, W.J.P.: Visual Traceability across Dynamic Ordered Hierarchies. Master's thesis, Eindhoven Univ. of Technology, The Netherlands (2011)
18. Sendall, S., Kozaczynski, W.: Model Transformation: The Heart and Soul of Model-Driven Software Development. IEEE Software 20(5), 42–45 (2003)
19. Sokolsky, O., Smolka, S.: Incremental Model Checking in the Modal μ-Calculus. In: Dill, D. (ed.) CAV 1994. LNCS, vol. 818, pp. 351–363. Springer, Heidelberg (1994)

Type-Safe Model Transformation Languages as Internal DSLs in Scala

Lars George, Arif Wider, and Markus Scheidgen

Department of Computer Science, Humboldt-Universität zu Berlin
Unter den Linden 6, 10099 Berlin, Germany
{george,wider,scheidge}@informatik.hu-berlin.de

Abstract. Although there are powerful model transformation languages (MTLs) like ATL, model-to-model transformations still are often implemented in general-purpose languages (GPLs) like Java, especially in EMF-based projects. Developers might hesitate to learn another language, use new tools, or they feel limited by the specific but less versatile constructs an MTL provides. However, model transformation code written in a GPL is less readable, contains redundancies or verbose expressions, and there are fewer possibilities for formal reasoning. Our approach combines some benefits of MTLs with GPL programming. We use the GPL Scala to realize MTLs similar to ATL as *internal domain-specific languages*. The benefits are seamless integration with EMF and state-of-the-art tool support as well as the possibility to extend MTLs and to mix MTL and GPL code. In contrast to similar approaches with dynamically typed languages like Ruby, Scala allows for *static type-safety* without adding syntactic clutter.

1 Introduction

Model transformations and their definition are still a prevalent research topic with no generally accepted single solution, but a multitude of transformation methods. In the beginning, model transformations have been regarded as a very general problem. But in fact, there are many specific applications for model transformations, each application with a distinct set of requirements. As a result, many transformation methods have been developed, each tailored for a specific set of similar model transformation applications. Examples for transformation methods are unidirectional imperative model transformations, graph transformations, declarative (uni- and bi-directional) approaches, specific methods for product lines, code generation, etc.

A commonality of all model transformation applications is the typed nature of transformation sources (and in many cases transformation targets). Metamodels for source (and target) models are used. Here, a metamodel defines a set of types (e.g., via classes), and each object in a model has at least one of these types.

To apply a model transformation method, one can either use a general purpose programming language (GPL) or (if available) a domain specific language (DSL) that realizes a transformation method. We call these DSLs *model transformation*

Z. Hu and J. de Lara (Eds.): ICMT 2012, LNCS 7307, pp. 160–175, 2012.
© Springer-Verlag Berlin Heidelberg 2012

languages (MTLs). Examples for MTLs are ATL[11], QVT Operational, ETL[1] from the Epsilon language family, and Tefkat[2].

In general, the DSL community distinguishes between two types of DSLs: external and internal DSLs. An external DSL is a language in its own right. External DSLs require explicit tool support (e.g., a parser, interpreter or compiler). An internal DSL is basically a library written in a so-called *host language* (usually a GPL). Internal DSLs use the existing tools of their host language. Some GPLs are more suitable host languages than others, since some languages offer more syntactical flexibility and hence more possibilities to create the desired DSL syntax. Examples for languages that are considered good host languages are Ruby, Smalltalk, Lisp (including dialects), Groovy, and Scala [7,8].

Most of these established host languages are dynamically typed with little or no compile time type checking. Here Scala is unique: it is a statically typed language and uses type inference to combine static safety with the clear syntax of dynamically typed languages. Due to the importance of types for model transformations, we consider Scala as an ideal host language to realize model transformation methods as internal DSLs.

In this paper, we evaluate Scala as a host language for MTLs, especially in comparison to dynamically typed languages. As an example, we present how to implement an internal DSL for rule-based, *unidirectional*, and hybrid *declarative* and *imperative* model-to-model transformations with *new target source-target relationship* in Scala that is similar to ATL (terminology taken from [6]).

The paper is organized as follows. The next section provides a background on internal DSL development with Scala. In section 3, we present a language similar to ATL and demonstrate it with the well-known Families2Persons[3] example. The next section discusses tool support. Section 5 describes the advantage of using Scala regarding static type-safety and type inference. In section 6, we extend the example MTL with further constructs and cover advanced topics like pattern matching. We end the paper with related work and conclusions.

2   Embedding DSLs in Scala

This section is about Scala features that are particularly helpful for developing internal DSLs.

Flexible Syntax

In general Scala's syntax resembles Java with three major exceptions. First, Scala permits omitting semicolons, dots in method invocations, and parentheses when a method is called with only one parameter. `"Hello".charAt(1);` can be written as `"Hello" charAt 1`. Suitable identifiers provided, statements can resemble natural language sentences. Secondly, type annotations are optional in

---

[1] http://www.eclipse.org/gmt/epsilon/doc/etl/

[2] http://tefkat.sourceforge.net

[3] http://wiki.eclipse.org/ATL/Tutorials_-_Create_a_simple_ATL_transformation

most cases and follow the identifier (as in UML): instead of their type, method definitions begin with def, immutable variable definitions with val, and mutable variable definitions with var. Their type can be inferred in most cases, while still providing static type-safety. Thirdly, type parameters of generic types are enclosed in square brackets and array (and list) items are accessed with normal parentheses.

Listing 1.1. Scala's syntax

```scala
class Container[T] { // type parameters are enclosed in square brackets
  val numbers = List(1,2,3) // type is inferred as List[Int]
  var content: List[T] = null // type cannot be inferred from null
  def access(i: Int): T = { return content(i) } // ': T' is optional
}
```

Implicit Conversions

Usually, you can only change or extend your own code. For example, adding a new method to the existing java.lang.String class is not possible. Languages like Ruby and Smalltalk circumvent this: they allow modifying a class for the whole application. Scala provides *implicit conversions* to change the perceived behavior of classes in each scope differently. Implicit conversions are methods annotated with the additional keyword implicit. The implicit method implicit def fromAToB(from : A) : B = new B(from) for example converts an object of type A to an object of type B. With this implicit conversion declared or imported, objects of type A can be used as objects of type B within the current scope.

Function Objects

In Scala, functions are treated as objects. Functions can be created anonymously, the syntax is (arg: T) => {block}. In Listing 1.2, the functions arithmeticMean or geometricMean can be passed for an easy to read invocation of calculate as demonstrated in line 8.

Listing 1.2. Simple example of an internal DSL

```scala
object Calculator {
  def calculate (fnc: (List[Int]) => Int) = { ...; this }
  def geometricMean(lst: List[Int]): Int = { ... }
  def arithmeticMean(lst: List[Int]): Int = { ... }
  def of(lst: List[Int]) = { ... }
}
// using the calculator DSL resembles natural language:
Calculator calculate arithmeticMean of List(1,2,3)
```

3   A Rule-Based Transformation Language in Scala

To demonstrate Scala as a host language for MTLs, we developed an internal DSL[4] for rule-based, *unidirectional,* and hybrid *declarative* and *imperative* model-to-model transformations with *new target source-target relationship.* This language is designed to resemble ATL. We demonstrate its usage with the help of ATL's Families2Persons example and compare the syntax with that of ATL.

A Simple Transformation

The basic example Families2Persons from the ATL tutorials is a model-to-model transformation. The family metamodel is shown in Fig. 1. Every family member is to be transformed into a person. A person can either be male or female; a person has only one (full) name. The person metamodel is shown in Fig. 2.

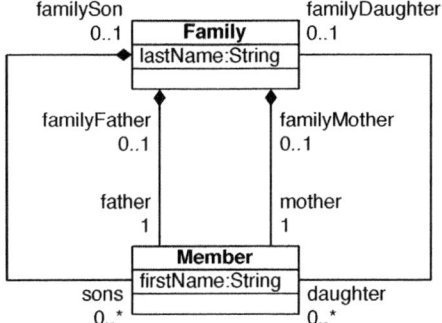

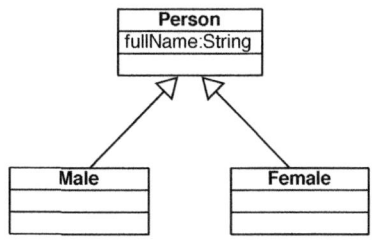

Fig. 1. Families metamodel   **Fig. 2.** Persons metamodel

The transformation creates a `Person` for each `Member` in the source model. The transformation determines the gender of each family member and creates a new male or female person, respectively. Finally, the `fullName` field is set according to the first name of the member and the last name of the family it belongs to.

Rule Definition

ATL transformation rules describe the transformation from a source to a target model element. Listing 1.3 shows an ATL rule that transforms a `Member` to a `Female` element. An ATL rule has different sections: two mandatory (*from* and *to*) and two optional (*using* and *do*) sections. A rule specifies types (i.e., metamodel classes) for its source and its targets within the *from* and *to* sections. In this example, a helper method `isFemale` (implemented in ATL's imperative function syntax; not shown here) is used to check the gender of the source `Member` element (line 3). This ensures that the rule `Member2Female` is only executed upon member elements that are either a `familyMother` or a `familyDaughter`. Within the *do* section, the newly created `Female` is assigned its `fullName` based on the

---

[4] accessible at `http://metrikforge.informatik.hu-berlin.de/projects/smtl`

source member's `firstName` and the `lastName` retrieved with another helper method called `familyName` (line 7).

Listing 1.3. Rule MemberToFemale using ATL

```
1  rule Member2Female {
2    from
3      s: Families!Member (s.isFemale())
4    to
5      t: Persons!Female
6    do {
7      t.fullName <- s.firstName + ' ' + s.familyName;
8    }
9  }
```

In our Scala MTL, rules are instances of the class `Rule` and source and target types are specified as type parameters. Listing 1.4 shows the `Member` to `Female` rule in our example MTL. The `Rule` class provides methods that—by omitting parenthesis, dots, and semicolons—act as the 'keywords' of our MTL. These methods ('keywords') are parametrized with functions. The types of these functions (and their parameters) are determined by the rule's type parameters (and due to type inference, types do not have to be specified again). The *when* method is used to define execution constraints that are passed as a function object (line 3). The passed function (defined in line 9) has to take an object of the rule's source type as input and has to return a boolean value. The actual transformation logic is passed as a function object to the **perform** method (do is already a keyword in Scala and cannot be used here). The passed function has to have two input parameters with types that correspond to the rule's source and target types. In the example, this function is anonymously defined (line 5-7) and the types of its parameters `s` and `t` (source and target) are implicitly inferred. Because of Scala's functional programming features and its concise syntax, it can also serve as a well-integrated alternative to OCL queries in many situations.

Listing 1.4. Rule MemberToFemale using the Scala MTL

```
1  new Rule[Member, Female]
2    when
3      isFemale
4    perform
5      ((s, t) => {
6        t.setFullName(s.getFirstName() + " " + getFamilyName(s))
7      })
8  // using Scala as a GPL for helper methods:
9  def isFemale(m: Member) = m.familyMother!=null ||
       m.familyDaughter!=null
```

Transformation Execution

No special tooling or plug-ins are needed for transformation execution. A transformation is an instance of the class `TransformationM2M` which manages transformation execution. It is parametrized (using 'keyword' methods) with source and target metamodels as shown in Listing 1.5 (line 4-5). One or many rules can be added to the transformation using the `addRule` method (line 7). Calling the `transform` method with the source model as argument starts the transformation. The source model can be provided as an XML file or as an Iterable of `EObjects`. To chain transformations, other transformations can be used as argument as well. The transformation result is returned as an EMF resource (default) or can be saved to the file system calling `export` (line 9).

Listing 1.5. Transformation execution example

```
1  val member2female = new Rule[Member, Female] ... // as in listing 1.4
2  ...
3  val transformation = new TransformationM2M
4    from "http://../Families"
5    to "http://../Persons"
6
7  transformation addRule member2female
8
9  transformation transform sourceModel export "output.xmi"
```

During transformation execution, the source model is traversed. Rules that are not marked as 'lazy' (using the `isLazy` 'keyword') are applied on source model elements directly. A rule is only executed, if the source type matches and if the `when` function returns true. This is comparable to 'matched' and 'lazy' rules in ATL or to the 'top' and 'normal' rule system in RubyTL [2]. Phasing [4] is also supported by calling the transformation's `nextPhase` method between adding multiple rules.

The transformation process keeps traces. The created target elements and the used rules are stored. Traces can be queried within a transformation. Therefore, rules do not necessarily create new objects. On the contrary, new target objects are only created, if not already created within the trace. A transformation rule can explicitly be declared to create new elements every time by calling the rule's `isNotUnique` method. This is similar to 'copy rules' in RubyTL.

4 Tool Support

The rationale for using internal DSLs and Scala as a host language is the better tool support and integration with existing modeling frameworks. Basic tool support is 'for free' for internal DSLs, since the host language's tools can be used. For external DSLs like ATL specific tools need to be developed and their quality directly depends on the efforts put into them. Compared to other internal DSLs

like RubyTL, our approach allows for better code assist and static checks based
on Scala's static type system. Furthermore, Scala is fully byte code compatible
with Java. Java-based modeling frameworks (e.g., EMF or Kermeta) can be used
effortlessly. Since Scala can access any Java class, content assist is also provided
for metamodel-based types, as long as there are corresponding Java classes (such
as in EMF).

```
new Rule[Member, Female] perform ((s, t) => {
    t.setFullName(s.getFirstName() + " " + getFamilyName(s))
})
```

Fig. 3. Code completion in the Scala MTL using Eclipse with the Scala IDE plug-in

To use the presented approach, a Scala compiler and EMF is needed. To get
the described tool support the *Scala IDE*[5] plug-in for Eclipse is recommended
(it includes a Scala compiler). This tooling provides syntax highlighting, wizards, templates, debugging and code completion (based on static types). The
transformation code can be debugged like any other Scala program and all attribute values can be observed at runtime. This includes the (Java-based) model
elements. Within the listings in this paper we also highlight the 'keywords' of
the internal DSL, although this would not be the case in an unmodified Scala
tooling, but could be provided by a separate plug-in.

A dynamically typed language (like Ruby) allows only limited code completion. Therefore, RubyTL for example offers an Eclipse plug-in called AGE[6]. It
provides a Ruby editor with syntax highlighting and code templates. The editor's code completion is limited to the keywords of RubyTL, since no static type
information is available. Errors based on wrong types can only be discovered at
runtime. For ATL (an external DSL) a specific rich-featured editor had to be
developed. In return, ATL's syntax could be perfectly tailored. However, ATL
uses only a small set of data types[7]. Therefore, full support in the editor can
only be offered for those types. Others will be presented as a default data type
named OclAny.

5   Static Type-Safety without Syntactic Clutter

Because of Scala's powerful type inference, the syntactic clutter is noticeably reduced in comparison to other statically typed languages like Java. Furthermore,
implicit conversions greatly improve flexibility; similar to the open class concept
in dynamically typed languages but with static type-safety.

---

[5] http://scala-ide.org/
[6] http://gts.inf.um.es/trac/age
[7] http://wiki.eclipse.org/ATL/User_Guide_-_The_ATL_Language#Data_types

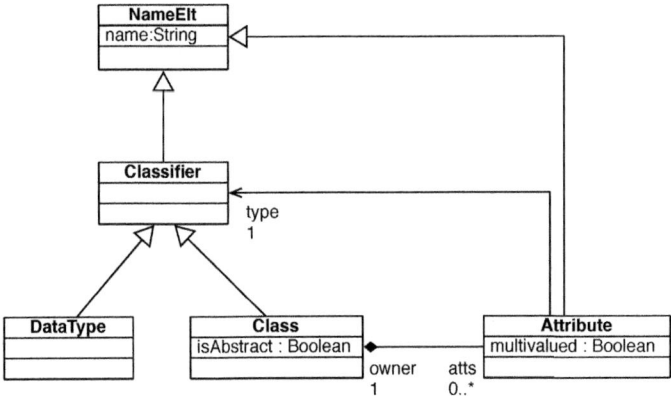

Fig. 4. Class metamodel

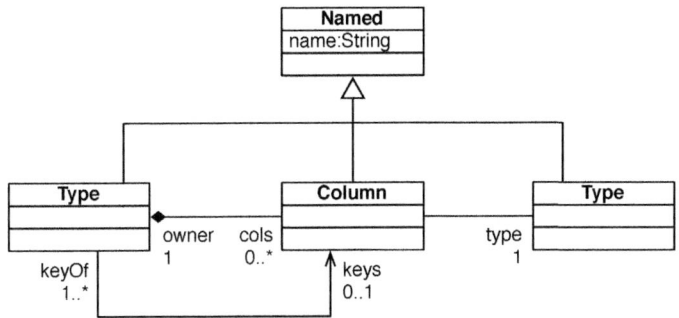

Fig. 5. Relation metamodel

The following listings are based on the ATL example Class2Relational[8]. The listings show a simplified transformation of a class schema model (metamodel in Fig. 4) to a relational database model (metamodel in Fig. 5). They cover a rule that generates a foreign key column in a table based on an attribute of type `Class`. Additionally, the source attribute has to be a single value and not a list (i.e., multivalued) to trigger the rule.

First, the newly created column gets its name. Secondly, the value of the column's attribute 'type' is retrieved with a helper method. Finally, the owner of the new column is set to the owning table. This table needs to be the same as the one generated when the owner of the source attribute was transformed.

The Scala MTL uses implicit conversions to provide a concise syntax: in line 8 of Listing 1.8, a value of type `Class` is passed to the `setOwner` method, which requires a value of type `Table`. Therefore another rule that transforms a `Class` to a `Table` is needed. The need for such a rule can be explicitly expressed with the MTL's method `as[ReturnType](inputObject)`. In this example an expression for explicit conversion would be `col.setOwner(as[Table](attr.getOwner))`. Similarly, a lazy rule (see Sec. 3) is also explicitly called like this. However, the

---

[8] http://www.eclipse.org/m2m/atl/atlTransformations/#Class2Relational

Listing 1.6. Rule Attribute2Column using ATL

```
1  rule classAttribute2Column {
2    from
3      attr : Class!Attribute (
4        attr.type.oclIsKindOf(Class!Class) and not attr.multivalued
5      )
6    to
7      col : Relational!Column (
8        name <- attr.name + 'Id',
9        type <- thisModule.objectIdType,
10       owner <- attr.owner
11     )
12 }
```

Listing 1.7. Rule Attribute2Column using RubyTL

```
1  top_rule 'classAttribute2Column' do
2    from Class::Attribute
3    to   Relation::Column
4
5    filter do |attr|
6      attr.type.kind_of? Class::Class and not attr.multivalued
7    end
8
9    mapping do |attr, col|
10     col.name = attr.name + 'Id'
11     col.type = objectIdType
12     col.owner = attr.owner
13   end
14 end
```

Listing 1.8. Rule Attribute2Column using the Scala MTL

```
1  new Rule[Attribute, Column]
2    when ((attr) => {
3      attr.getType.isInstanceOf[Class] && !attr.isMultivalued
4    })
5    perform ((attr, col) => {
6      col.setName(attr.getName + "Id")
7      col.setType(objectIdType)
8      col.setOwner(attr.getOwner)
9    })
```

explicit call can be omitted if the needed rule was declared to be implicitly available: `implicit val classToTable = new Rule[Class, Table] perform (...)`.

This is possible, because the `Rule` class in the Scala MTL extends the predefined `Function1` type. As a result, a rule can be used like a function (with one parameter). The signature of this function is determined by the rule's type parameters. An invocation of the `classToTable` function therefore needs a parameter of type `Class` and returns a `Table`. The Scala compiler inserts invocations of these 'rule functions' automatically to convert objects implicitly as long as the required rules are marked as implicit and are in scope.

In the example, a conversion from the attribute's owner (of type `Class`) to the type that is needed for the column's owner (of type `Table`) is necessary. The Scala compiler solves this type problem by automatically calling the `classToTable` rule. If no appropriate rule is available, the developer gets an error message at compile time stating that no suitable conversion could be found or why available conversions do not fit.

This example shows how Scala's type inference and implicit mechanisms can be used to create a syntax that is as concise as in ATL or RubyTL but still preserves static type-safety. An exemplary calculation of the Halstead metrics [9] for the Families2Persons and Class2Relational example using ATL, QVTo, RubyTL, and our Scala MTL confirmed this. In fact, Scala's implicit mechanism is a rule-based system itself and therefore plays well into the implementation of rule-based transformation languages. However, as the insertion of an implicit conversions is only based on the source and target types, the inserted rule can still fail at runtime because its value-based constraint is not satisfied.

6 Extending the Internal DSL

One of the main advantages of internal DSLs is their easy extensibility in contrast to external DSLs where DSL-specific tools have to be adapted accordingly. In this section, we demonstrate how to add functionality to the Scala MTL.

Multiple Target Model Elements

In the simple example of Listing 1.4 one object of type `Member` is transformed into one object of type `Female`. Other transformation languages allow to create more than one target object per rule. This can be a list of objects of the same type (often called one-to-many rule) or objects of different types.

Regarding different target types, a drawback of using type parameters to define a rule's source and target type is the fixed number of type parameters. Scala's type system does not allow to overload types with a different number of type parameters. In order to allow a rule to have more than one target type, we could define different rule types with a different number of type parameters like `Rule2[S,T1,T2]`, `Rule3[S,T1,T2,T3]` etc. This leads to duplication of code. To avoid this, one could use *heterogeneously typed lists* (HLists [12]). However, this increases code complexity considerably.

We propose a more lightweight solution. We use a statically available Scala object that enhances the syntax of the Scala MTL with a `create` method. With this method, one creates additional output objects without changing the rule's 'signature'. There is a type parameter to determine the target object's type (refer to Listing 1.9, line 5). All attributes and methods of the returned model object are accessible as usual. However, these additional target objects are not defined within the rule's signature and are created as a side effect. This becomes important, if a rule is used implicitly as a function (see section 5).

Listing 1.9. Creating more target model elements

```
1  new Rule[Member, Female]
2    perform ((s, t) => {
3      t.setFullName(s.getFirstName() + " " + getFamilyName(s))
4
5      var newFemale = create[Female]
6      newFemale.setFullName("...")
7    })
```

Another obvious solution to target object creation is the standard **new** operator. But this solution is inadequate for two reasons. First, EMF objects should never be created directly. EMF objects should only be exposed via (generated) interfaces. Instances based on concrete implementations are created by factories. This allows EMF models to function in different contexts transparently (e.g., EMF models stored in multiple resources, and EMF models stored in a CDO database). Secondly, new target objects need to be registered within the transformation trace; the create method does this automatically.

The ability to create multiple target model elements in a rule also allows one-to-many transformations, i.e., rules with multiple target objects of the same type. To further support this, the 'keyword' toMany can be used instead of `perform`. The **toMany** method expects a function as argument similar to `perform`. But the second argument that is given to this passed function is a reference to an empty list and not the target object. This list can be filled with an arbitrary number of target objects.

Declarative Element Creation Using Case Classes

Many transformation rules only create a target model element and set its attributes. Therefore MTLs often provide features to create objects and pass their attributes along. ATL for example supports this in its declarative 'to' section.

In Scala, a similar way for a more declarative object creation is the use of *case classes*. Scala case classes are declared with the **case** keyword. The compiler generates a bunch of instance and static methods for each case class. These methods (among other things) allow creating instances without the **new** operator and allow passing constructor arguments directly behind the class' name. This is especially useful, if the fields of a case class are also implemented as case classes.

Nested object structures can be created in a single line of code, e.g.: val x = ClassA(ClassB(1),ClassC(ClassD())).

However, since EMF does not allow to use concrete model element implementations directly, we cannot use case classes directly. Instead, we generate corresponding case classes for all types of the target metamodel. Furthermore, we generate implicit conversions that convert instances of these generated case classes to their corresponding target model objects (using EMF's factories). The required code can be generated explicitly with a Scala script or with an Eclipse plug-in. By default, the case classes are named like their corresponding metaclass but with a 'CC' postfix.

Syntax can be further shortened by overloading the perform method. The overloaded method expects a function with a single parameter as the argument: only the source model element is passed and the target object is expected as the return value. Listing 1.10 shows a simplified version of the Member2Female rule that uses case classes and the simplified perform method.

Listing 1.10. Object creation using case classes

```
1  new Rule[Member, Female]
2    perform ((s) => {
3      FemaleCC(s.getFirstName() + " " + getFamilyName(s))
4    })
```

Here, it is particularly helpful that Scala supports *named* and *default parameters*: In the example above, the target element creation in line 3 could alternatively be written as FemaleCC(*fullName* = s.getFirstName() + ...). This makes the instantiation of classes with many constructor parameters easier. Case classes can also be used for rules with multiple targets. In summary, case classes allow declarative-like object creation in an otherwise imperative MTL.

Pattern Matching

Pattern matching is a powerful Scala feature, particularly useful for transformation code with a lot of alternatives or null checks. Pattern matching in Scala is only available on instances of case classes because here, the Scala compiler automatically provides the needed instance methods. Therefore, if we not only generate case classes and corresponding implicit conversions for the target metamodel but also for the source metamodel, we can seamlessly integrate Scala's pattern matching: source model elements are implicitly converted to case class instances and patterns can be matched on those instances.

Listing 1.11 shows an example. It uses the relation metamodel (Fig. 5). Whereas the attributes in a metamodel class are unordered, the constructor parameters of the corresponding case class are ordered. We implemented this order to be alphabetic by the attribute's name. This results in the following case class constructors: ColumnCC(keyOf, name, owner, type), TypeCC(name), TableCC(cols, keys, name).

Pattern matching is made available by using `use.matching` instead of the `perform` keyword. This way, the implicit conversion is automatically triggered. The case statement compares a specified case class instance (the pattern) with the source model element that was converted to a case class instance of the same type, here a `MaleCC` instance as the pattern with a (converted) Male source model element. The case statement's body has to return a target model element (or an instance of the corresponding case class).

Listing 1.11. A pattern matching example using generated case classes

```
new Rule[Column, Type].use.matching {
  case ColumnCC(_, _, TableCC(_, _, name), t@TypeCC(_))
    if name.startsWith("m2m_") => t
  case _ => TypeCC("unknown") // the default case
}
```

The presented pattern matching uses constructor patterns. Such patterns consist of a case class instance, its attributes, and potentially nested case class instances. An attribute that does not need to match any pattern can be specified with the '_' character. Matching on patterns with nested case class instances is called deep matching: not only the top level objects are checked, but also the content of the contained (case class) instances. The '_' can also be used as a top level pattern to define the default case. The example pattern in Listing 1.11 matches if the `Column` has an `owner` and a `type` set (line 2). Furthermore the `owner`'s `name` needs to start with 'm2m_': the pattern does only match columns in tables called 'm2m_...'. On a match the `type` of the `Column` is returned.

Scala's pattern matching makes null checks on attributes unnecessary. The example of Listing 1.11 will not fail, even if the source `Column`'s type is null. The '_' pattern matches anything including null; null is only matched by null (explicit null check) or '_'. Beyond that, a fine grained error handling is possible with pattern matching: Each unsatisfying attribute occurrence can be addressed explicitly with a case statement and the appropriate error handling code. This allows for effective separation of actual logic (triggered by the desired input pattern) from error handling code. A `MatchError` will only be thrown, if no case matched at all. This can be prevented by providing a default case as the last case statement.

Listing 1.12 demonstrates a possible rule written in ATL covering the pattern match of Listing 1.11. First, null checks for the needed attributes are done (line 7), and then the name of the owner is tested. The readability can suffer from several nested `If` statements, and more complex pattern structure can easily lead to missing cases: The rule in Listing 1.12 for example does something for all cases, except if `type` and `owner` are defined and the `owner name` is incorrect.

However, there is an issue with our approach: pattern matching with implicit conversions to case classes is designed for trees and many models are graphs (i.e., with references); a naive implementation of the implicit conversions to case classes can lead to circular dependencies so that the implicit mechanism prevents

Listing 1.12. An ATL version of the pattern matching example

```
1   rule ColumnTypeSelectTest {
2     from
3       c : Relational!Column
4     to
5       type : Relational!Type
6     do {
7       if(not c.type.oclIsUndefined() and not c.owner.oclIsUndefined()) {
8         if(c.owner.name.startsWith('Family')) {
9           type.name <- c.type.name;
10        }
11      } else {
12        type.name <- 'unknown';
13      }
14  }  }
```

the compile process from terminating. A solution is to traverse the containment hierarchy (a spanning tree existing in each EMF model) twice: one time for creating new objects and a second time for setting the cross references.

7   Related Work and Discussion

The general idea and best practices of internal DSLs have been extensively discussed by Martin Fowler on his blog (finally edited into a book [7]). A set of patterns for internal DSLs in several languages has been published by Günther and Cleenewerck in [8]. Scala's potential as a host language for general DSLs has been evaluated in [14]. Hofer et al. showed the extensibility of DSLs written in Scala in [10]. Scala has already been used as a host language for a variety of internal DSLs, e.g., in [16] and [1]. Sloane showed how the term-based transformation language *Stratego* can be implemented as an internal DSL in Scala [15].

Picard showed how to use Scala for EMF model transformations [13]. However, no domain-specific model transformation constructs or syntax elements were implemented. The work basically shows how to parse an EMF model from its XMI serialization, create Scala objects from it, and how Scala as a GPL can be used to implement transformations. Therefore, the fact that Scala is JVM-based is not leveraged and there is no integration with EMF tooling.

Cuadrado and Molina used Ruby as host language for their MTL called RubyTL [3,2]. Similar to the work presented in this paper, they designed RubyTL to be a hybrid transformation language that uses declarative constructs to realize pattern matching and rule selection and an imperative style to realize rule actions. Furthermore, RubyTL is designed as an extendable MTL with the goal to efficiently implement and evaluate new transformation techniques. In [5,4] the authors facilitated these characteristics to research rule factorization and composition techniques based on rule phasing (i.e., assign rules to consecutively executed phases).

Ruby is a dynamically typed language with fast prototyping capabilities but also a lack of static type-safety. This is in direct contrast to our work with Scala. Scala uses static type inference, which allows for a similar programming style as in dynamically types language but preserves static type-safety. Furthermore, Scala as a host language provides better tool support than RubyTL (or any other dynamically typed language) due to the annotation of static errors and superior content assist based on programming-time knowledge about the type of any variable. Furthermore, RubyTL is not based on EMF or any other comparable technology, and works directly on an in-memory representation of XMI. RubyTL therefore does not use any metamodel information and even if Ruby was statically typed, RubyTL would have no types to work with. Additionally, most existing modeling APIs are written in Java and as such can be used from within Scala, but cannot (at least not without limitations) be used in Ruby.

However, our approach also shares some of the general disadvantages of internal DSLs: In contrast to external DSLs, code completion and error messages are not tailored for the DSL. Therefore, some knowledge of the host language is needed when using an internal DSL. Internal DSLs are easier extensible than external DSLs (because no DSL-specific tools have to be adapted) but often advanced features of the host language are used in order to achieve a desired DSL syntax. Therefore, the DSL's implementation is sometimes hard to understand which makes extensions to the DSL less straightforward. The complexity of the DSL's implementation can also show through in complex error messages. Finally, the ability to mix MTL constructs with GPL code is also a disadvantage because arbitrary GPL code seriously limits possibilities for formal reasoning.

8   Conclusions

In this paper, we used Scala to implement model transformation languages as internal DSLs. We showed that Scala can be used as a host language for model transformation languages and is flexible enough to create syntaxes that resemble known model transformation languages, for example ATL. In [17], we also implemented an MTL for *bidirectional model transformations* that makes more use of advanced features of Scala's type system. Since we use an internal DSL approach, model transformation languages are easy to extend: language features can be added and existing behaviour can be adopted to the specific needs of one's current transformation task. Furthermore, Scala is rooted in the Java programming environment and existing modeling APIs (which are mostly written in Java, e.g., EMF and anything written for it) can be used immediately. Furthermore, transformations or helper methods that were already written in Java can be reused, integrated and gradually migrated to the more concise means that the internal DSL provides. Compared to other internal DSL approaches to model transformations, Scala is statically typed based on type inference: it provides a clean syntax similar to dynamically typed languages, but still provides the benefits of static type-safety. These benefits are (among others) compile type warnings and errors as well as better code completion based on type information. Compared to external languages, powerful tool support (including full debugging support) does already exist, even though not tailored for each specific internal DSL: especially error messages can be confusing. An IDE plug-in could be provided to improve error messages or to provide templates and syntax highlighting for the internal DSL. However, this would eliminate the advantage of being independent from DSL-specific tools and their development.

References

1. Barringer, H., Havelund, K.: TRACECONTRACT: A Scala DSL for Trace Analysis. In: Butler, M., Schulte, W. (eds.) FM 2011. LNCS, vol. 6664, pp. 57–72. Springer, Heidelberg (2011)
2. Cuadrado, J.S., Molina, J.G., Tortosa, M.M.: RubyTL: A Practical, Extensible Transformation Language. In: Rensink, A., Warmer, J. (eds.) ECMDA-FA 2006. LNCS, vol. 4066, pp. 158–172. Springer, Heidelberg (2006)
3. Cuadrado, J.S., Molina, J.G.: A Plugin-Based Language to Experiment with Model Transformation. In: Wang, J., Whittle, J., Harel, D., Reggio, G. (eds.) MoDELS 2006. LNCS, vol. 4199, pp. 336–350. Springer, Heidelberg (2006)
4. Cuadrado, J.S., Molina, J.G.: Modularization of Model Transformations Through a Phasing Mechanism. Software and Systems Modeling 8(3), 325–345 (2009)
5. Cuadrado, J.S., Molina, J.G.: Approaches for Model Transformation Reuse: Factorization and Composition. In: Vallecillo, A., Gray, J., Pierantonio, A. (eds.) ICMT 2008. LNCS, vol. 5063, pp. 168–182. Springer, Heidelberg (2008)
6. Czarnecki, K., Helsen, S.: Classification of model transformation approaches. In: 2nd Workshop on Generative Techniques in the Context of MDA, pp. 1–17 (2003)
7. Fowler, M.: Domain-Specific Languages. Addison-Wesley (October 2010)
8. Günther, S., Cleenewerck, T.: Design Principles for Internal Domain-Specific Languages: A Pattern Catalog illustrated by Ruby. In: 17th Conference on Pattern Languages of Programs, PLoP (2010)
9. Halstead, M.: Elements of Software Science. Elsevier Science Inc. (1977)
10. Hofer, C., Ostermann, K., Rendel, T., Moors, A.: Polymorphic Embedding of DSLs. In: 7th Conference on Generative Programming and Component Engineering, GPCE 2008, pp. 137–148. ACM (2008)
11. Jouault, F., Kurtev, I.: Transforming Models with ATL. In: Bruel, J.-M. (ed.) MoDELS 2005. LNCS, vol. 3844, pp. 128–138. Springer, Heidelberg (2006)
12. Kiselyov, O., Lämmel, R., Schupke, K.: Strongly typed heterogeneous collections. In: Haskell 2004: ACM SIGPLAN Workshop on Haskell, pp. 96–107. ACM (2004)
13. Picard, C.: Model Transformation with Scala. Master's thesis, Universitat Politècnica de Catalunya (2008)
14. Pointner, R.: An Evaluation of Scala as a host language for DSLs. Tech. rep., University of Central Lancashire (June 2010)
15. Sloane, A.M.: Lightweight Language Processing in Kiama. In: Fernandes, J.M., Lämmel, R., Visser, J., Saraiva, J. (eds.) GTTSE 2009. LNCS, vol. 6491, pp. 408–425. Springer, Heidelberg (2011)
16. Spiewak, D., Zhao, T.: ScalaQL: Language-Integrated Database Queries for Scala. In: van den Brand, M., Gašević, D., Gray, J. (eds.) SLE 2009. LNCS, vol. 5969, pp. 154–163. Springer, Heidelberg (2010)
17. Wider, A.: Towards Combinators for Bidirectional Model Transformations in Scala. In: 4th Conference on Software Language Engineering (SLE 2011). Springer (2012)

Towards a Family
of Model Transformation Languages

Jesús Sánchez Cuadrado

Universidad Autónoma de Madrid, Spain
jesus.sanchez.cuadrado@um.es

Abstract. Many model transformation languages of different nature have been proposed during the last years, each one of them suitable for a certain kind of transformation task. However, a complex transformation problem may not fall into a single transformation category, making the solution written in the chosen transformation language suboptimal, as some concerns cannot be handled naturally.

To tackle this issue, we propose to define a model transformation tool as a family of model transformation languages. Each member of the family is a simple language intended to deal with a particular kind of transformation task. In this paper we discuss the different issues involved, such as design decisions, interoperability among languages, and composability. We illustrate the paper with a transformation from UML and OCL to Java, in which languages for pattern matching, mapping, attribution and target-oriented transformations are used. Finally, the approach is validated with a proof-of-concept implementation.

1 Introduction

Model transformation is one of the key elements in Model Driven Engineering (MDE). Hence, in the last years a number of model transformation languages of different nature have been proposed. As acknowledged by the classifications of model transformation languages given in [4] and [13], each language provides a series of features that make it more suitable to address a certain kind of transformation problems. So far, two paths have been taken by transformation language designers: a) keep the language focused or b) add more features to the language in order to widen its scope. The first approach limits the applicability of the language, while the second one tends to pollute the original design.

We have been working on an alternative design, in which a model transformation language is made up of smaller languages. Each language is focused on a specific kind of transformation task, and altogether form a so-called *family of model transformation languages*. In this way, a complex transformation problem could be split into smaller tasks using the most appropriate language for each one of them, with the additional advantage of enhanced declarativeness and intentionality, as languages are really tailored for the problem being solved. Realizing this approach requires a way to make the languages interoperable, as

Z. Hu and J. de Lara (Eds.): ICMT 2012, LNCS 7307, pp. 176–191, 2012.
© Springer-Verlag Berlin Heidelberg 2012

well as composition mechanisms in order to specify how the results provided by each language contribute to the global transformation result.

This paper reports the initial results of our work building a family of model transformation languages, named Eclectic. We focus on two aspects: the main design decisions which span all languages in the family, and language interoperability and composability mechanisms which are the foundation of our approach. To show the feasibility of the approach the paper is illustrated with a transformation from UML and OCL to Java, which is addressed using four languages of Eclectic, for pattern matching, mapping, attribution and target-oriented transformations. The paper also reports on a proof-of-concept implementation, which includes textual editors and a compiler for the Java Virtual Machine (JVM), that is freely available at [5]. The architecture of the tool is extensible so that it would allow us to integrate domain specific transformation languages (DSTL).

Paper Organization. Section 2 explains the main design decisions, presents the running example, and introduces four languages of Eclectic. Section 3 describes the interoperability and composition mechanisms. Section 4 reviews some related work. Section 5 gives some conclusions and outlines the future work.

2   Design of the Family

The design of a family of transformation languages must take into account two main concerns: the design of the different languages that build it up, and how to compose them which in turn will require making them interoperable.

In this section we will outline the design principles of Eclectic and illustrate some of its member languages by means of a running example. Section 3 will explain the interoperability and composability mechanisms for them.

2.1   Design Principles

The aim of our approach is to tame complex model transformations by promoting intentionality. As a motivating example, ATL is well-known for being a simple, declarative transformation language, but as the transformation problem at hand moves away from being a mapping task, intentionality blurs. This is so because non-declarative constructs such as lazy rules, imperative rules or complex navigation code must be tangled with declarative code. Our design tries to solve this issue by providing separate languages for different kinds of transformation task, following a series of design principles:

- **Few Features.** The number of constructs of each language should be kept as low as possible, including only those that are important to tackle the task each language is intended to.
- **Orthogonality.** Each language should only be useful for a few tasks, avoiding redundancy with respect to the ones addressed by other languages. This will facilitate users choosing a language for each particular task.

- **Simple Syntax.** The fact that few features will be included in each language will facilitate the definition of a simpler and cleaner syntax than complex languages (e.g, no need for statement separators). In addition, syntax should highlight those constructs that are the essence of each language, while "hiding" constructs that are more accidental.
- **Lightweight Type Information.** The amount of type information should be low. This can be achieved with type inference (which would be facilitated because languages are simple) or by relying on dynamic typing. Our current implementation uses dynamic typing, but we plan to support type inference.
- **Eclecticism.** As a major design principle, we believe that each style of model transformation has its own value to tackle certain problems, so we do not restrict Eclectic to the languages considered so far, but we are looking into other possible languages, and we are willing to contributions in this sense.

2.2 Running Example

The rest of the paper will be illustrated with a transformation that takes a UML model plus an OCL model with invariants and preconditions, and generates a Java model. We have used the UML meta-model of the Eclipse UML2 plug-in, an OCL meta-model based on the ATL implementation, and the Java meta-metamodel of MoDisco. Figure 1 shows relevant excerpts of them.

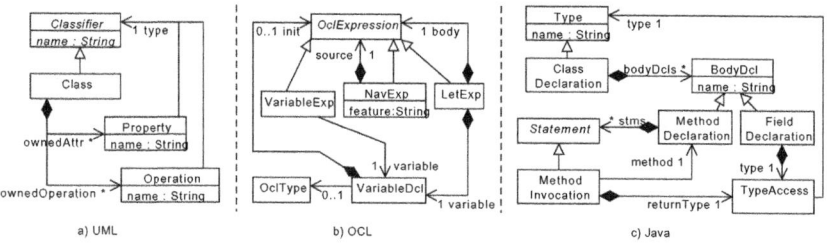

Fig. 1. Excerpts of the meta-models involved in the running example. They have been slightly simplified for the sake of clarity (e.g., renamings, hierarchy flattening).

In this transformation (uml2java) there are several aspects to consider. The mapping between UML class models and Java classes is more or less straightforward, except for some cases that requires detecting concrete patterns. On the other hand, translating OCL expressions to Java can be (partially) done with templates that generate pieces of Java code, but it requires computing type information for an accurate translation. Finally, we restrict UML models to use single inheritance.

In the rest of the section the languages are briefly introduced by showing an excerpt of the running example. The examples in this section do not consider language composition, but it is added in the next one. In addition, the development of the languages that compose Eclectic is being inspired by existing languages of different nature. Thus, we will briefly comment on related work as the languages are explained, although more details are given in Section 4.

2.3 Mapping Language

Establishing correspondences among meta-model elements in order to fix heterogeneities between semantically equivalent meta-models is the kind of transformation task that rule-based model-to-model transformation languages in the style of declarative ATL can handle naturally. For our initial prototype we have chosen to create a simplification of ATL, called **SMaps**.

Listing 1 shows an excerpt of the mapping between UML and Java using this language. The input and output models are indicated between parenthesis in the transformation header (line 1). This style is followed in the rest of the languages of Eclectic. As can be seen, mappings among source and target metaclasses are specified using from - to (lines 4 and 14), and they may include modifiers such as linking (stating how to relate both target elements). Mappings between structural features are specified using ←. We allow the same reference to be mapped more than once if it is multivalued (lines 10-11).

Conversions between datatypes and explicit transformations are done with the notion of converter. A converter is basically a function that is offered as a library (line 2 performs the importation) or can be specified as a mapping between datatypes (lines 20-23). It is implicitly invoked using the the convert modifier (line 7 and 8). The rationale for this notation instead of plain function calls is to enhance text clarity, so that the reader clearly identifies the left part and the right part of mapping (i.e., with a function call the right part of the mapping is "wrapped" into the actual parameters). In our implementation, libraries of converters are provided as Java classes.

```
1    mapping struct(uml) −> (java)
2      uses java_conventions
3
4      from uml!Class to cd: java!ClassDeclaration, cu : java!CompilationUnit
5        linking cd.originalCompilationUnit = cu
6
7        cd.name <− name convert java_conventions.camelCase
8        cd.visilibity <− visibility convert mapVisibility
9
10       cd.bodyDeclaration <− ownedAttribute
11       cd.bodyDeclaration <− ownedOperation
12     end
13
14     from uml!Property to m: java!MethodDeclaration, ta: java!TypeAccess
15       linking get.typeAccess = ta
16       m.name <− name convert java_conventions.getterName
17       ta.type <− type
18     end
19
20     converter mapVisibility: uml!Visibility −> String
21       #vk_public −> 'public'
22       #vk_private −> 'private'
23     end
```

Listing 1. Excerpt of the mapping from UML to Java

The execution semantics is a mix between ATL and RubyTL. Mappings correspond to ATL rules, and the ← construct is a form of ATL binding. However, binding resolution works as in RubyTL, taking into account the conformance relationship of both the source and the target element, but, so far, only the first target element is resolved as in ATL.

With the aim reducing the cluttering of the transformation text, we have decided to do without navigation language (e.g., OCL), so that model navigation must be done in separate navigation modules.

As can be observed the language is intended neither to extract implicit information or to perform one-to-many transformations (when the number of target elements of a mapping is not known before hand). Instead, it is focused on resolving structural heterogeneities between semantically equivalent models. In this line, we expect to evolve SMaps by considering constructs in the style of the Mapping Operators (MOPs) proposed in [23].

2.4 Target-Oriented Language

This language, called Tao, is intended to address transformations mainly driven by the structure of the target model. This roughly corresponds to the style of model-to-text template languages, where fixed pieces of text are parameterized with expressions that fill in the holes. This kind of transformations usually has a high-degree of nesting, thus a design decision has been to consider object syntax as a way to specify large instantiation sequences (similar to QVT syntax).

Listing 2 generates a Java class from an OCL specification, with one method per OCL invariant. Templates are specified with template, and take one or more parameters, being polymorphic on the first parameter (e.g., lines 15 and 18). We have chosen this syntax because we plan to experiment with multiple-dispatch templates. The instantiation of a new object is specified with model!Metaclass { assignments }, where assignments initialize attributes and references. For

```
1    tao gen_java(uml, ocl) −> (java)
2      uses java_conventions, string
3
4      template mapProgram(p : ocl!Program)
5        java!ClassDeclaration {
6          name = p.name convert java_conventions.camelCase, string.concat("Check")
7          bodyDeclarations = p.invariants to java!Method {
8            name = name
9            visibility = "public"
10           expressions = p.body with mapExpression
11         }
12       }
13     end
14
15     template mapExpression(expr : ocl!NavigationExpr)
16       ...
17     end
18
19     template mapExpression(expr : ocl!LetExpr)
20       ...
21     end
```

Listing 2. Generator from OCL to Java

example, line 6 sets the name of the Java class by combining two converters, where the result of the first one is the input of the second one. Then, line 7 initializes bodyDeclarations by creating one method per invariant defined in the OCL program. Finally, line 10 invokes mapExpression explicitly.

As can be seen, this language has only a few elements yet it simplifies a task that is sometimes cumbersome. Please note that, although some constructs such as object-syntax or template invocation resemble parts of QVT, there is the important difference that we have avoided complex mechanisms, for instance QVT Operational initialization rules.

Finally, it could be possible to consider template languages that use the syntax of the target language (e.g., Java) to make the transformation text more fluent. In fact, these kind of languages could be integrated in Eclectic as libraries contributed by third-parties, acting as domain-specific transformation languages.

2.5   Attribute Computation

Attribute grammars are a well known technique for specifying how to compute properties of language constructs, called attributes, by defining their values in terms of the attribute values of related constructs [1]. Attribute computation is defined by rules (or equations), and the attribution system is in charge of performing the evaluation by associating attribute values to syntax tree nodes, propagating values through the nodes as needed. Transformation problems that require propagating values top-down or bottom-up are typically difficult to express with some transformation languages (e.g., ATL, TGGs). In QVT it is possible to use *when* and *where* clauses to propagate values, but "propagation code" gets tangled with mapping code.

In this way, a simple language for attribute computation has been defined. It is called SAttr. It supports synthesized attributes that propagate information bottom-up, and inherited attributes that propagate top-down. It also includes a simple expression language, as the essence of this kind of transformations is to perform computations based on previously computed values. An attribution transformation is composed of attribution rules. Each attribution rule matches an element of a given metaclass and computes attribute values. To this end, there are two basic constructs: attribute initialization and attribute access. The expr[attr] ← right-part construct is used to initialize an attribute, and it has the effect of associating the value of right-part to attr for the element referred by expr. In the case of synthesized attributes, self is used to refer to the element matched by the rule. Similarly, attribute values can be accessed with the expr[attr] construct (in this case using self[attr] to access an inherited attribute). Our implementation schedules the execution according to the dependencies.

Listing 3 shows a piece of transformation that associates type information to the elements of the OCL abstract syntax model (it is partially inspired in the OCL specification). In this example, the type of each sub-expression is propagated from the leaves using the type synthesized attribute (line 2), while contextual type information is provided by the env inherited attribute (line 3, an immutable map that associates a variable declaration with its type). In line 7, the type of a navigation expression (NavExpr) is gathered by getting the type of the receptor object (self.source[type]), and then looking up the type of the navigated feature (the feature operation is a helper defined in a navigation library). As another example, to deal with *let* expressions, line 11 adds a variable declaration to the inherited env, computing the variable type as the type of the initialization expression (so performing weak type inference). Then, it is propagated as an inherited attribute so that it is available for the body expressions. Note that, in line 16 the env attribute is used to gather the type of a variable reference that points to a variable declaration. Finally, even though it is not shown in this example, it is possible to create target elements if needed by interoperating with languages with this capacity, as is the case of Tao (see Section 3.2).

```
1   attribution typing(uml, ocl) −> ()
2     syn type : uml!Classifier
3     inh env : _!Map
4
5     rule ocl!NavExp
6       receptor[env] <− self[env]
7       self[type] <− self.source[type].feature(self.name).type
8     end
9
10    rule ocl!LetExp
11      body[env] <− self[env].put(self.varDcl, self.init[type])
12      self[type] <− self.body[type]
13    end
14
15    rule ocl!VariableExp
16      self[type] <− self[env].get(self.variable)
17    end
```

Listing 3. Collecting type information from OCL expressions

Our current design only considers basic features of attribution systems. Other systems such as Kiama or Silver implement more complex features, and we want to explore which ones are more useful in a model transformation setting. Nevertheless, in its current state, we have found this language particularly useful to compile expression languages to a low-level representation (see Section 3.1).

2.6   Pattern Matching

The languages shown so far just match a single model element. In order to address transformations where more complex patterns have to be found, we have included a simple pattern matching language in the family, named SPat.

Here we just briefly introduce the language, by means of the example shown in Listing 4, which is completed in the next section. The GettableProperty pattern gathers all public UML Property whose owning class does not contain an operation whose name would collide with the corresponding Java getter method. As can be seen this language is in the style of Tefkat, although other styles such as the one of VIATRA2 one could be possible. Interestingly, the getterName converter can be reused as a function call (line 4).

```
1   pattern GettableProperty
2     forall p: uml!Property [ p.visibility = #pk_public ] and
3       not exists o: uml!Operation [ p.owner.includes(o) and p.name = java_conventions.getterName(a.name) ]
4   end
```

Listing 4. Matching properties that do not collide with an existing "get"

In this section we have commented on four languages that illustrates the design of Eclectic. They have have been presented without taking into account how to make them work together. Next section discusses the issues involved.

3 Language Composition

Our design based on a family of model transformation languages allows us to decompose a transformation problem into subproblems, where each subproblem is tackled with the most appropriate language. However, this poses two main concerns: language composition and interoperability.

In this context, *composition* is the ability of combining different languages to achieve a common task, while *interoperability* is the ability of two or more components (transformation languages in this case) to exchange information and to use it. There are two types of transformation composition: internal and external. Internal transformation composition refers to the composition of transformation constructs of a single language, while external transformation composition must take into account how to compose heterogeneous constructs belonging to different languages. Indeed, in this setting we are dealing with external transformation composition which requires interoperability.

3.1 Interoperability

Transformation language interoperability has been regarded as an important topic in model transformation [10]. However, so far, only limited forms of interoperability has been achieved [22].

Figure 2 shows the architecture of our solution. Our approach to interoperability is based on a common intermediate language, called Intermediate Dependency Code (IDC), so that each member of the family compiles down to it. IDC is composed of a few basic instructions (some of them specialized for model manipulation), which use a simplified form of Static Single Assignment (SSA) to represent data dependencies between instructions [3]. IDC does not force any particular transformation style (e.g., rule-based transformations) as it does not provide any notion of rule, but lower-level mechanisms. IDC is compiled to the JVM, and it uses a runtime library to deal with different modeling frameworks (we currently support EMF).

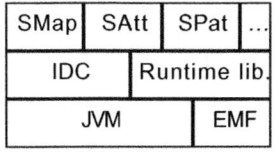

Fig. 2. Eclectic architecture

The key element of IDC is its ability to schedule the execution of several transformations based on their data-dependences using continuations. A continuation reifies the concept of "the rest of the computation", so that execution state of a given program can be saved into a continuation and restored later. This concept is supported in some programming languages, for instance Scheme or Scala [16]. Basically, we use continuations to enable a transformation execution to be suspended until a required piece of data (provided by another transformation) is available. This provides a means to integrate heterogeneous languages. Additionally, as all transformations run over the JVM, method-level interoperability is also supported. Due to space reasons we do not give a full explanation of IDC, but more details are given in [17].

3.2   Composition

Our strategy to tackle transformation composition has been to identify abstract relationships among transformation constructs, and reify them in each language as a composition mechanism. The definition of each composition mechanism has two aspects: how one language publishes or makes available the data it handles, and how another language requires and consumes this data.

Based on our experience with model transformations, and during the process of building Eclectic, we have identified four types of composition: *feeding* a transformation rule or a pattern with some value(s), *resolving* a reference from a source element to a target element, *decorating* model elements with virtual properties or operations, and *configuring* transformation definitions for execution.

Please note that this list is not exclusive, but others means of composition are possible, for instance inheritance, if one can make a transformation written in one language be extended by of another one written in another language, as proposed in [22]. In the following we discuss these forms of composition, showing how they are integrated in Eclectic.

Feeding Transformation Constructs. Transformation constructs such as transformation rules and patterns, are normally fed with model elements in order to start processing them. Sometimes a rule embeds the pattern (e.g., ATL), while sometimes both constructs are separated (e.g., VIATRA2). In any case, this can be seen as an abstract relationship where the transformation engine feeds some language construct with model elements (e.g., a pattern) or a transformation construct feeds another transformation construct (e.g., a pattern feeding a rule).

Both SMap and SAttr use simple patterns based on the name of a metaclass (i.e., model!Metaclass syntax) but if we want more complex patterns (such as the one shown in Listing 4) we need to include in each language a way to specify patterns or filter expressions. Hence, we would like a composition mechanism that does not require to change these languages to refer to a pattern expressed with SPat. Our solution has been to make the result of a pattern available as a new type, that is instantiated for each match of the pattern. Listing 5 shows the GettableProperty pattern, that uses the providing keyword to initialize the result of the match (line 7). We have included the possibility of assigning self to some object, so that the result is referentially equal to such object (p in this case), but extended with additional properties. Now, Listing 6 uses the pattern as if it were a normal metaclass (line 5) but it is actually refering to the result of a pattern, that may be the result of performing a complex search.

```
1  spat umlext(uml)
2
3  pattern GettableProperty
4    forall p: uml!Property [p.visibility= #pk_public] and
5    not exists o: uml!Operation [ p.owner.includes(o) ...]
6
7    providing self = p,
8              self.isPublic = true
9  end
```

Listing 5. Publishing a pattern in SPat

```
smap struct(uml) −> (java)
  uses java_conventions
  uses umlext

from umlext!GettableProperty to
     get: java!MethodDeclaration,
     ta: java!TypeAccess
     // Same as as original transformation
end
```

Listing 6. Using a pattern in SMap

What is distinctive of this approach is that it separates patterns from rules without requiring any special syntax, but there is a seamless integration making a pattern result looks like a type. Likewise, recursive patterns in SPat are allowed using the same strategy.

Resolving References. Resolving a target element from a source element that is pointed by a reference is a primary element of model-to-model transformation languages. ATL, for instance, performs this task implicitly through a binding construct (←). In this case, we need to resolve relationships established by different languages in their own manner.

We take inspiration from Tefkat's tracking classes and the proposal of [12] for our mechanism to make source-target correspondences available. In these works, the underlying idea is to establish an interface between transformation rules by means of an intermediate model (which can be considered a trace model, where a tracking class is a type of trace link), so that there is a layer of indirection through this model to refer to the data produced by another rule. We generalize this mechanism to span several languages, making the intermediate model implicit.

In the case of SMap, the strategy is to tag each mapping, so that the set of tags of a transformation is the interface of the transformation. From an external program, a tag is as a new type of trace link that keeps a correspondence, and the interface is the set of trace link types that can be instantiated by a given transformation execution. Listing 7 shows how the SMap transformation uses a tag (classifier in line 4) to make a mapping resolvable from other transformations. Implicitly, a trace link called classifier is created, that has a source reference pointing to a UML class, and two target references pointing to the cd and cu elements (lines 5-6). Please note that in SMap a tag is not the same as a typical rule name, since the same tag may be contributed by different mappings, for instance the classifier tag applies both to mappings from UML classes to Java classes as from UML primitive types to Java.

In SAttr, however, there is no need to explicitly set the interface with tags, but it is automatically derived. In the example of Listing 3 the interface consists of two trace links, one for each declared attribute, type and env. An external transformation would refer to an attribute value as if it were a trace link.

Transformations written in other languages, like Tao, may need to refer to these trace links. To this end, the following syntax can be used: transf!trace_link(.target)? where transf is the name of a external transformation, trace_link is the name of the trace link that will be used to resolve the source element to a target element, and target is the name of the target element to be gathered (if not given the first target element is used).

Listing 8 shows how the gen_java Tao transformation interoperates with SAttr and SMap (struct and typing transformations). Line 6 shows how to obtain the Java getter method that corresponds to a given UML property (the uml_property helper is part of a navigation library that links the OCL and UML models). This mapping between properties and methods has been performed by the struct transformation (Listing 7, lines 10-14). A more complex example is also shown in line 12, where the UML type of a *let* expression that has been computed in the typing transformation is gathered (expr[typ!type]), and from obtained type, the corresponding Java is next obtained.

```
1   mapping struct (uml) −> (java)
2   uses java_conventions
3
4   [classifier]
5   from uml!Class to cd: java!ClassDeclaration,
6                     cu : java!CompilationUnit
7      ...
8   end
9
10  [ get ]
11  from uml!Property to
12     m: java!MethodDeclaration,
13     ta: java!TypeAccess
14  end
```

```
tao gen_java(uml, ocl) −> (java)
uses typing, struct

template mapExpression(expr : ocl!NavExp)
   java!MethodInvocation {
      method = expr.uml_property[struct!get]
   }
end

template mapExpression(expr : ocl!LetExp)
java!VariableDeclarationExpression {
   type = expr[typing!type][struct!classifier.cd]
}
end
```

Listing 7. Tagging mappings   **Listing 8.** Resolving references

Decorating Model Elements. The possibility of adding virtual properties and operations (sometimes known as helpers) to model elements has been used so far as a way to enable navigation libraries in model transformation languages [9][15][2]. Although we have not shown any example of this scenario, Eclectic supports navigation libraries as well.

Nevertheless, we have also used this feature as a way to enable invocation of Tao templates. The interface of a Tao transformation is just the set of rules seen as operations, which return target elements. This form of interface is independent of how these operations have been implemented (with Tao in this case), but those transformations requiring explicit creation of elements simply invoke one operation expecting one or more target elemens as a result.

Listing 9 shows a piece of an SMap transformation that invokes the mapPrecondition operation to generate an assert expression from an operation precondition (lines 9-10). From the point of view of SMap, Tao templates are converters. This means that at the SMap level the way to use languages that require explicit rule invocation is through a converter.

The SAttr language is also allowed to use Tao transformations. This is important as SAttr does not have any construct to create new elements. As we explained, SAttr provides a simple expression language, so we just rely on normal method calls as the composition mechanism.

Configuring Transformation Definitions. The mechanisms discussed so far allow us to use the results produced by a transformation in another one written in a different language. To this end, the uses keyword establishes a dependency with an external transformation. However, we still need to configure the composite transformation which consists of the smaller transformation programs. We also would like to consider the configuration of transformation chains (i.e., feeding a transformation with the output of another).

We have devised a simple language to specify composite transformation programs and transformation chains. It basically treats transformations as functions with zero or more parameters, and with zero or more result models. There is a special construct, composite, which performs all the necessary plumbing, at the IDC level, to schedule two or more transformations to be executed together as a unit. Listing 10 shows the complete transformation chain for the UML2Java example. First of all, a new composite transformation (uml2java_si, lines 2-7), which uses the four transformations previously presented, is defined. When the same model name is used as output (lines 3-4) it means that both transformations contribute to it. If a transformation program has no output models, it is indicated using an underscore (lines 5-7). Note that we use the term transformation to refer to a piece of program that just contributes to a global result, although it does not perform any actual transformation (e.g., umlext that find patterns).

The composite transformation uml2java_si, however, does deal with multiple inheritance, so the first step in the chain would be to rewrite the UML model to remove multiple inheritance (e.g., introducing interfaces). Line 9 invokes the rewriting transformation, remove_multiple obtaining a UML model with only single inheritance (uml_sing_inh)[1]. Afterwards, the composite transformation is invoked normally (line 12), obtaining the target model. Please note that our composition mechanism is able of dealing with transformations that depend on one another. This is possible because our engine is based on continuations as explained in Section 3.1.

```
1   mapping struct (uml) −> (java)
2     uses gen_java
3
4     from uml!Operation to m: java!Method
5       m.name <− name
6       // The mapPrecondition template will create
7       // a java!MethodInvocation to assert the
8       // precondition (if it exists)
9       m.bodyDeclaration <− m.pre
10        convert gen_java.mapPrecondition
11    end
```

Listing 9. Invoking a template from SMap

```
1   chain uml2java(uml, ocl) −> (java)
2     composite uml2java_si(uml, ocl) −> (java)
3       java = struct(uml)
4       java = gen_java(uml, ocl)
5       _ = typing(uml, ocl)
6       _ = umlext(uml)
7     end
8
9     uml_single_inh = remove_multiple(uml)
10    java = uml2java_si(uml_single_inh, ocl)
11  end
```

Listing 10. Configuring UML2Java

---

[1] We have not implemented an in-place language in Eclectic yet, but it can be simulated with a copy transformation.

The task of writing transformation chains has been typically addressed with build scripts or Java programs. In our case, this language simplify writing transformation chains considering transformations as functions with a number of input/output models (in the style of MCC [11]). Besides, it provides facilities to compose transformations, and we are working on giving support to higher-order transformations.

4   Related Work

The notion of family of languages has been used both to refer to independent DSLs that share common implementation artefacts [7][21] and to refer to a set of related languages that must be composed to achieve a common goal [18]. For instance, UML [14] can be considered as a family of related modeling languages, each one intended to address some concern (e.g., structure, behaviour, deployment) of object-oriented modeling, which share a common core.

In the context of model transformations there are some examples of families of languages. First of all, the QVT architecture is similar to our proposal. However, there is a significant difference, as we advocate for simple languages, while QVT Relational and Operational are complex languages. Epsilon [6] is a family of model management languages, where each language is intended for a model management task, such as validation, migration or model transformation. There is a base language, EOL, that is common to all of them. These languages, however, work independently, and they are composed by means of ANT scripts that feed one language with the ouput of another. TransML [8] is a family of languages for modeling model transformations. It is organized as a stack, with lower languages refinining the upper ones. Finally, ATL has two basic execution modes: normal mode that corresponds to model-to-model transformations, and refining mode that corresponds to in-place transformations, thus ATL can be considered as a family with two languages.

In the context of program transformation, Kiama [19] is a Scala library for language processing that provides several internal languages for describing attribute grammars, tree rewriting, abstract state machines, and pretty printing.

Composition of heterogeneous rule-based transformation languages is studied in [22], where a common virtual machine, called EMFTVM, is used to implement ATL and a rewriting language. Our approach is also based on an intermediate language, but with different characteristics. EMFTVM provides a common notion of module and transformation rules, which enables a common semantics for module import and rule inheritance. In our case, the intermediate language provides more general composition services (see Section 3), so that each language of Eclectic is allowed to have its own semantics. For instance, this would allow us to integrate a purely imperative language in the family (i.e., with no rules).

Creating chains of model transformations is a widely used technique to split complex transformation problems [20][11]. In these chains the input of one transformation is just fed with the output of a previous one, which has the disadvantage that the transformation execution context is lost so actual interoperability is not possible (unless complex architectures are used [24]).

5 Assessment and Future Work

In this paper we have presented a family of model transformation languages, called Eclectic. Our aim is to address complex transformation problems by splitting them into smaller problems that can be solved with simple languages. However, this requires a careful design of the different languages as well as taking into consideration interoperability and composability issues. Here, we have presented the design and composition mechanisms of Eclectic, which have been illustrated by means of a running example. Besides, the feasibility of the approach is demonstrated by means of an proof-of-concept implementation available at http://sanchezcuadrado.es/projects/eclectic.

As a summary, Table 1 relates the languages and the composition mechanisms presented in this paper. An entry with *Use* means that a language can interoperate with another language that has an entry with *Enable* for the same composition mechanism. For instance, all languages (except Chain) are able to invoke operations that decorate models elements, but Tao is the only language that currently decorates models. The SPat/Decorating entry is special because in this language only methods without side-effects are allowed, which means that it could interoperate with a navigation library but not with a Tao program.

Table 1. Summary of mechanisms and languages

	Feeding	Resolving	Decorating	Composing
Mapping (SMap)	Use	Use, Enable	Use	Enable
Attribution (SAttr)	Use	Use, Enable	Use	Enable
Target-oriented (Tao)	-	Use	Use, Enable	Enable
Pattern Maching (SPat)	Use,Enable	Use	Use*	Enable
Configuration (SChain)	-	-	-	Use, Enable

As can be observed in the table, an important aspect of our design is that languages are loosely coupled, so that it is possible to evolve members of the family without affecting the other languages. In fact, new languages could be added seamlessly. This is particularly important to enable interoperability with *domain-specific transformation languages* (DSTL). We envision an scenario where part of a complex transformation is written with Eclectic, and it is completed and extended by means of a DSTL that addresses variable parts.

In principle, one possible drawback of this approach is learning facility. However, we have tried to keep the languages small and with a similar syntax so that one can learn how to use them just looking a few examples. In addition, we believe that this approach enhances intentionality of the transformation text, which favours comprehensibility.

One concern that we would like to address is the fragmentation of the transformation code. Being languages with few features, sometimes one has to rely on other languages to perform simple operations. Thus, further evaluation is needed to assess the real possibilities of the approach. One such evaluation would be to apply Eclectic to transform other UML models apart from the class diagram, testing the transformations with large models to benchmark the performance of our engine.

Finally, we have presented several proof-of-concept components of Eclectic, but further experiments are need to find out simpler and even more compact constructs. Additionally, we are looking into how to integrate other transformation styles such as in-place and bidirectional transformations.

Acknowledgements. Work partially funded by the Spanish Ministry of Economy and Competitivity (TIN2011-24139), and the R&D programme of Madrid Region (S2009/TIC-1650). This work is based on the initial results obtained during project TIN2009-11555 funded by Spanish Ministry of Economy and Competitivity. I also thank the referees for their valuable comments. Last but not least, I thank Jesús Perera Aracil for implementing the first version of the JVM compiler for Eclectic.

References

1. Aho, A.V., Sethi, R., Ullman, J.D.: Compilers: principles, techniques, and tools. Addison-Wesley Longman Publishing Co., Inc., Boston (1986)
2. Cuadrado, J.S., Jouault, F., Molina, J.G., Bézivin, J.: Experiments with a High-Level Navigation Language. In: Paige, R.F. (ed.) ICMT 2009. LNCS, vol. 5563, pp. 229–238. Springer, Heidelberg (2009)
3. Cytron, R., Ferrante, J., Rosen, B.K., Wegman, M.N., Zadeck, F.K.: Efficiently computing static single assignment form and the control dependence graph. ACM Trans. Program. Lang. Syst. 13, 451–490 (1991)
4. Czarnecki, K., Helsen, S.: Feature-based survey of model transformation approaches. IBM Syst. J. 45, 621–645 (2006)
5. Eclectic website, http://sanchezcuadrado.es/projects/eclectic
6. Epsilon, http://www.eclipse.org/gmt/epsilon
7. Greenfield, J., Short, K., Cook, S., Kent, S.: Software Factories: Assembling Applications with Patterns, Models, Frameworks, and Tools. Wiley (2004)
8. Guerra, E., de Lara, J., Kolovos, D., Paige, R., dos Santos, O.: Engineering model transformations with transML. Software and Systems Modeling, 1–23 (2011)
9. Jouault, F., Allilaire, F., Bézivin, J., Kurtev, I.: ATL: A model transformation tool. Science of Computer Programming 72(1-2), 31–39 (2008), http://www.emn.fr/z-info/atlanmod/index.php/Main_Page (last accessed: November 2010)
10. Jouault, F., Kurtev, I.: On the interoperability of model-to-model transformation languages. Sci. Comput. Program. 68, 114–137 (2007)
11. Kleppe, A.: MCC: A Model Transformation Environment. In: Rensink, A., Warmer, J. (eds.) ECMDA-FA 2006. LNCS, vol. 4066, pp. 173–187. Springer, Heidelberg (2006)
12. Kurtev, I., van den Berg, K., Jouault, F.: Rule-based modularization in model transformation languages illustrated with atl. Science of Computer Programming 68(3), 138–154 (2007); Special Issue on Model Transformation
13. Mens, T., Van Gorp, P.: A taxonomy of model transformation. Electron. Notes Theor. Comput. Sci. 152, 125–142 (2006)
14. OMG. UML 2.3 specification, http://www.omg.org/spec/UML/2.3/
15. OMG. Final adopted specification for MOF 2.0 Query/View/Transformation (2005), www.omg.org/docs/ptc/05-11-01.pdf

16. Rompf, T., Maier, I., Odersky, M.: Implementing first-class polymorphic delimited continuations by a type-directed selective cps-transform. In: Proceedings of the 14th International Conference on Functional Programming, pp. 317–328 (2009)
17. Sanchez Cuadrado, J.: Compiling ATL with Continuations. In: Proc. of 3rd International Workshop on Model Transformation with ATL (MtATL 2011), pp. 10–19. CEUR-WS (2011)
18. Sanchez Cuadrado, J., Molina, J.G.: A model-based approach to families of embedded domain-specific languages. IEEE Trans. Softw. Eng. 35, 825–840 (2009)
19. Sloane, A.M., Kats, L.C., Visser, E.: A pure embedding of attribute grammars. Science of Computer Programming (2011)
20. Vanhooff, B., Ayed, D., Van Baelen, S., Joosen, W., Berbers, Y.: UniTI: A Unified Transformation Infrastructure. In: Engels, G., Opdyke, B., Schmidt, D.C., Weil, F. (eds.) MODELS 2007. LNCS, vol. 4735, pp. 31–45. Springer, Heidelberg (2007)
21. Voelter, M.: A family of languages for architecture description. In: 8th OOPSLA Workshop on Domain-Specific Modeling, DSM 2008 (October 2008)
22. Wagelaar, D., Tisi, M., Cabot, J., Jouault, F.: Towards a General Composition Semantics for Rule-Based Model Transformation. In: Whittle, J., Clark, T., Kühne, T. (eds.) MODELS 2011. LNCS, vol. 6981, pp. 623–637. Springer, Heidelberg (2011)
23. Wimmer, M., Kappel, G., Kusel, A., Retschitzegger, W., Schönböck, J., Schwinger, W.: Surviving the Heterogeneity Jungle with Composite Mapping Operators. In: Tratt, L., Gogolla, M. (eds.) ICMT 2010. LNCS, vol. 6142, pp. 260–275. Springer, Heidelberg (2010)
24. Yie, A., Casallas, R., Deridder, D., Wagelaar, D.: Realizing model transformation chain interoperability. Software and Systems Modeling, 1–21 (2010)

Translational Semantics of a Co-evolution Specific Language with the EMF Transformation Virtual Machine

Dennis Wagelaar[1,*], Ludovico Iovino[2], Davide Di Ruscio[2], and Alfonso Pierantonio[2]

[1] Department of Computer Science,
Vrije Universiteit Brussel,
B-1050 Brussels, Belgium
dennis.wagelaar@vub.ac.be
[2] Dipartimento di Informatica
Università degli Studi dell'Aquila
I-67010 L'Aquila, Italy
{ludovico.iovino,davide.diruscio,alfonso.pierantonio}@univaq.it

Abstract. Model-to-model transformations are often employed to establish translational semantics of Domain-Specific Languages (DSLs) by mapping high-level models into more concrete ones. Such semantics are also executable when there exists a target platform able to execute the target models. Conceiving a transformation that targets a low-level language still remains arduous due to the large semantic gap between the DSL and the corresponding target language. In this respect, depending on the domain of the DSL, this task can be made easier by reusing an existing platform and bytecode language for that domain, as for instance the EMF Transformation Virtual Machine (EMFTVM) for the domain of model transformation. This paper defines executable semantics for EMFMigrate, a model transformation language specifically designed for managing the coupled evolution in model-driven development. To this end, the approach considers EMFTVM as the runtime engine targeted by the proposed semantic mappings.

1 Introduction

Domain-specific languages [1] (DSLs) are software languages which allow the designer to express problems in terms of concepts proper to a given application domain. A DSL essentially consists of an *abstract syntax*, the set of language concepts and their relationships usually given in terms of a metamodel; a *concrete syntax*, the (textual or graphical) notation that the end user will use to specify programs conforming to the abstract syntax; and a *semantics*, the meaning of the language constructs by means of corresponding semantic mappings [2]. Among the available approaches to semantics, model-to-model transformations are often employed to establish translational semantics [3] of a DSL by mapping its metamodel to those of low-level programming languages and platforms. When there exists a target platform able to execute the target

---

* The author's work is funded by a postdoctoral research grant provided by the Institute for the Promotion of Innovation by Science and Technology in Flanders (IWT-Flanders).

Z. Hu and J. de Lara (Eds.): ICMT 2012, LNCS 7307, pp. 192–207, 2012.

models, then the semantics of the DSL is executable too. However, writing a transformation that targets a general-purpose programming language, or a bytecode/assembly language, still remains arduous due to the large semantic gap between DSL and target language [4].

Depending on the domain of the DSL, the task of providing executable semantics can be made easier by reusing an existing runtime engine for that domain. Such a runtime engine would support DSL primitives to reduce the semantic gap between the DSL and the runtime engine, and provides dedicated facilities for source languages, such as source-level debugging. Also, the fact that multiple languages reuse a common runtime engine means that the runtime engine receives more exposure to real-world usage. This translates into a more stable and mature runtime engine.

This paper proposes to use the EMF Transformation Virtual Machine (EMFTVM) as a reusable runtime engine for the model transformation domain. EMFTVM was conceived as a common runtime engine for heterogeneous model transformation languages [5], and provides a high-level bytecode metamodel to express transformations. Domain-specific language primitives are part of the bytecode metamodel, such as explicit transformation modules and rules, including their composition mechanisms, and model manipulation instructions. Since the bytecode is represented as a metamodel, as aforementioned the DSL semantics may be given in terms of a model transformation, e.g. in ATL [6], and executed in EMFTVM itself. Moreover, the bytecode metamodel implementation enables runtime performance optimizations by automatically collecting low-level information, such as local variable slot assignment, maximum stack usage, and instruction branch offsets. A discussion on such aspects can be found in [5].

We show how EMFTVM can be used for providing executable semantics to EMFMigrate [7], a DSL dedicated to the management of the coupled evolution[1] of metamodel and a wide range of related artifacts (e.g., models, transformations, and concrete syntax specifications). Whenever a metamodel undergoes modifications such artifacts might become invalid and demand specific adaptation in order to consistently recover their validity. In this respect, EMFMigrate provides a way to express metamodel changes, and use those changes to drive the adaptations of the affected artifacts. The EMFMigrate language itself is a domain-specific transformation language, that allows one to specify how a model should be transformed in order to work with the new metamodel. As an example, an ATL transformation is automatically adapted by means of an EMFMigrate program as a consequence of the modifications operated on the source metamodel.

By leveraging the primitive semantics of EMFTVM, it is possible to give EMFMigrate a corresponding semantics and implementation. In this respect, an ATL transformation has been developed to translate EMFMigrate programs to EMFTVM bytecode. The development of the transformation took ten days for two people, which is very efficient if compared with alternative methods as discussed later in the paper. Because EMFMigrate is a rule-based language the built-in rule construct of EMFTVM can be used directly, and the general rule matching/application behaviour does not need to be programmed explicitly. For instance, the OCL implementation provided by EMFTVM can be completely reused to implement the filter mechanisms of EMFMigrate.

---

[1] The terms *coupled evolution* and *co-evolution* are used interchangebly throughout this paper, in the literature the term *co-adaptation* (e.g., [8]) is also used.

Alternatively, additional efforts would have been required to implement a parser for OCL and its semantics which is known to be complex. As a result, EMFMigrate leverages the proven functionality of EMFTVM, its debugging facilities, and integrates with other languages that target EMFTVM, such as ATL and SimpleGT[2].

The remainder of this paper is structured as follows: In section 2 we introduce the EMFTVM bytecode language, and in section 3, we introduce the EMFMigrate language. In section 4, we demonstrate how EMFMigrate is implemented on top of EMF-TVM. In section 5, we discuss related work. Finally, section 6 concludes this paper.

2 EMFTVM

The EMF Tranformation Virtual Machine (EMFTVM) is a virtual machine for model transformation/manipulation on top of the Eclipse Modeling Framework (EMF). It is a stack-based VM (i.e. instructions communicate values via a stack), and uses a low-level bytecode language to describe model transformations. Three important features of this bytecode language are *i)* that it is represented as an EMF model, and that it supports *ii) first-class rules* and *iii) closures*. Closures are nested, nameless functions that can be invoked or passed as parameters to other functions. These features make it easier to compile a source language to EMFTVM bytecode: any model transformation language with EMF support can be used to implement the compiler, rule-based source languages may directly target the *rule* construct, and each (executable) source language expression element can be translated to a single EMFTVM closure (one-on-one mapping).

EMFTVM bytecode is organised into *modules*, which represent self-contained units of execution. Each module consists of a number of *fields*, *operations*, and *rules*. Fields and operations can be static or dynamic, similar to Java fields and methods. Modules also specify a number of *input*, *in/out*, and *output* models. This distinction allows one to enforce read-only or write-only constraints at run-time: input models are read-only, output models write-only, and in/out models can be read and written. Finally, modules may *import* other modules.

Instructions are organised into *code blocks*. Fig. 1 shows the structure of code blocks. Code blocks are executable lists of instructions, and have a number of local variables. Code blocks are used to represent operation bodies and field initialisers. Code blocks may also have nested code blocks, which effectively represent *closures* in EMFTVM.

EMFTVM's nameless function support makes it an implementation of lambda calculus, and hence a Turing-complete language. Although this may be considered a necessary precondition for any common runtime language, the real value of EMFTVM of course lies in its domain-specific primitives. The notion of *modules* and *rules*, and their configurable semantics, allow a language designer to map their source language module and rules directly to their EMFTVM counterparts.

EMFTVM rules consist of input elements, output elements, a *matcher* code block, *applier* code block, and *post-apply* code block. This distinction between *matcher, applier*, and *post-apply* allows one to execute rules in stages: the *matcher* filters potential input element matches, the *applier* assigns element properties and deletes elements, and the *post-apply* block contains code that should be run after a rule has been applied.

---
[2] http://soft.vub.ac.be/soft/research/mdd/simplegt

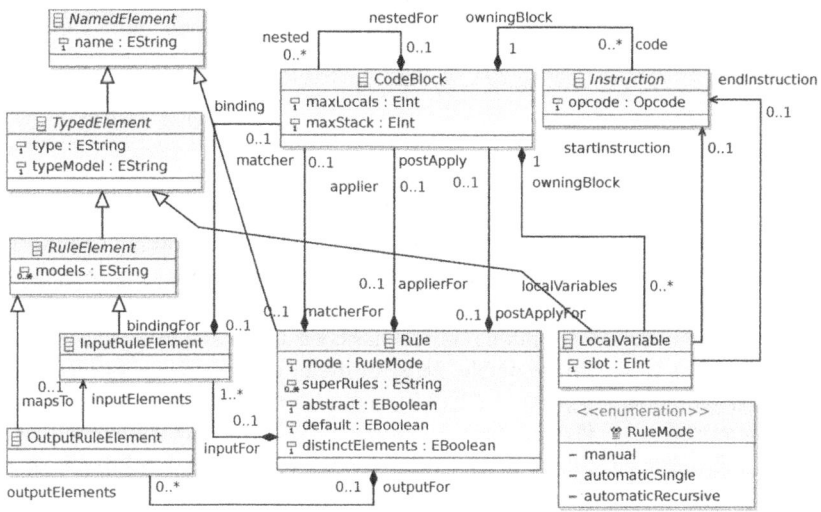

Fig. 1. Structure of EMFTVM rules and code blocks

EMFTVM provides a framework for automatic matching and tracing, which invokes these three different code blocks at specific stages.

Input elements can have a *binding* code block. This allows EMFTVM to apply a *search plan* strategy [9] in its automatic matcher. Each *binding* block calculates the valid values for an input element, given the values of the input elements that have already been bound (either by iteration or by another *binding*). Furthermore, rules have a name that is unique within its module, and can have a number of super-rules. These super-rules are stored as names only, and are resolved at load-time, when rules are composed. This is done to facilitate interaction with the module import mechanism. Module import, super-rules and rule inheritance are further explained in [5].

Rules can be *abstract*, which means that they are only applied in combination with a non-abstract sub-rule. A rule may create *default* traces, which allows the transformation module to *resolve* target elements from a (list of) source element(s). Default traces have as consequence that the same input pattern may not be matched by another rule that creates default traces, as this would result in ambiguous source-target value resolution. Rules may also match against *distinct elements*, which means that no two elements in a single input pattern match can be equal.

Finally, rules have an execution *mode*, which can be either *manual, automatic single*, or *automatic recursive*. *Manual* rules have to be explicitly invoked. *Automatic single* rules are matched once, then applied once by the automatic matching framework. *Automatic recursive* rules are matched and applied by the automatic matching framework until there are no more matches.

EMFTVM provides two built-in composition mechanisms: *module import* and *rule inheritance*. As these composition mechanisms work at the level of the bytecode, it is possible to compose heterogeous transformation languages, as long as they compile to EMFTVM bytecode. A detailed description of the composition semantics can be found in [5].

3   EMFMigrate

Similarly to any software artefacts, metamodels can evolve over time [8]. Since models, transformations and other modelling artefacts are defined according to corresponding metamodels, proper co-evolution techniques have to be provided to propagate the changes operated on a given metamodel to those modelling elements depending on it. The metamodel/model co-evolution problem has been already intensively explored and a research corpus is already available. Most of the efforts are focusing on metamodel/model co-evolution (e.g., [10,11,12,13,7]), but also transformations (e.g., [14]) and supporting tools (e.g., [15]) have been taken into account too. At the moment, a comprehensive approach capable of dealing with the co-evolution of the different artefacts in a homogeneous manner, i.e., adopting the same tools and techniques for the different kinds of artefact adaptations, is still missing. As a consequence, keeping the metamodels and all the artefacts depending on it in a consistent state requires the modeller to deal with a wide range of techniques, languages, and tools.

EMFMigrate [7] is an attempt aiming at supporting the coupled evolution in general, in the sense that it is not restricted to specific kinds of artefacts. It allows the declarative specification of migration strategies for different modelling artefacts affected by the same metamodel modifications. The approach consists of a DSL which provides modellers with dedicated constructs for *i)* specifying migration libraries, which aim to embody and enable the reuse of recurrent artefact adaptations; *ii)* customizing migrations already available in libraries; and *iii)* managing those migrations which are not fully automated and that require user intervention. In other words, the metamodels refactorings originate different adaptations depending on the kind of artefact to be kept consistent, each adaptation is formalized in a library. Recurrent adaptations are specified in default libraries which can be in turn customized in order to address ad-hoc needs. An EMFMigrate specification is given as follows:

```
1 migration migrationID;
2 include library;
3 migrate A : MM with Delta {
4   rule mr₁
5     [guard₁]  rewritingRule*
6   rule mr₂
7     [guard₂]  rewritingRule*
8   ...
9   rule mrₙ
10     [guardₙ]  rewritingRule*
11 }
```

List. 1.1. Simple EMFMigrate migration program

In particular, a migration program is able to migrate the artefact A, conforming to the metamodel MM, according to the differences in the model $Delta$, conforming to the difference metamodel proposed in [16] already applied to other co-evolution cases (e.g., [10,17]). The parameters A, MM, and $Delta$ are either specified within the program

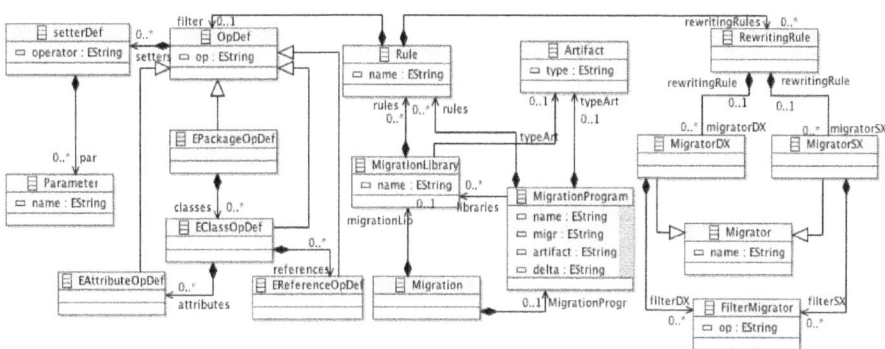

Fig. 2. EMFMigrate language metamodel

(as for the transformation and delta models in List. 1.3) or in a launch configuration dialog. The migration is specified in terms of migration rules mr_i. Each rule is applied on the artefact A if the corresponding $guard_i$ evaluated on the difference model $Delta$ holds. The body of a migration rule consists of a sequence of rewriting rules like the following

$$s[guard] \rightarrow t_1[assign_1]; t_2[assign_2]; \ldots t_n[assign_n]$$

where s, t_1, ..., t_n refer to metaclasses of MM, and $guard$ is a boolean expression which has to be *true* in order to rewrite s with t_1, t_2, and t_n. It is possible to specify the values of the target term properties by means of assignment operations (see $assign_i$).

As said above, an EMFMigrate specification can include libraries consisting of recurrent migration rules. For example, the EMFMigrate specification in List. 1.1 includes the library 'myLib.mig'. The definition of a migration library for managing the adaptation of artefacts conforming to a metamodel MM is given as follows:

```
1 library 'myLib.mig' : MM{
2   rule mr₁
3     [guard₁] rewritingRule*
4   rule mr₂
5     [guard₂] rewritingRule*
6   ...
7   rule mrₘ
8     [guardₘ] rewritingRule*
9 }
```

List. 1.2. Simple EMFMigrate migration library

Fig. 2 shows the metamodel of EMFMigrate. *Migration* is the root metaclass of the metamodel. By means of *Migration* instances, it is possible to specify a library or a migration program (see the metaclasses *MigrationLibrary* and *MigrationProgram*, respectively). Both of them consist of migration rules (see *Rule*) whose application depends on a guard (see the reference *filter* typed *OpDef*). If the guard evaluated on the delta model holds, the corresponding migration rules (see *RewritingRule*) are applied. Each rewriting rule consists of a left hand side (see *MigratorSX*) specifying the elements which have to be migrated as specified in the right hand side of the rule (see *MigratorDX*).

```
1 migration migrationExample;
2 migrate "petriNet2PNML.atl": ATL with "PN1_PN2.delta" {
3   rule renameClass[
4       class c=changeClass(c1:class) where {
5           set name= %newName;
6       }
7   ]
8   {
9       o1 :OclModelElement where[
10          name= c1.name
11      ] ->
12      o2 :OclModelElement [
13          name = newName
14      ]
15  }
16 }
```

List. 1.3. EMFMigrate program for adapting a sample ATL transformation

To discuss an exemplar application of EMFMigrate, List. 1.3 shows a migration program defined for managing the adaptation of the simple petriNet2PNML ATL transformations with respect to the changes represented in the PN1_PN2.delta delta model. The considered ATL transformation is in the left-hand side of Fig. 3 together with its representation as a model conforming to the ATL metamodel (see the right-hand side of the same figure). Such a transformation is able to generate a Petri Net Markup Language (PNML) model from a PetriNet one. The delta model, which is not shown here due to space limitations, represents the differences between the two versions of the PetriNet metamodel shown in Fig. 4. The new version of the metamodel has been produced by operating a number of changes, such as: *i)* the metaclasses TransitionToPlace and PlaceToTransition have been added; *ii)* the new metaclass Arc has been added as a superclass of TransitionToPlace and PlaceToTransition, *iii)* the metaclass Net has been renamed as PetriNet, and *iv)* the the old references places and transitions in the old Net metaclass have been merged in the elements reference

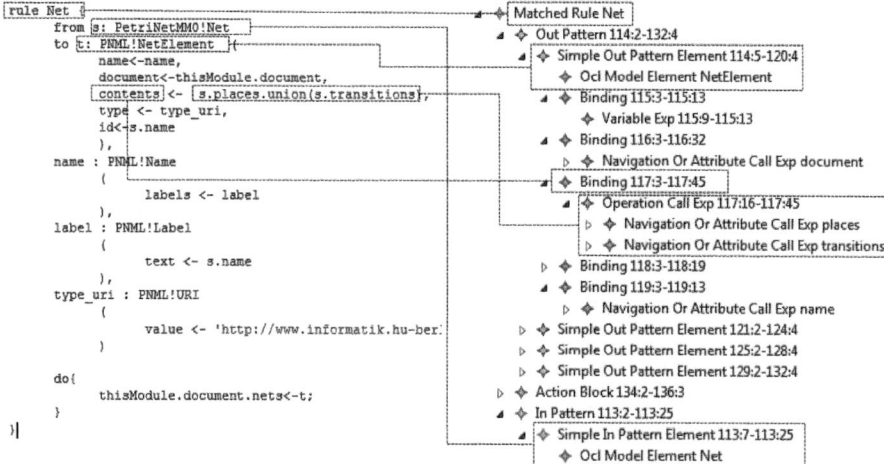

Fig. 3. Sample ATL transformation rule and its abstract syntax

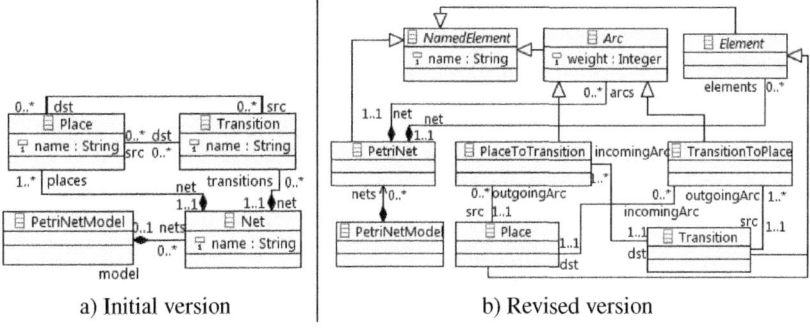

a) Initial version | b) Revised version

Fig. 4. Different versions of the source PetriNet metamodel

of the new `PetriNet` metacalass. The dashed lines in Fig. 3 denote the ATL transformation elements, which have been affected by the operated metamodel changes.

The migration in List. 1.3 is able to adapt the transformation in Fig. 3 with respect to the renaming operation performed on the initial metaclass `Net` in Fig. 4.a. In fact, the guard of the migration program in List. 1.3, lines 3–7 matches with the renaming operation performed on the metaclass `Net` to obtain the final `PetriNet` in Fig. 4.b. In case of metaclass renaming the considered ATL transformation can be adapted by replacing all the occurrences of the old metaelement with the new one (see List. 1.3, lines 9–14). By considering the transformation in Fig. 3, the execution of the migration program in List. 1.3 adapts the input pattern `PetriNetMM0!Net` of the rule `Net` by replacing it with the new `PetriNetMM0!PetriNet`.

More complex scenarios can be managed by means of EMFMigrate. However, the EMFMigrate validation is beyond the purpose of this paper, which instead focuses on the adoption of EMFTVM for providing semantics of DSLs in the model transformation domain, such as EMFMigrate. In this respect, the next section shows how the semantics of EMFMigrate are established by harnessing EMFTVM.

4   Implementing EMFMigrate with EMFTVM

In this section we describe how the semantics of EMFMigrate are defined by means of model-to-model transformations targeting the EMFTVM bytecode metamodel. In particular, Section 4.1 describes the mappings between EMFMigrate and EMFTVM, and how these mappings are implemented in ATL. Finally, section 4.2 discusses the benefits of adopting EMFTVM to implement EMFMigrate instead of using more traditional languages, such as Java.

4.1   Mapping of EMFMigrate Constructs to EMFTVM Constructs

As the EMFTVM bytecode language is based on a metamodel, model transformations can be used to translate EMFMigrate specifications to EMFTVM modules. Fig. 5(a) shows how an ATL transformation module translates EMFMigrate into EMFTVM. Fig. 5(b) shows how the generated EMFTVM module is then executed on top of the

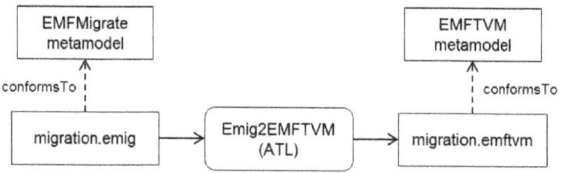

(a) Generation of an EMFTVM module out of an EMFMigrate specification

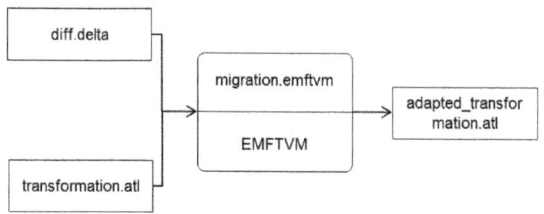

(b) Execution of the generated EMFTVM module

Fig. 5. Executing EMFMigrate specifications

existing runtime engine infrastructure to obtain a migrated version of the modelling artefact – in this case an ATL transformation module. In the remainder of this section, we will discuss how the EMFMigrate to EMFTVM transformation[3] is organised, and what patterns are used to perform the translation.

Table 1 lists the correspondences between the EMFMigrate and EMFTVM constructs. The top part refers to constructs which are used to form migration programs, while the bottom part refers to simple and nested expressions, as discussed later in this section. Additionally, at the end of the section details about debugging aspects are provided.

Migration Programs. By referring to the metamodels illustrated in Fig. 1 and Fig. 2, EMFMigrate programs are contained in *migrations*, which are mapped to EMFTVM *modules*. Migrations can *include* libraries, which maps to EMFTVM module *import*. The model-to-be-migrated is mapped to an *in/out* model in EMFTVM, while the delta

Table 1. Mapping of EMFMigrate constructs to EMFTVM constructs

EMFMigrate construct → EMFTVM construct
migration → module
include → imports
migrated model → in/out model
delta model → input model
rule → automatic single default rule
rewriting rule → input and output elements for containing rule
rewriting rule output bindings → code in rule applier block
rule filter expression → code in rule matcher block
rewriting rule "where" expression → code in rule matcher block

---

[3] http://tinyurl.com/emig2EMFTVM-atl

model is mapped to an *input* model (i.e. read-only). This is done because the model-to-be-migrated is transformed in-place. EMFMigrate *rules* consist of a *filter* expression and a number of *rewriting rules*. The rule itself is mapped to an *automatic single default rule* in EMFTVM, just like ATL matched rules. The filter expression becomes part of the EMFTVM rule's matcher code while the list of rewriting rules provide the input and output elements for the EMFTVM rule. Each rewriting rule can also provide a "where" clause, which becomes part of the EMFTVM rule's *matcher* code block, and can provide output bindings, which become part of the EMFTVM rule's *applier* code block.

```
1 rule MigrationProgram {
2  from s: EMig!MigrationProgram in IN
3  to t: EMFTVM!Module (
4    name <- s.name,
5    sourceName <- s.name + '.emig',
6    rules <- s.rules,
7    features <- Sequence{main},
8    inputModels <- Sequence{delta},
9    inoutModels <- Sequence{inmodel}),
10 ... ,
11 ln: EMFTVM!LineNumber (
12   startLine <- s.line,
13   endLine <- s.endline,
14   startChar <- s.offset,
15   endChar <- s.endoffset)
16 }
```

List. 1.4. EMFMigrate MigrationProgram rule excerpt

List. 1.4 shows an excerpt of the "MigrationProgram" transformation rule, which transforms an EMFMigrate migration into an EMFTVM module. Lines 5 and 11–15 are related to debugging, which is discussed later in the section.

Simple Expressions. The pattern used for translating expressions is different, as shown in List. 1.5. Expressions are mapped to EMFTVM code blocks, which can be invoked from the surrounding source model element using the *InvokeCb* instruction (see List. 1.5, line 13). In this way, the compiler can abstract from the specific expression type used, as all expression types are mapped to a code block. In the case of nested expressions, the current expression can specify the nested expression as a nested code block, and invoke that code block at the desired time using *InvokeCb*.

```
1 rule FilterMigratorDX {
2  from s: EMig!FilterMigrator in IN (not s.isSX)
3  to t: EMFTVM!CodeBlock (
4    code <- Sequence{invokecb_value, invoke_resolve, load_migrDXObj, setFeature},
5    nested <- Sequence{s.value},
6    lineNumbers <- Sequence{ln}),
7  ln: EMFTVM!LineNumber (
8    instructions <- Sequence{invokecb_value, invoke_resolve, load_migrDXObj,
         setFeature},
9    startLine <- s.line,
10   endLine <- s.endline,
11   startChar <- s.offset,
12   endChar <- s.endoffset),
13   invokecb_value: EMFTVM!InvokeCb (
14     codeBlock <- s.value),
15   invoke_resolve: EMFTVM!Invoke (
```

```
16   opname <- 'resolve'),
17 load_migrDXObj: EMFTVM!Load (
18   localVariable <- s.refImmediateComposite().localVar(false)),
19 setFeature: EMFTVM!"Set" (
20   fieldname <- s.featureSX.obj.name)
21 }
```

List. 1.5. EMFMigrate-to-EMFTVM compiler FilterMigratorDX rule (in ATL)

This specific rule transforms a FilterMigrator expression that is used in the right-hand-side of a rewrite rule (e.g, see List. 1.3, lines 12–14). This expression is composed of a feature and a value to assign to it. The value takes the form of a nested expression, and is therefore listed as a nested code block. In List. 1.5, line 8, the value code block is invoked, which leaves the resulting value on the stack. The *resolve()* operation is invoked on this value (List. 1.5, lines 15–16), which applies the implicit tracing mechanism of EMFTVM, and translates any transformed source elements into their target element counterparts. Next, the rewriting rule's target element is loaded onto the stack, and the specified feature is set to the resolved value (List. 1.5, lines 17–20).

Nested Expressions. List. 1.6 shows the rule that transforms the nested value expression of the FilterMigrator expression. It is again transformed into a code block, this time containing only a *load* instruction that loads the referred element onto the stack. As the FilterMigratorDX rule does not have to refer to the internals of the nested code block, it is possible to insert any sequence of instructions into the nested code block. That suffices to represent any kind of nested expression.

```
1 rule DotNavigationObjDXTarget {
2 from s: EMig!DotNavigationObjDX in IN (not s.isSX)
3 to t: EMFTVM!CodeBlock (
4    code <- Sequence{load},
5    lineNumbers <- Sequence{ln}),
6  ln: EMFTVM!LineNumber (
7    startLine <- s.line,
8    endLine <- s.endline,
9    startChar <- s.offset,
10   endChar <- s.endoffset,
11   instructions <- Sequence{load}),
12 load: EMFTVM!Load (
13 localVariable <- s.obj.localVar(false))
14 }
```

List. 1.6. EMFMigrate-to-EMFTVM compiler DotNavigationObjDXTarget rule (in ATL)

The bytecode resulting from this compiler strategy typically contains many nested code blocks that are simply invoked without parameters from their surrounding code block. In order to improve runtime performance, such nested code blocks can simply be in-lined into the surrounding code block. EMFTVM provides a reusable in-lining transformation for this purpose[4], which can be applied after the compiler. This transformation moves the contents of the code block to be inlined (transitively) upward to the first surrounding code block that is not being in-lined. The result is a "flattened" bytecode structure, with minimal code block nesting.

---

[4] http://tinyurl.com/InlineCodeblocks-atl

Debugging. An interesting aspect of implementing transformation languages with EMFTVM is related to the debugging features which are natively offered by the VM. Firstly, the file name of the source file is stored in the EMFTVM module (List. 1.4, line 5), such that the VM can point to the right file in a debugging session. Secondly, EMFTVM code blocks allow for line number information to be attached to instructions (List. 1.4, lines 11–15). The line number information can include the start line, start column, end line, end column, start character offset, and end character offset. Depending on the EMF-based parser generator used (TCS, EMFText, or xText), a specific subset of these is specified. As EMFMigrate is based on xText, the start line, end line, start character offset, and end character offset can be provided (the other information is not available through xText). During a debugging session, the VM can provide the line number information based on the current instruction offset. Together with the source file name information, this allows the Eclipse-based debugger to highlight the corresponding text passage in the source file.

4.2   Discussion

The EMFMigrate-to-EMFTVM compiler was implemented over the course of ten days by two people (pair programming). The resulting ATL program contains 650 lines of code (LOC). If we compare this to generating Java code for EMFMigrate, this might be considered very efficient: that Java code would contain a lot of the code that is already part of EMFTVM, which took four months of development by one person. It is more difficult to compare EMFTVM to the ATL virtual machine as a compilation target, as we have no data on that. The main difference lies in programming the semantics of transformation rules, which is already built into EMFTVM, but not in the ATL VM. The size of the code of the ATL-to-ATL VM compiler is 1975 LOC against 1555 LOC for the ATL-to-EMFTVM compiler. This is not a significant difference, considering that the first compiler needs to implement rule behaviour from scratch. Note that the first compiler is written in ACG, a DSL for generating ATL VM bytecode, which is more concise than ATL (but can only target ATL VM bytecode).

By using EMFTVM, EMFMigrate also leverages the exposure EMFTVM has already received as a runtime for ATL and SimpleGT: many improvements have already been made to EMFTVM since its first release. As mentioned before, all work on performance optimisation done on EMFTVM is available to EMFMigrate as well. EMFTVM currently has similar performance as the production ATL VM for EMF, which was heavily optimised by Obeo[5].

Regarding the correctness and the completeness of the translation of EMFMigrate to EMFTVM, our approach does not provide additional advantages over regular compiler development. As the EMFMigrate-to-EMFTVM transformation serves as the official semantics specification, it is by definition the correct specification. Completeness of the transformation depends on whether the mapping represented by the EMFMigrate-to-EMFTVM transformation is *total*. This can be determined for (a subset of) ATL by analysing the coverage of the source metamodel in the transformation rules [18].

---

[5] http://www.obeo.fr

In any case, the completeness analysis for the EMFMigrate-to-EMFTVM compiler has not been done: the EMFMigrate language is still under development, and additional language semantics are likely required.

5   Related Work

Defining Semantics for DSLs. For the definition of the executable – or runtime – semantics, often a transformation language/engine is provided, which allows for defining a compiler that translates the DSL into a general-purpose programming language or bytecode/assembly code. Translational semantics or semantic anchoring (e.g., [3]) is used in similar situations where the target models are not necessarily given in an executable notation. The Spoofax language workbench [19] includes the Stratego transformation language [20] for this purpose. Stratego offers the option of specifying transformation rules in terms of concrete (textual) syntax, which helps to bridge the semantic gap between DSL and target language. Our approach to reduce the actual gap is by providing a high-level domain-specific virtual machine that includes language constructs semantically close to the DSL domain. Recent work on Spoofax has added the option to define debuggers for DSLs using *syntax event linking* specifications [21]. EMFTVM comes with a DSL-aware debugger that can be reused for any DSL that compiles to EMFTVM bytecode. Most other language workbenches, such as MetaEdit+ [22] and JetBrains MPS [23], come with a template language for code generation, which also aims to bridge the semantic gap between DSL and target language by using concrete syntax. The Whole Platform [24] uses a framework approach to define DSL executable semantics, which requires the developer to write transformations in Java. AToM3 [25] uses graph transformation to rewrite a DSL into terms of another language for which executable semantics have been defined, such as Petri Nets and State Charts. If a DSL's semantics lie sufficiently close to one of these provided language implementations, this approach is very effective. The EMFTVM bytecode language also provides such a language implementation for the domain of model transformation. Rascal can represent grammars and interpreters within the same language: an interpreter is written as a series of `eval()` functions that take a specific DSL expression as input. The way an `eval()` function can just call other `eval()` functions without knowing the nested DSL expression types is similar to how an EMFTVM compiler can just invoke a nested code block without knowing what expression type produced that code block.

Runtime Engines for Model Transformation. In the domain of model transformation, there have been two efforts to provide common runtime engine for multiple model transformation languages. One of these concerns the alignment of ATL and QVT Operational [26]. The executable semantics are provided by the ATL virtual machine in this case. Another such effort is the ATC VM[6], which aims to provide a common execution framework for languages such as QVT or RubyTL. In both cases, high-level constructs, such as transformation rules, are compiled away into low-level primitives. EMFTVM leaves a smaller semantic gap, and requires less transformation effort than these VMs. In the case of the ATL VM, the bytecode uses a proprietary XML format

---

[6] http://sourceforge.net/projects/atc/

(ASM), and compilers are specified in a bytecode-specific compiler language (ACG). ACG specifies ASM bytecode very concisely, which helps to bridge the semantic gap between ATL and ASM.

T-core [27] is a language that provides a set of model transformation primitives, focused on graph transformation. T-core appears to be for graphs what EMFTVM is for EMF, and provides an implementation of low-level model transformation primitives, including transformation rules. The implementation – and choice of graph representation framework – is ongoing work.

Co-evolution. The application scenario of EMFTVM in this paper pertains to the EMFMigrate language, which is related to the research on coupled evolution in model-driven engineering. The definition of the EMFMigrate approach has been inspired by existing techniques and tools like [13] and [11]. COPE [13] is an approach that permits the migration of models in response to an evolving metamodel, combining metamodel changes and model migration using coupled-transactions. Flock [11] is a metamodel/-model co-evolution approach that permits the specification of model migration strategies. When a migration strategy is evaluated, some parts of the migrated model are derived from the original model and other parts from user specification rules. In [14] the authors deal with the metamodel/transformation co-evolution problem. The approach is based on Higher Order Transformations which are able to support the adaptation of existing transformations developed in the GME/GReAT toolset. In [15] the authors provide a solution to adapt GMF models with respect to the changes applied to the metamodel underpinning the overall definition of the editor. In particular, specific GMF model adapters are provided to automate the propagation of domain-model changes.

Differently to the previous works, EMFMigrate aims at supporting the coupled evolution in general, in the sense that it is not restricted to a particular kind of co-evolution. Interestingly, the approach permits to specify libraries of migrations, which can be reused and in case customized.

6   Conclusions and Future Work

This paper presents how to give a "useful" semantics to EMFMigrate, a model transformation language specifically tailored for the management of coupled evolution. In particular, we have established an executable semantics by adopting EMFTVM, a reusable runtime engine for the domain of model transformation. EMFTVM provides a high-level bytecode language that has specific constructs for model manipulation, such as transformation rules and model element manipulation instructions. The bytecode language is defined as a metamodel, which allows the semantics to be translational, i.e., the translator is given as a model-to-model transformation. Low-level information for performance optimization, such as local variable slot assignment and maximum stack usage, is computed automatically by EMFTVM. Bytecode verification helps track down compiler bugs. As to productivity, the EMFMigrate-to-EMFTVM translator was implemented in ten days by two people. If we compare this to generating Java code for EMFMigrate, this may be considered very efficient as the Java code would contain a lot of the code that is already part of EMFTVM, which took four months of development by one person. In the future, we plan to embed OCL in EMFMigrate to allow for richer model navigation. As the ATL-to-EMFTVM compiler already includes an OCL module, this module can be reused in the definition of the EMFMigrate-to-EMFTVM compiler.

References

1. van Deursen, A., Klint, P., Visser, J.: Domain-Specific Languages: An Annotated Bibliography. ACM SIGPLAN Notices 35, 26–36 (2000)
2. Cuadrado, J., Molina, J.: A model-based approach to families of embedded domain-specific languages. IEEE Transactions on Software Engineering 35, 825–840 (2009)
3. Chen, K., Sztipanovits, J., Abdelwalhed, S., Jackson, E.: Semantic Anchoring with Model Transformations. In: Hartman, A., Kreische, D. (eds.) ECMDA-FA 2005. LNCS, vol. 3748, pp. 115–129. Springer, Heidelberg (2005)
4. Bryant, B.R., Gray, J., Mernik, M., Clarke, P.J., France, R.B., Karsai, G.: Challenges and Directions in Formalizing the Semantics of Modeling Languages. Computer Science and Information Systems 8, 225–253 (2011)
5. Wagelaar, D., Tisi, M., Cabot, J., Jouault, F.: Towards a General Composition Semantics for Rule-Based Model Transformation. In: Whittle, J., Clark, T., Kühne, T. (eds.) MODELS 2011. LNCS, vol. 6981, pp. 623–637. Springer, Heidelberg (2011)
6. Jouault, F., Allilaire, F., Bézivin, J., Kurtev, I.: ATL: A model transformation tool. Science of Computer Programming 72, 31–39 (2008)
7. Di Ruscio, D., Iovino, L., Pierantonio, A.: What is needed for managing co-evolution in MDE? In: Proceedings of the 2nd IWMCP 2011, pp. 30–38. ACM (2011)
8. Wachsmuth, G.: Metamodel Adaptation and Model Co-adaptation. In: Ernst, E. (ed.) ECOOP 2007. LNCS, vol. 4609, pp. 600–624. Springer, Heidelberg (2007)
9. Varró, G., Friedl, K., Varró, D.: Adaptive Graph Pattern Matching for Model Transformations using Model-sensitive Search Plans. Electr. Notes Theor. Comput. Sci. 152, 191–205 (2006)
10. Cicchetti, A., Di Ruscio, D., Eramo, R., Pierantonio, A.: Automating Co-evolution in Model-Driven Engineering. In: 12th International IEEE ECOC 2008, Munich, Germany, September 15-19, pp. 222–231. IEEE Computer Society (2008)
11. Rose, L.M., Kolovos, D.S., Paige, R.F., Polack, F.A.C.: Model Migration with Epsilon Flock. In: Tratt, L., Gogolla, M. (eds.) ICMT 2010. LNCS, vol. 6142, pp. 184–198. Springer, Heidelberg (2010)
12. Garcés, K., Jouault, F., Cointe, P., Bézivin, J.: Managing Model Adaptation by Precise Detection of Metamodel Changes. In: Paige, R.F., Hartman, A., Rensink, A. (eds.) ECMDA-FA 2009. LNCS, vol. 5562, pp. 34–49. Springer, Heidelberg (2009)
13. Herrmannsdoerfer, M., Benz, S., Juergens, E.: COPE - Automating Coupled Evolution of Metamodels and Models. In: Drossopoulou, S. (ed.) ECOOP 2009. LNCS, vol. 5653, pp. 52–76. Springer, Heidelberg (2009)
14. Levendovszky, T., Balasubramanian, D., Narayanan, A., Karsai, G.: A Novel Approach to Semi-automated Evolution of DSML Model Transformation. In: van den Brand, M., Gašević, D., Gray, J. (eds.) SLE 2009. LNCS, vol. 5969, pp. 23–41. Springer, Heidelberg (2010)
15. Di Ruscio, D., Lämmel, R., Pierantonio, A.: Automated Co-evolution of GMF Editor Models. In: Malloy, B., Staab, S., van den Brand, M. (eds.) SLE 2010. LNCS, vol. 6563, pp. 143–162. Springer, Heidelberg (2011)
16. Cicchetti, A., Di Ruscio, D., Pierantonio, A.: A Metamodel Independent Approach to Difference Representation. Journal of Object Technology 6, 165–185 (2007)
17. Cicchetti, A., Di Ruscio, D., Pierantonio, A.: Managing Model Conflicts in Distributed Development. In: Czarnecki, K., Ober, I., Bruel, J.-M., Uhl, A., Völter, M. (eds.) MODELS 2008. LNCS, vol. 5301, pp. 311–325. Springer, Heidelberg (2008)

18. Planas, E., Cabot, J., Gómez, C.: Two Basic Correctness Properties for ATL Transformations: Executability and Coverage. In: MtATL 2011. CEUR-WS, vol. 742, pp. 1–9 (2011)
19. Kats, L.C.L., Visser, E.: The Spoofax Language Workbench. Rules for Declarative Specification of Languages and IDEs. In: Proceedings of the 25th Annual ACM SIGPLAN OOPSLA 2010, pp. 444–463. ACM Press (2010)
20. Bravenboer, M., Kalleberg, K.T., Vermaas, R., Visser, E.: Stratego/XT 0.17. A language and toolset for program transformation. Science of Computer Programming 72, 52–70 (2008)
21. Lindeman, R.T., Kats, L.C.L., Visser, E.: Declaratively defining domain-specific language debuggers. In: Proceedings of the 10th ACM GPCE 2011, pp. 127–136. ACM Press (2011)
22. Tolvanen, J.P., Rossi, M.: MetaEdit+: defining and using domain-specific modeling languages and code generators. In: Companion of the 18th Annual ACM SIGPLAN OOPSLA 2003, Anaheim, CA, USA, pp. 92–93. ACM Press (2003)
23. Dmitriev, S.: Language Oriented Programming: The Next Programming Paradigm (2004), http://www.jetbrains.com/mps/docs/ Language_Oriented_Programming.pdf
24. Solmi, R.: Whole Platform. Phd thesis, Università di Bologna e Padova (2005)
25. de Lara, J., Vangheluwe, H., Alfonseca, M.: Meta-Modelling and Graph Grammars for Multi-Paradigm Modelling in AToM^3. Software and Systems Modeling (SoSyM) 3, 194–209 (2004)
26. Jouault, F., Kurtev, I.: On the Architectural Alignment of ATL and QVT. In: Proceedings of the 21st Annual ACM SAC 2006, Dijon, France (2006)
27. Syriani, E., Vangheluwe, H.: De-/Re-constructing Model Transformation Languages. In: GT-VMT. Electronic Communications of the EASST, vol. 29 (2010)

Towards Multi-level Aware Model Transformations

Colin Atkinson, Ralph Gerbig*, and Christian Tunjic

University of Mannheim, Mannheim, Germany
{atkinson,gerbig,tunjic}@informatik.uni-mannheim.de

Abstract. As practical tools for disciplined multi-level modeling have begun to emerge, the problem of supporting simple and efficient transformations to-and-from multi-level model content has started to assume growing importance. The problem is not only to support efficient transformations between multi-level models, but also between multi-level and traditional two-level model content represented in traditional modeling infrastructures such as the UML and programming languages. This is not only important to facilitate interoperability between multi-level modeling tools and traditional tools, but also to extend the benefits of multi-level modeling to transformations. Multi-level model content can already be accessed by traditional transformation languages such as ATL and QVT, but in a way that is blind to the ontological classification information they contain. In this paper we present an approach for making rule-based transformation languages "multi-level aware" so that the semantics of ontological instantiation can be exploited when writing transformations.

Keywords: multi-level transformation, orthogonal classification architecture, ontological classification, linguistic classification.

1 Introduction

Transformations are one of the key pillars of model-driven software engineering [11] and are key to the productivity and flexibility advantages that make model-driven development so attractive. This is reflected in the rapidly growing interest in model transformations in academia and the increasing sophistication of the transformation capabilities offered by leading modeling environments. However, most contemporary model transformation technologies suffer from the same fundamental weakness as the modeling languages they are based-on, the restriction to a linear modeling architecture that usually accommodates only one pair of classification levels (types and instances). This makes it difficult for models, and transformation languages that build on them, to handle deep classification scenarios (when there are more than just two classification levels in a domain of interest) without introducing additional accidental complexity [3] into models.

---

* Ralph Gerbig was supported by Deutsche Forschungsgemeinschaft (DFG) as part of SPP 1496 "Reliably Secure Software Systems".

Z. Hu and J. de Lara (Eds.): ICMT 2012, LNCS 7307, pp. 208–223, 2012.
© Springer-Verlag Berlin Heidelberg 2012

Over the last few years a new architecture for model organization has emerged that accommodates deep classification in a much simpler and uniform way, whilst retaining all the advantages of traditional modeling architectures. The key idea behind this so called Orthogonal Classification Architecture (OCA) [2], is to recognize two fundamentally distinct forms of classification relationship and represent them in two completely orthogonal dimensions. One form, so called linguistic classification, captures a model element's role from a linguistic perspective, while the other form, ontological classification, captures a model element's role in the domain of interest. Disentangling linguistic classification from ontological classification in this way allows modelers to represent multiple ontological classification levels in a uniform and natural way without having to worry about the difference between classes and objects or how to use ad hoc modeling concepts like stereotypes or power-types.

Because of these advantages several research groups have developed prototype realizations and applications of OCA-based modeling environments in recent years [10][1]. However, these have all focused on the core structural models rather than on transformations between them. To date no transformation language has been developed to specifically support transformations between multi-level models represented using an OCA modeling framework. While the structural models themselves may be multi-level aware, therefore, the transformations that go with them are often not, or require the use of numerous workarounds and ad hoc techniques to operate on multi-level model content. Again, the end result is greater accidental complexity and lower efficiency in the resulting transformations.

In this paper we present an approach that makes the first steps towards addressing this problem by making a traditional "two-level" transformation language "multi-level aware". The work is presented in the context of the Melanie (Multi-level modeling and ontology engineering environment) [12] prototype modeling environment developed by the Software Engineering Group at the University of Mannheim based on the OCA principles. This environment is based on the Eclipse EMF and GMF frameworks, thereby allowing the rich set of EMF languages and tools to be applied to multi-level models represented using Melanie's default general purpose modeling language - the Level-Agnostic Modeling Language (LML). This includes several well known model transformation frameworks such as ATL and QVT. The work presented in this paper is actually based on ATL (ATLAS Transformation Language) and can be understood as a technique for making ATL "multi-level aware". Technically this is achieved by writing a so-called ATL adapter that provides several additional capabilities beyond those in the core language. For example, the ATL adapter allows transformations to explicitly distinguish between ontological instances of a model element and linguistic instances.

The remainder of the paper is organized as follows. In the next section we provide a brief introduction to the OCA and the deep instantiation approach to multi-level modeling upon which the Melanie framework is based. After that we outline the different classes of transformations that make sense between multi-level and two-level model content. Section 4 then describes the ATL adapter that we have implemented to support the identified transformation types. Since using ATL this adapter supports the EMF environment upon which Melanie is built. Once the transformation approach has been defined, section 5 shows how it has been applied in the context of another prototype modeling tool [5]. Section 6 continues with a discussion of current limitations and the future evolution paths for the technology. Finally section 7 concludes with some closing observations.

2   Orthogonal Classification Architecture

In this section we give a brief introduction to the Orthogonal Classification Architecture (OCA) upon which multi-level models are based. This serves as the basis for the description of multi-level aware model transformations in the following sections. The goal of multi-level modeling is to allow users to create models containing as many classification levels as needed to best model the domain in hand. If the domain features deep classification, which is more often than not the case, three or more levels are needed. The aim of the OCA is to allow deep classification scenarios to be modeled in a "uniform" and "adaptable" way. "Uniform" means that all model content at all levels is represented using the same basic set of concepts and symbols, while "adaptable" means that all model content, across all levels, is "soft" (i.e. treated as editable data) and can be changed interactively. Changes to a model element at any level therefore take place immediately and have an immediate effect on all the other model elements that depend on them. In contrast, current meta-modeling technologies such as the Eclipse Modeling Framework (EMF) only make two classification levels available to modelers to capture a problem domain (the meta-model and the instance of the meta-model). Moreover, only one level is available for editing at a time. Either the meta-model is available for editing without the model or the model is available for editing with a fixed meta-model. Since there is no explicit support for more than two levels we characterize such approaches as "two-level" in this paper.

By clearly separating ontological classification from linguistic classification, and using the same set of linguistic classifiers across all levels, the OCA allows an arbitrary number of ontological classification levels to be visualized and edited at all times. All model elements in the OCA typically have two direct types - an ontological type (horizontally dashed arrows in Figure 1) which characterizes its domain properties and a linguistic type (vertically dotted arrows in Figure 1) which characterizes its linguistic properties (i.e. what kind of model element it is). The ontological type of a model element is contained in the higher (i.e. more abstract) ontological level while the linguistic classifier is contained in the linguistic meta-model. This is referred to as the Pan-level Model (PLM) since it spans all the ontological levels. For example, in Figure 1 "Collie" is the ontological type of "Lassie" and "Shep" while "Breed" is the ontological type of "Collie". All four model elements are linguistic instances of the linguistic type "clabject" defined in the PLM. The term clabject is a contraction of the concatenation of the words "class" and "object" and is used as a neutral term to avoid characterizing a model element as either a class or an object (since most elements usually combine the properties of both). As illustrated in Figure 1, the name "Orthogonal Classification Architecture" reflects the fact that the two forms of classification are arranged in separate (i.e. orthogonal) dimensions.

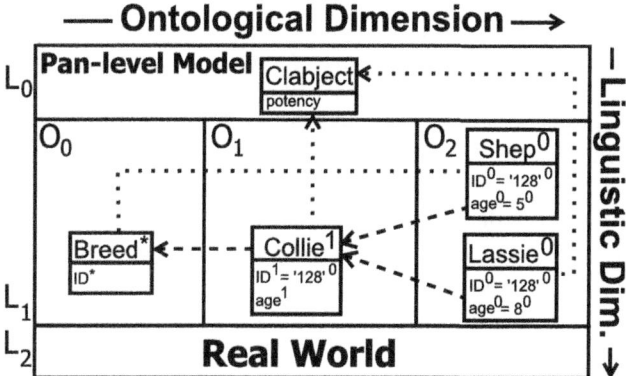

Fig. 1. The Orthogonal Classification Architecture and the resulting two transformation dimensions

The precise type/instance properties of model elements are captured using the notion of potency as part of the so called "deep instantiation" mechanism. Potency is a non-negative integer that states how many subsequent levels a model element can influence. The potency of an instance is always one less than the potency of its type. Thus, clabjects with potency 0 cannot have any instances, while clabjects of potency 1 can have instances one level below but no further, and so on. In other words, if a model element has a potency of 1 it only influences the following level, with 2 it influences the following two levels and so on. A special value for potency is the "*" value. This value states that a model element influences all following model levels.

Melanie is the first fully-fledged graphical editor supporting the OCA approach with multi-level modeling based on deep instantiation. It is an Eclipse plug-in built on the EMF modeling environment and GMF graphical editing environment. From the point of view of EMF, the L_0 (PLM) is a regular Ecore meta-model and L_1 is a regular meta-model instance. All standard EMF based technologies such as OCL and ATL can be used on the L_1 model content, but they are oblivious to the multi-level interpretation of the content in terms of ontological classification levels. Awareness of ontological levels has to be built into a special multi-level aware interpretation of the model content. For example, one of the main components of Melanie is a GMF based graphical editor which is aware of ontological classification and supports the usual (and enhanced) instantiation semantics and services between them.

3   Multi-level Aware Transformations

Model transformations are normally used in environments like EMF to transform between content represented in different technology spaces. An important example is the transformation of information from the EMF technology space (in UML models) to the Java technology space (for execution on Java virtual machines) [9]. In the long run, there may come a time when all relevant information in software information is represented in a multi-level way based on the OCA. When that time comes, all transformations will essentially map multi-level model content to multi-level information. However, until then, for practical purposes it will also be important to be able to map multi-level model content to traditional 2-level content and vice versa. In general, therefore, there is a need for three forms (or modes) of transformations, as illustrated in Figure 2 - multi-level to multi-level, multi-level to 2-level and 2-level to multi-level. The latter two transformation modes involving 2-level models are particularly important in making Melanie compatible with existing 2-level based tools. These distinct transformation modes are described more fully below. Due to space restrictions multi-level is marked as *-level in Figure 2.

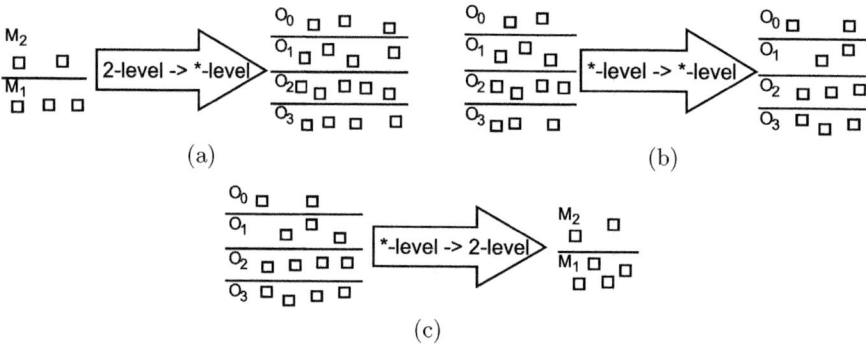

Fig. 2. The three supported transformation modes: (a) 2-level to *-level, (b) *-level to *-level, (c) *-level to 2-level

2-level to Multi-level mode. This mode, which focuses on transforming 2-level (e.g. Ecore) models to multi-level models, has two main areas of application. The first is to migrate existing meta-models to multi-level models. This enables language engineers to import already existing meta-models when migrating to multi-level modeling. For example, a language engineer can use a model transformation to transform a meta-model, defined in Ecore, to the O_0 level of a new Ontology. The instances of this model can then be transformed to the O_1 level of the model as instances of the types at O_0. The second area of application is to achieve interoperability with existing two-level model technology based tools. For instance, data that is stored in a two-level model can be automatically imported into a multi-level model. In short, this form of transformation makes all two-level models suitable input for multi-level models.

Multi-level to 2-level mode. This mode is used to make multi-level models available as input to existing two-level tools. For example a transformation that translates multi-level BPMN models into a two-level format understood by a workflow or simulation engine can be developed to allow a multi-level modeling tool such as Melanie to be used together with an existing ecosystem of tools.

Multi-level to Multi-level mode. The multi-level to multi-level mode serves the same purpose as ATL in connection with Ecore models nowadays. Example use cases include model-to-model transformations between two models conforming to two different meta-models or model refactoring by applying refinement transformations.

3.1   Impact of Multi-level Modeling on Transformations

The main difference between OCA-based model content and regular Ecore/MOF based model content from the point of view of defining transformations is that the OCA defines two classification dimensions on which transformation rules can be defined whereas Ecore/MOF is only aware of one classification dimension. Figures 1 and 3 show these distinct linguistic and ontological dimensions. The key challenge in making rule-based transformations "multi-level aware" is to enable to be explicitly distinguished between ontological instances and linguistic instances when identifying sets of objects to transform. Rules intended to operate in the linguistic dimension should affect all model elements that match a linguistic type no matter which ontological level they reside in, while rules defined to operate in the ontological dimension should only affect the model elements that are ontological instances of a particular model element.

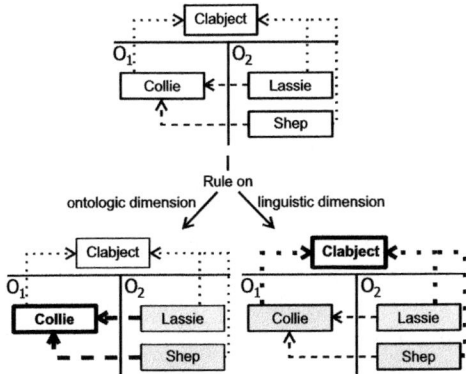

Fig. 3. An example of the two transformation dimensions arising in the OCA

Figure 3 shows a simple example of the difference between the definition of rules on the linguistic and ontological dimension. The top part of the figure shows the bottom two levels of the ontology featured in Figure 1 with "Collie" at O_0 and two instances of "Collie", "Lassie" and "Shep", at O_1. All these three model elements are instance of the linguistic type clabject. The lower left shows all ontological instances of "Collie" marked with a grey background. Notice that only the ontological instances of Collie are identified as instances, and these all exist in the ontological level below Collie. The lower right side, on the other hand, shows all linguistic instances of the linguistic type clabject. In contrast to the ontological dimension, all model elements that are clabjects, are identified, regardless of their ontological level. A key feature of "multi-level aware" transformation rules, therefore, is the explicit distinction between ontological and linguistic instances.

The definition of transformation rules on clabjects has to take deep instantiation into account, particularly the possibility to have an arbitrary number of levels in a multi-level model. If one defines a transformation on an ontological model element the question arises as to how to treat the instances on distinct levels, especially when transforming from more than two levels to an 2-level model which only supports two levels (the meta-model and the meta-model instances). For simplicity, the current implementation always focuses on the instances which exist at the lowest level in the instantiation hierarchy. For example if an ontology contains three levels and a transformation is defined on the highest level, only the model elements on the lowest level are transformed into the target model. All model elements in between are currently ignored. However, there are use cases in which it is desirable to take particular pairs of levels into account. This is a topic for further research. The problem does not arise when transforming from 2-level to multi-level models because the number of levels in a multi-level model is sufficient to accommodate all the classification levels in an 2-level model.

In the domain of 2-level to multi-level transformations a problem arises when transforming model elements to the highest ontological level. Here model elements have no ontological type at a higher level which can be specified during transformation definition. A workaround for this could be to specify the linguistic type of the element to be created and by convention place the by the transformation created model element at the highest ontological level available.

4   Multi-level Aware ATL

To make ATL multi-level aware, its syntax must be extended to facilitate explicit differentiation between the linguistic and ontological dimensions. In our current implementation, linguistic classification is handled using the standard in-built ATL notation. Thus, to identify a linguistic type the default ATL syntax ($MetaModelName!MetaModelElement$) is used. To identify an ontological type, on the other hand, this syntax is changed to $MetaModelName!"Level ::$ $OntologicalTypeName"$. The rule "ComponentClass2Class" in Listing 1 line 18 gives an example of a source pattern defined on an ontological type and a target pattern defined on a linguistic type. The helper called "createName" in line 6 is an example of a helper defined on an ontological type.

The second change to the ATL language occurs in the way features of types are accessed. In a multi-level aware transformation a user can specify whether a linguistic attribute or an ontological attribute is intended. To access a linguistic feature the syntax "_l_.featureName" is used. To access an ontological feature the syntax "_o_.featureName" is used. This syntax is used heavily in the former mentioned "createName" helper.

The ATL Regular VM's adapter concept is used to realize the required ATL dialect. The Regular VM architecture, which allows different adapters to be plugged in according to the model used in a transformation, is shown in Figure 4. Further technical details about the Regular VM are described in the "ATL Developer Guide" [6]. The following paragraphs explain the role of the adapter's three components named "ASMPLMModel", "AtlPLMModelHandler" and "ASMPLMModelElement".

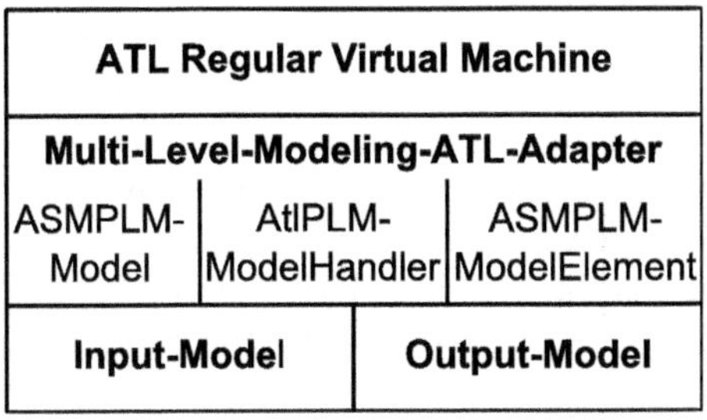

ATL Regular Virtual Machine		
Multi-Level-Modeling-ATL-Adapter		
ASMPLM- Model	AtlPLM- ModelHandler	ASMPLM- ModelElement
Input-Model	Output-Model	

Fig. 4. The architecture implemented by the multi-level aware ATL adapter

ASMPLMModel - implements the default behavior of ASMEMFModel extended by the loading of ontological model elements as meta-model types. This enables a user to define rules which operate on, or create, ontological model elements as well as linguistic model elements, by using the multi-level aware ATL syntax extensions.

AtlPLMModelHandler - is the default handler implementation provided by ATLEMFModelHandler. Its main functionality is to delegate function calls to the multi-level specific implementations of ASMModel and ASMModelElement.

ASMPLMModelElement - enables access to linguistic/ontological model elements and their attributes. The default implementation is extended to support the reading and writing of linguistic and ontological features. This is achieved by overriding the default implementation of operations when they are defined on the ontological layer. Linguistic requests are passed on to the base class functionality which is provided through ASMEMFModelElement.

5   Multi-level Aware Transformation Example

This section presents an example of the use of the aforementioned ATL adapter in the context of another Eclipse tool called nAOMi (opeN, Adaptable, Orthographic Modeling Environment) [13]. This is a prototype tool developed at the University of Mannheim to support the notion of Orthographic Software Modeling (OSM). The example shows how multi-level aware ATL transformations can be used to generate system views on-the-fly from a Single Underlying Model (SUM).

5.1   Orthographic Software Modeling

The OSM approach aims to provide a flexible and intuitive way of organizing multiple views of software systems and components. It achieves this by integrating three main innovations - on-demand view generation, dimension-based navigation and an inherently view-based method.

On-demand View Generation - ensures that views are kept synchronized and consistent with an underlying database of information about the system under development. This database is known as the Single Underlying Model (SUM) [5]. It contains all available information about the modeled system but is never directly seen by end users. Instead views tailored to specific stakeholders are generated as projections from the SUM on-demand. Consistency is achieved by ensuring that all views are up-to-date with the SUM rather than between the different views themselves.

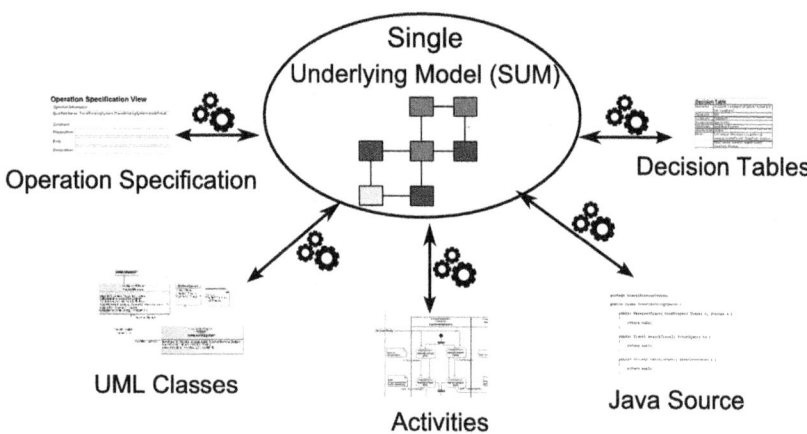

Fig. 5. Orthographic Software Modeling Overview

Dimension-based Navigation - offers an intuitive and platform independent way for users to navigate around views. It achieves this by regarding views as cells in a multidimensional cube. Each dimension of the underlying methodology forms a different dimension of the cube and each independently selectable aspect of that dimension represents a dimension element. Selecting a view therefore corresponds to selecting a single cell within the cube. Figure 6 shows the nAOMi Eclipse plug-in that supports this metaphor for navigating around views. The left part of Figure 6 shows the dimension explorer which modelers can use to select a view by picking an element from each dimension. In the example screenshot the dimensions for the KobrA [4] approach are available, which are "Abstraction", "Version", "Component", "Encapsulation", "Projection", "Granularity" and "Operation". However, these dimensions can be tailored to the needs of the software development methodology used. The right part shows the view that has been generated for the selected dimension elements.

An Inherently View-based Method - defines the dimensions and dimension elements used to represent a system and the contents of the different views.

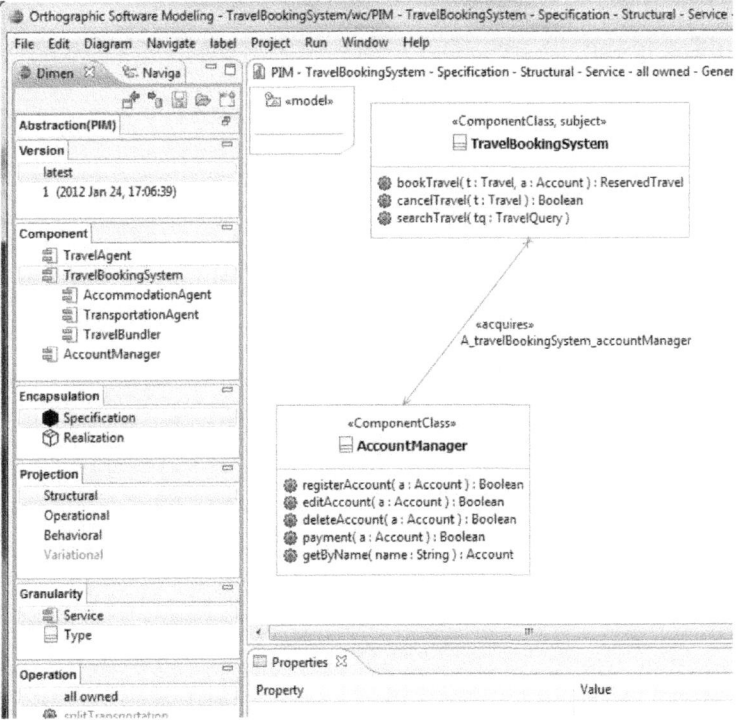

Fig. 6. Orthographic Software Modeling Tool

In our current nAOMi prototype the KobrA method is used, but it is also possible to use other methods. To provide a concrete example of the application of this approach we use a Travel Booking System.

5.2   View Generation by ATL Transformations

The ATL dialect described in this paper is an ideal vehicle for describing transformations to create UML-views of a system on-the-fly from a multi-level SUM. We demonstrate a multi-level to 2-level transformation which generates an Ecore based component diagram UML view from a multi-level SUM. This transformation uses the multi-level aware ATL adapter. Figure 7 shows a part of the SUM, related to component diagram modeling, modeled with our multi-level modeling environment. The model contains two ontological types on level O_0 which are "ComponentClass" and the "Acquires" relationship. The "Acquires" model element is connected via a source and a target connection to two "Component-Classes". On level O_1, two instances of the "ComponentClass" type exist which are called "TravelBookingSystem" and "AccountManager". These are connected to an instance of "Acquires" which is called "TBS_AM".

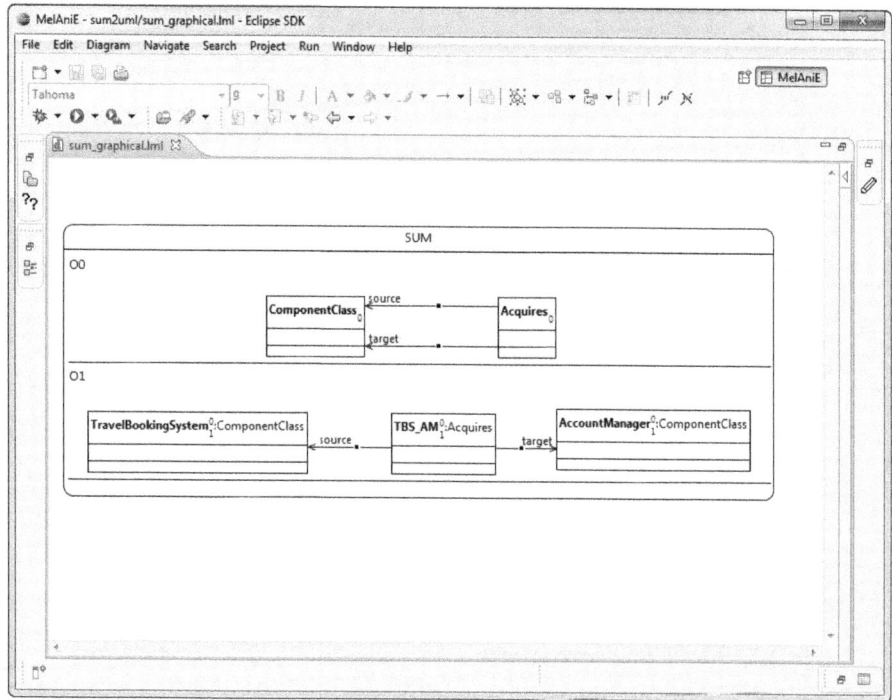

Fig. 7. The small part of the SUM which is used for evaluation modeled in Melanie

```
    module kobra;
    create OUT : UML from IN : PLM;

    helper def : umlModel : UML!Model = OclUndefined;
5
    helper context PLM!"O0::Acquires" def : createName:String = '
        From' + self._o_.source._l_.name + 'To' + self._o_.target.
        _l_.name;

    rule Ontology2Model {
      from s : PLM!Ontology
10    to t : UML!Model (
        name <- s.name
        )
        do {
        thisModule.umlModel <- t;
15    }
    }

    rule ComponentClass2Class {
      from s : PLM!"O0::ComponentClass"
20    to t : UML!Class (
        name <- s._l_.name
        )
        do {
        thisModule.umlModel.packagedElement <- thisModule.umlModel.
            packagedElement->append(t);
25    }
    }

    rule Acquires2Association {
      from s : PLM!"O0::Acquires"
30    to t : UML!Association (
        name <- s.createName,
        ownedEnd <- Sequence{ thisModule.createMemberEnds(s._o_.
            source), thisModule.createMemberEnds(s._o_.target) }
        )
        do {
35      thisModule.umlModel.packagedElement <- thisModule.umlModel.
            packagedElement->append(t);
        }
    }

    lazy rule createMemberEnds {
40    from s : PLM!"O0::ComponentClass"
      to t : UML!Property (
        type <- s
        )
    }
```

Listing 1. The complete ATL transformation used for evaluation

The multi-level aware transformations on the model are defined by employing the standard ATL transformation editor. Listing 1 shows the fully functional transformation that is used for the transformation of the use case presented here. ATL's support for adding source pattern elements in quotation marks is used to make an ATL rule aware of ontological levels. By putting the expressions in quotation marks these are not checked by the syntax checker which allows customizations of the ATL syntax. The non-standard pattern PLM!"O0::ComponentClass" in Listing 1 line 19 is used to define a transformation on the ontological instances of the "ComponentClass" model element which reside at level O_1. If a transformation developer wishes to create an ontological multi-level model element he can use this style of describing a pattern for target patterns. To define rules on linguistic model elements one can specify source and target patterns using the standard ATL syntax. When accessing the attributes of model elements, switching between ontological and linguistic mode is achieved without any extensive modifications to the ATL syntax. The user can use the keywords ".$\_$l$\_$." or ".$\_$o$\_$." to switch between linguistic traits and ontological attributes. In Listing 1 line 21 the linguistic attribute name is accessed by "$\_$l$\_$.name". To access the ontological attribute "name" if one is specified by the ontological type, the statement would have the form "$\_$o$\_$.name". The same also works for the left hand side of the assignment statement when one wants to write values to ontological instead of linguistic attributes.

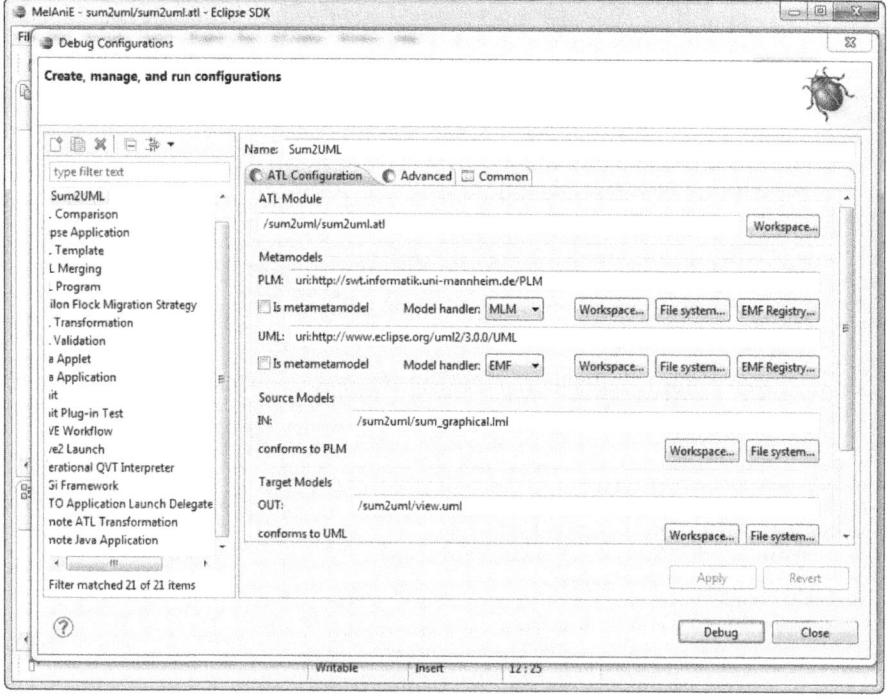

Fig. 8. Setup of a multi-level to Ecore transformation

Figure 8 shows how the transformation from the 2-level to the multi-level modeling technology space is set up. Transformations are run by the "Regular ATL Virtual Machine" instead of the EMF tailored "EMF-specific Virtual Machine". This is necessary as only the regular virtual machine supports the concept of adapters. In the run dialog the path to the PLM registered in the Eclipse workbench, "uri:http://swt.informatik.uni-mannheim.de/PLM", must be entered as meta-model in order to make ATL aware of the linguistic meta-model. Additionally, "MLM" must be selected for all multi-level meta-models. The path to the multi-level model itself is put into the corresponding source and target model text boxes. At runtime the ontological levels are extracted as meta-model information by the ATL adapter, which enables ATL to match rules on ontological model elements.

After running the transformation, the two-level UML component diagram view is created as shown in Figure 9. It contains the "TravelBookingSystem" and "AccountManager" components connected via an acquires relationship, called "FromTravelBookingSystemToAccountManager", as described in the multi-level SUM. The created views can be opened with the standard Ecore tools since they are standard Ecore models created through the EMF ATL adapter used as the target model adapter. In this figure the resulting model is opened in the generic EMF model editor.

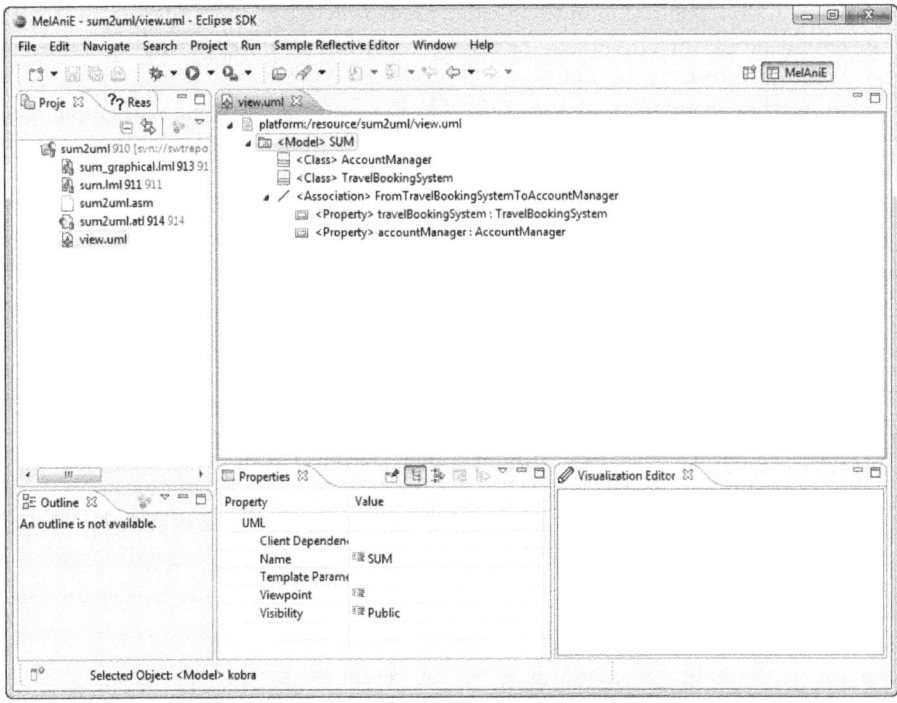

Fig. 9. The result of the executed transformation in the generic EMF model editor

6 Limitations and Future Work

Even though the example has deliberately been kept small it contains all the important concepts of the ATL transformation language [7]. These are: matched rules, lazy or called rules, imperative statements and helpers. Hence, this small example gives a good insight into the functionality of the adapter. In particular, it shows that multi-level aware ATL transformations look and feel very much like 2-level ATL transformations. The definition of rules on ontological types and switching between the linguistic dimension can be realized without requiring any additional knowledge on the part of a transformation developer.

It is a further research topic to refine the syntax used for realizing such transformations. At the time of writing we are developing and evaluating 2-level to multi-level transformation capabilities and have so far not encountered any transformation requirements which are not easily describable using current ATL syntax. In order to distinguish between setting linguistic or ontological properties, the prefixes "_l_." and "_o_." can be used, as demonstrated for multi-level to 2-level transformations.

The approach described in this paper is just the first step in the direction of multi-level aware transformations, and there are many open research questions. These include the question of how to bridge the mismatch in classification level numbers between 2-level and multi-level models and the question of how to specify transformation rules that create model elements on the highest ontological level which do not have an ontological type. However, we are confident that with the evolution of our implementation and extension of our experiments solutions to these questions can be found.

For further evaluation of the ATL adapter approach we plan to implement a second case study using this technology within the "Reliably Secure Software Systems" priority program funded by the Deutsche Forschungsgemeinschaft (DFG). This case study will import business processes from EMF based tools and will annotate them with security requirements for the imported process. These security annotated processes will then again be translated back into the business process modeling tool understandable format and will additionally configure enforcement engines to enforce the security requirements.

Our long term goal is it do define an integrated, multi-level aware textual transformation language that supports all classes of services applicable to multi-level models in a uniform and level-agnostic way. This language, known as TREACLE (Transformation, Rule, Enquiry, Action and Constraint Language), will also provide support for OCL-like constraints, EOL [8] like action operations, data-base like queries and ontology-like inference rules as well as ATL/QVT like transformation rules in a unified, multi-level aware form.

7 Conclusion

In this paper we have presented an approach that takes the first steps towards supporting model transformations that fully exploit the multi-level modeling environments that are starting to emerge from the research community. The paper provides insights into the topic of multi-level aware transformation definition and the problems that need to be solved when implementing transformations on top of a multi-level modeling platform. Rather than focusing on multi-level to multi-level transformations, which are likely to be the long term use case for multi-level aware model transformations, the paper focuses on the more important near term goal of supporting interoperability with existing two-level modeling platforms such as those based on the EMF and the MOF. For this use case, multi-level to 2-level transformations and vice versa are more important. Finally, the work shows that a simple working prototype can be developed by extending existing technologies and the theories on which state-of-the art modeling technologies are based. This significantly lowers the learning curve and adoption barrier for development organizations intending to migrate to multi-level modeling technology. Even though we only showed how to extend ATL we are confident that the same approach will work for other transformation technologies such as QVT. For instance, we indirectly created a multi-level aware OCL implementation in order to support ATL's assignment statement.

References

1. Asikainen, T., Männistö, T.: Nivel:a metamodelling language with a formal semantics. Software and Systems Modeling (2009)
2. Atkinson, C., Kühne, T.: Rearchitecting the UML infrastructure. ACM Trans. Model. Comput. Simul. (2002)
3. Atkinson, C., Kühne, T.: Reducing Accidental Complexity in Domain Models. Software and Systems Modeling (2007)
4. Atkinson, C., Muthig, D.: Component-Based Product-line Engineering with the UML. In: Gacek, C. (ed.) ICSR 2002. LNCS, vol. 2319, pp. 343–344. Springer, Heidelberg (2002)
5. Atkinson, C., Stoll, D.: Orthographic Modeling Environment. In: Fiadeiro, J.L., Inverardi, P. (eds.) FASE 2008. LNCS, vol. 4961, pp. 93–96. Springer, Heidelberg (2008)
6. Eclipse Foundation: ATL Developer Guide - Regular VM, http://wiki.eclipse.org/ATL/Developer_Guide#Regular_VM
7. Eclipse Foundation: ATL/User Guide - The ATL Language, http://wiki.eclipse.org/ATL/User_Guide_-_The_ATL_Language
8. Kolovos, D., Paige, R., Polack, F.: The Epsilon Object Language (EOL) (2006)
9. Kurtev, I., Bézivin, J., Aksit, M.: Technological spaces: An initial appraisal. In: CoopIS, DOA 2002 Federated Conferences, Industrial track, Irvine (2002)
10. de Lara, J., Guerra, E.: Deep Meta-modelling with METADEPTH. In: Vitek, J. (ed.) TOOLS 2010. LNCS, vol. 6141, pp. 1–20. Springer, Heidelberg (2010)
11. Sendall, S., Kozaczynski, W.: Model transformation: the heart and soul of model-driven software development. IEEE Software 20(5) (2003)
12. University of Mannheim - Software Engineering Group: MelAniE - Multi-levEl modeLing And oNtology engIneering Environment, http://www.eclipselabls.org/p/melanie
13. University of Mannheim - Software Engineering Group: nAOMi - opeN, Adaptable, Orthographic Modeling Environment, http://eclipselabs.org/p/naomi

An Algorithm for Generating
Model-Sensitive Search Plans for EMF Models

Gergely Varró[*], Frederik Deckwerth, Martin Wieber, and Andy Schürr

Technische Universität Darmstadt,
Real-Time Systems Lab,
D-64283 Merckstraße 25, Darmstadt, Germany
gergely.varro@es.tu-darmstadt.de, f.deckwerth@stud.tu-darmstadt.de,
{martin.wieber,andy.schuerr}@es.tu-darmstadt.de

Abstract. In this paper, we propose a new model-sensitive search plan generation algorithm to speed up the process of graph pattern matching. This dynamic programming based algorithm, which is able to handle general n-ary constraints in an integrated manner, collects statistical data from the underlying EMF model, and uses this information for optimization purposes. Additionally, runtime performance measurements have been carried out to quantitatively evaluate the effects of the search plan generation algorithm on the pattern matching engine.

Keywords: graph pattern matching, search plan generation algorithm, model-sensitive search plan.

1 Introduction

Efficient, scalable, and standard compliant techniques and tools are still undoubtedly needed to promote the spread of model-driven technologies in an industrial context. As numerous scenarios in the model-based domain, such as (i) checking the application conditions in rule-based model transformation tools [1,2], (ii) bidirectional model synchronization, or (iii) on-the-fly consistency validation, can be described as a general pattern matching problem, its efficient implementation is undisputedly an important task.

In this general pattern matching context, a pattern consists of constraints, which place restrictions on variables, and the number of variables involved in a constraint is referred as its arity. The pattern matching process determines a mapping of variables to the elements of the underlying model in such a way that the assigned model elements must fulfill all constraints. Structural constraints can be checked by using the services of the modelling layer (e.g., type checks, navigation along links), while non-structural constraints are handled by some other means (e.g., integer or textual comparison).

As non-structural constraints are easily manageable [3], the current paper only focuses on structural constraints, which correspond to the graph pattern

---

[*] Supported by the Postdoctoral Fellowship of the Alexander von Humboldt Foundation and associated with the Center for Advanced Security Research Darmstadt.

Z. Hu and J. de Lara (Eds.): ICMT 2012, LNCS 7307, pp. 224–239, 2012.

matching problem [4]. Although available pattern matching engines support type checks and link navigations as unary and binary structural constraints, respectively, practical model-driven scenarios additionally require the handling of n-ary constraints to express ordered references or pattern composition [5].

When building a pattern matching engine, its performance highly depends on the order in which the constraints of a pattern are evaluated (cf. the impact of the variable ordering in general backtracking). This rationale motivates the construction of heuristics-based algorithms for generating constraint sequences or search plans [6], which can be efficiently evaluated.

While the majority of state-of-the-art search plan generation algorithms [1,7,8] exploits only type and multiplicity restrictions derived from the metamodel of the problem domain, two novel *model-sensitive* approaches [9,10] take, for optimization purposes, the potential structure of *instance models* into account as further domain-specific knowledge. Although the inherent performance advantages of model-sensitive search plan generation techniques have already been clearly shown [11], the applicability of the tools themselves in a more general modeling context is hindered by the fact that both engines (i) operate on non-standard (tool specific) model representations, and (ii) apply graph-based algorithms for search plan generation, which can handle only unary and binary constraints in an integrated manner.

In this paper, we propose a completely new model-sensitive search plan generation algorithm, based on dynamic programming, to enable the integrated handling of general n-ary constraints. The algorithm collects statistical data from the model under transformation via an extensible framework and uses this information for optimization purposes. The pluggable collection of statistical data is exemplified on Eclipse Modeling Framework (EMF) compliant models. Finally, the effects of the search plan generation algorithm on the performance of pattern matching are quantitatively evaluated by using runtime measurements.

The remainder of the paper is structured as follows: Section 2 introduces basic modeling and pattern specification concepts. The general pattern matching process is sketched in Sec. 3, while Sec. 4 presents the new search plan generation algorithm. Section 5 gives a quantitative assessment and performance comparison. Related work is discussed in Sec. 6, and Sec. 7 concludes our paper.

2 Metamodel, Model and Pattern Specification

2.1 Metamodels and Models

A *metamodel* represents the core concepts of a domain. In this paper, our approach is demonstrated on a real-world running example from the railway domain [12] (developed in the MOGENTES project [13]), whose metamodel is depicted in Fig. 1(a). *Classes* are the nodes in the metamodel: Routes, Sensors, Signals, SwitchPositions, and TrackElements, which can either be Switches or Segments. *References* are the edges between classes, which can be uni- or bidirectionally navigable as indicated by the arrows at the end points. A navigable end is labelled with a *role name* and a *multiplicity*, which restricts the number of target objects that can be reached via the given reference. In our example, a Route has at least 2 Sensors (as shown by the unidirectional reference hasSensors), and defines an arbitrary number of SwitchPositions, which is a bidirectional reference. *Attributes* (depicted in the lower part of classes) store values of primitive or enumerated types, e.g., the length integer in a Segment, or the actualState of a Switch whose possible values are listed in the *enumeration* SwitchStateKind. Figures 1(b) and 1(c) depict two *models* from the domain, whose nodes and edges are called *objects* and *links*, respectively.

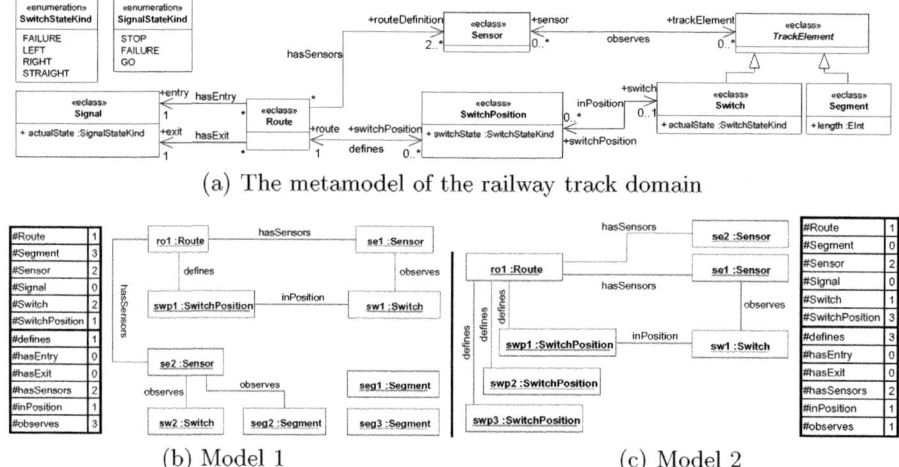

(a) The metamodel of the railway track domain

(b) Model 1   (c) Model 2

Fig. 1. Metamodel of the railway track domain and two sample models

EMF-Specific Issues: References and attributes are collectively referred to as *structural features* and handled uniformly in EMF. Each navigable direction of a structural feature is represented by an indexed List in the source class, which stores corresponding target objects.

Our approach collects statistical data from the model at runtime via EMF adapters. An *object* and *link counter* is introduced for each class and structural feature, which stores the number of type conforming objects and links, respectively, as shown by the tables in Figures 1(b) and 1(c).

2.2   Pattern Specification

As defined in [5,14], a *pattern* is a set of constraints over a set of variables. A *variable* is a placeholder for an object in a model, and it has a reference to a class from the metamodel, which defines the type of the objects that can be assigned to the variable during pattern matching. A *constraint* specifies a condition on a set of variables (which are also referred to as *parameters* in this context) that must be fulfilled by the objects, which are assigned to the parameters.

EMF-Specific Issues: Although the pattern matcher has a pluggable infrastructure for the constraints that can be used for specifying patterns, only one kind of constraints is used throughout the paper.[1] In the following, a constraint maintains a reference to a structural feature, and it prescribes the existence of a link, which (i) conforms to the referenced structural feature and (ii) connects the source and the target object assigned to the first and last parameter, respectively.

An *ordered* or *unordered* structural feature can be modeled by a *binary* constraint in the pattern specification, when *the order information is irrelevant* in the pattern matching process. In contrast, *ternary* constraints should be used for *ordered unidirectional* structural features, where the second parameter is an integer index, which prescribes the location of the target object in the list of the source object containing links that conform to the structural feature.

Example. Pattern `routeSensor` (Fig. 2) expresses a sample requirement defined by railway domain experts, which has been slightly simplified for presentation purposes. It states that a route must have a sensor observing a switch, and the observed switch itself must be part of the route. The pattern has 5 variables (`RO`, `IDX`, `SE`, `SW` and `SWP`), 1 ternary and 3 binary constraints, which prescribe the existence of an ordered unidirectional and 3 bidirectional references, respectively.

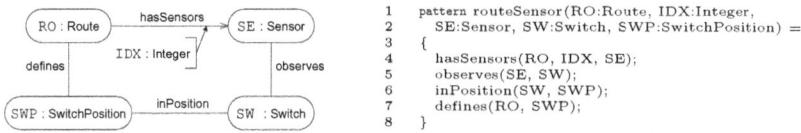

```
1    pattern routeSensor(RO:Route, IDX:Integer,
2        SE:Sensor, SW:Switch, SWP:SwitchPosition) =
3    {
4        hasSensors(RO, IDX, SE);
5        observes(SE, SW);
6        inPosition(SW, SWP);
7        defines(RO, SWP);
8    }
```

Fig. 2. Pattern `routeSensor` in a graphical and textual representation

3   Pattern Matching Process at Runtime

As [14] states, *pattern matching* is the process of determining mappings for all variables in a given pattern, such that all constraints in the pattern are fulfilled. The mappings of variables to objects are collectively called a *match*, which can be a *complete match* when all the variables are mapped, or a *partial match* in all other cases. The overall process of pattern matching is as follows:

Section 3.1. Operations representing atomic steps in the pattern matching process are created from the pattern specification.

Section 3.2. The operations are filtered and sorted by a *search plan generation algorithm* (for the details see Sec. 4) to produce efficient search plans.

Section 3.3. The search plan is then used by an interpreter to control the actual execution of pattern matching, which is carried out as a depth-first traversal.

---

[1] Type restrictions for variables are going to be represented as constraints only in a future version of the pattern matcher.

3.1   Creating Operations

This subsection, which reuses some definitions from [5,14], describes the process of creating operations from the constraints in the pattern specification. In the following, it is assumed that an (arbitrary) order is fixed for the variables in the pattern, and the notation v_p denotes the pth variable according to this order.

An *adornment* [5] represents binding information for *all variables* in the pattern by a corresponding character sequence consisting of letters B or F, which indicate that the variable in that position is *bound* or *free*, respectively.

An *operation* represents a single atomic step in the matching process. It consists of a constraint, an operation adornment, and a mask, which is derived from the operation adornment. An *operation adornment* prescribes which *parameters* must be bound when the operation is executed, while a *mask* represents the same binding information, but projected on *all variables* in the pattern. A *check operation* has only bound parameters. An *extension operation* has free parameters, which get bound when the operation is executed.

Setting operation adornments. For presentation purposes, we assume that operations use the standard EMF services, which restricts the set of operations created for a constraint in the following manner.

For each *binary constraint referring to a bidirectional structural feature*, 3 operations with the corresponding BB, BF, and FB adornments are created. The check operation (BB) verifies the existence of a link, while the other two, adorned by BF and FB, denote forward and backward navigations, respectively. Analogously, for each *binary constraint referring to a unidirectional structural feature*, 2 operations with the corresponding BB and BF adornments are prepared.

For each *ternary constraint (referring to an ordered unidirectional structural feature)*, operations adorned by BBB, BBF, and BFF are prepared (adornment BFB is disallowed for presentation purposes). The check operation (BBB) verifies that (i) a link connects the source and the target object mapped to the first and the third parameter, respectively, and (ii) the target object is stored in the appropriate List of the source object at the index assigned to the second parameter. The operation with the BBF adornment is a forward navigation along the *single* link, which is stored at the index assigned to the second parameter. Finally, the operation adorned by BFF is a forward navigation along *all* links that conform to the structural feature of the constraint, and that retain the source object mapped to the first parameter.

Mask derivation. A *mask* m_o is a sequence of *, B, and F characters. Character * at position p means that the binding of variable v_p is irrelevant, while letters B or F at position p explicitly prescribe the corresponding variable v_p to be bound or free, respectively. For each letter B (F) in the adornment, the position p of the corresponding parameter v_p is looked up by using the fixed variable order, and position p is set to B (F) in the mask. All other locations of the mask are set to *.

Categorizing and applying operations. In the context of an adornment, operations can be categorized. An operation o is a *present* (or applicable) *operation* with respect to an adornment a, if the following conditions hold:

1. **General operation applicability.** Each variable v_p, that must be *free* according to the mask m_o of operation o, is also *free* in adornment a.
2. **Immediate operation applicability.** Each variable v_p, that must be *bound* according to the mask m_o of operation o, is also *bound* in adornment a.

An operation o is a *past operation*, if the first condition on general operation applicability is violated. An operation o is a *future operation*, if only the second condition on immediate operation applicability is violated.

If an operation o is a present (or applicable) operation w.r.t. adornment a, then *applying the operation o on adornment a resulting in an adornment a'* (denoted by $a \xrightarrow{o} a'$) (i) binds all free variables indicated by mask m_o of operation o, and (ii) leaves the binding of all other variables unaltered.

Example. Figure 3(a) lists the operations derived from the routeSensor pattern. In the following, we suppose that variables RO, IDX, SE, SW and SWP are ordered in this specific sequence. For instance, operation observes(SE,SW) adorned by BF is an extension operation, and it is only applicable if variable SE is bound, and variable SW is free, which is also reflected in mask **BF* as SE and SW are the third and fourth variable, respectively. This operation can be categorized as a future operation with respect to adornment BFFFF, as it violates the immediate operation applicability condition at the third position.

Constraint	Op. Adornm.	Mask	Applic.	Type
hasSensors(RO,IDX,SE)	BBB	BBB**	future	check
hasSensors(RO,IDX,SE)	BBF	BBF**	future	extension
hasSensors(RO,IDX,SE)	BFF	BFF**	present	extension
observes(SE,SW)	BB	**BB*	future	check
observes(SE,SW)	BF	**BF*	future	extension
observes(SE,SW)	FB	**FB*	future	extension
inPosition(SW,SWP)	BB	***BB	future	check
inPosition(SW,SWP)	BF	***BF	future	extension
inPosition(SW,SWP)	FB	***FB	future	extension
defines(RO,SWP)	BB	B***B	future	check
defines(RO,SWP)	BF	B***F	present	extension
defines(RO,SWP)	FB	F***B	past	extension

(a) Operations

Search plan	Step	Constraint	Op. Adornm.	Mask	Adornm. a_i (a_0 = BFFFF)
Search plan 1 (derived from model 1)	(1)	defines(RO,SWP)	BF	B***F	BFFFB
	(2)	inPosition(SW,SWP)	FB	***FB	BFFBB
	(3)	hasSensors(RO,IDX,SE)	BFF	BFF**	BBBBB
	(4)	observes(SE,SW)	BB	**BB*	BBBBB
Search plan 2 (derived from model 2)	(1)	hasSensors(RO,IDX,SE)	BFF	BFF**	BBBFF
	(2)	observes(SE,SW)	BF	**BF*	BBBBF
	(3)	inPosition(SW,SWP)	BF	***BF	BBBBB
	(4)	defines(RO,SWP)	BB	B***B	BBBBB

(b) Search plans as sequence of operations

Fig. 3. Operations and search plans for the routeSensor pattern

3.2 Search Plan Generation

When pattern matching is invoked, variables can already be bound to objects to restrict the search. The corresponding binding information of all variables is called *initial adornment* a_0. By using the initial adornment, a search plan generation algorithm filters and sorts the operations to produce a search plan. The current search plan formalism is a precise and extended variant of [5].

A *search plan* $SP = \langle o_1, o_2, \ldots, o_l \rangle$, *starting from an initial adornment* a_0, is a sequence of operations satisfying the following conditions:

1. **No multiple constraint checks.** Each constraint in the pattern has *at most one* corresponding operation in the search plan.
2. **Valid adornment sequence.** An adornment sequence $a_0, a_1, \ldots, a_l$ can be derived in such a way that $a_0 \overset{o_1}{\Longrightarrow} a_1 \overset{o_2}{\Longrightarrow} \ldots \overset{o_l}{\Longrightarrow} a_l$. The last element a_l in this adornment sequence is referred as the *adornment of the search plan.*

A search plan is *complete*, if each constraint is represented by *exactly one* operation in the sequence, and its adornment has only B characters.

Example. Figure 3(b) depicts two search plans generated by our algorithm for Models 1 and 2, when variable RO is initially bound and, thus, the initial adornment is BFFFF. The rightmost column presents the adornment *after* applying the operation in the same line. SP1 extends the partial match along two separate directions before joining the branches with the last (check) operation, while SP2 employs a clockwise navigation along the references in the pattern.

3.3   Search Plan Execution by a Pattern Matcher Interpreter

By conceptually following the corresponding part of [14], the interpreter uses a *match array* for storing the matches, and the search plan for guiding the pattern matching process. The size of the match array is determined by the number of variables in the pattern. Each operation has a mapping, which identifies the slots in the match array that correspond to the parameters of the operation.

When pattern matching is invoked, the initial match array is filled in by the objects that are initially assigned to the variables, and it is passed on to the first operation in the search plan. When an extension operation is executed, the structural feature of its constraint is navigated in forward (BF, BBF, BFF) or backward (FB) direction depending on the operation adornment, then each accessed object is type checked and bound to the corresponding free variable, and the execution is passed on to the following operation for subsequent processing together with the extended match array. A check operation simply passes on the unchanged match array, if the actual check succeeded, and stops triggering further processing steps otherwise. If a match array passes beyond the last operation, then it represents a complete match, which is copied and stored in the result set.

This pattern matching (PM) process implements a depth-first traversal of a PM state space, where a *PM state* represents a partial match that is produced by an extension operation during pattern matching. The PM state space can be described by a tree, whose root is the initial match, while internal nodes and leaves correspond to partial and complete matches, respectively. Note that each tree level is produced by a corresponding extension operation, and check operations do not influence the tree structure as they do not bind any variables.

Example. Figure 4 depicts two PM state spaces, which are generated by performing search plans SP1 and SP2 on Model 2, respectively. E.g., the second level of Fig. 4(a) represents the partial matches that are prepared when navigating along defines links from route ro1 to switch positions swp1, swp2, and swp3, as prescribed by operation defines(RO,SWP) with adornment BF. Framed leaves represent those complete matches that pass beyond the last check operation (only shown in Fig. 3(b)), while unframed ones fail this check. It is obvious from Fig. 4 that SP2 is better than SP1, as SP2 traverses less PM states.

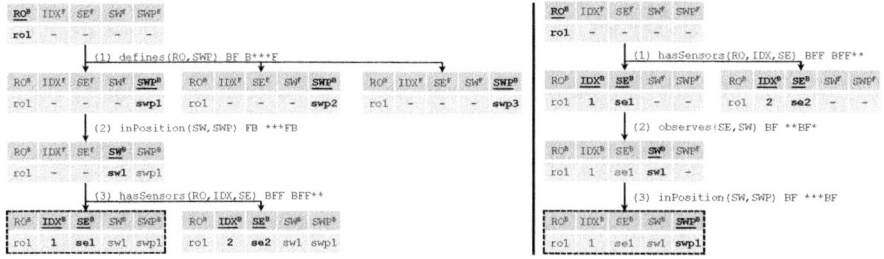

(a) PM state space by performing SP1 on Model 2   (b) PM state space by performing SP2 on Model 2

Fig. 4. Sample PM state spaces for Model 2

4   Dynamic Programming Based Search Plan Generation

As demonstrated in Fig. 4, the search plan has a large impact on the number of produced matches, and consequently, on the performance of pattern matching. As such, the production of a good search plan is an essential issue, and that is why a quantitative characterization of operations and search plans is introduced for optimization purposes by means of weights and costs. Note that a cost function should ideally have a strong correlation with the size of the PM state space.

Operation weight calculation. An extension operation o is augmented by a *weight* w_o, which denotes the cost of performing the operation. In our approach, a weight is defined as an average *branching factor* for that level of the PM state space tree, which represents the operation execution, and is calculated using the statistical data collected from the underlying model. The weights of ternary operations with the BBF adornment are set to 1 (irrespective of the model), as these operations never induce any branching in the matching process. For binary and ternary operations with the corresponding BF and BFF adornments (forward navigation), the structural feature referenced by the constraint of the operation is determined, and the weight is the ratio of the link and object counters defined for this structural feature and its *source* class, respectively. For binary operations with FB adornment (backward navigation), the link counter of the structural feature is divided by the object counter of the *target* class to define the weight.

Search plan costs. The search plan cost c_l used in this paper estimates the size of the PM state space tree via the $c_l = \sum_{j=1}^{l} \prod_{i=1}^{j} w_{o_i}$ expression [10], which sums up the estimated number of PM states on a level-by-level basis (excluding the root). To support an iterative search plan cost calculation, the cost c_l is complemented by a product value π_l and the calculation is rearranged as $(c_l, \pi_l) = f(c_{l-1}, \pi_{l-1}, w_{o_l})$, where $c_0 = 0$, $\pi_0 = 1$, $\pi_l = \pi_{l-1} w_{o_l}$, and

$$c_l = \sum_{j=1}^{l} \prod_{i=1}^{j} w_{o_i} = \overbrace{w_{o_1} + \ldots + w_{o_1} w_{o_2} \cdots w_{o_{l-1}}}^{c_{l-1}} + \overbrace{\underbrace{w_{o_1} \cdots w_{o_{l-1}}}_{\pi_{l-1}} \cdot \underbrace{w_{o_l}}_{w_{o_l}}}^{\pi_l} = c_{l-1} + \pi_l.$$

Algorithm data structures. To avoid unnecessary recalculations in our approach, a state stores only the best of those search plans that share the same adornment. A *state* S contains a *search plan* SP_S with its *adornment* a_S and costs (c_S, π_S); and sequences of *present extension* O_S^{pe}, *future extension* O_S^{fe}, and *future check* O_S^{fc} operations[2] (w.r.t. adornment a_S), which are (i) pairwise disjoint by definition, and (ii) ordered based on their weights. Two states are *adornment disjoint*, if they have different adornments.

The initial state S_0 has an empty operation sequence as its search plan, the initial adornment a_0 as its adornment, and its cost values are set as $c_{S_0} := c_0$, $\pi_{S_0} := \pi_0$. Its operations are categorized w.r.t. the initial adornment a_0.

Algorithm. An efficient search plan is generated by a dynamic programming based algorithm (see Algorithm 1), which iteratively fills states into an initially empty table T with $n + 1$ columns and k rows, where n is the number of free variables $|a_{S_0}|$ in the adornment a_{S_0} of the initial state S_0 and $k \geq 1$ is a user-defined parameter that influences the trade-off between efficiency and optimality of the algorithm. In general, the column $T[i]$ stores the best k adornment disjoint states (in an increasing cost order), which have i free variables in their adornment, while $T[i][j]$ is the jth best from these adornment disjoint states.

The two key features of the algorithm can be summarized as follows. (i) The table *only stores adornment disjoint states* with the consequence of keeping only the best search plan from those ones that share a common prefix. (ii) Additionally, the table *only stores a constant number* of adornment disjoint states *in each column*, immediately discarding costly search plans, which are not among the best k solutions, and implicitly all their possible continuations. This avoids the production of all search plans, which could alone result in the same (exponential) complexity as the match calculation process.

First, the algorithm determines the number of free variables $n = |a_{S_0}|$ in the adornment a_{S_0} of the initial state S_0 (line 1), and stores this state S_0 in $T[n][1]$ (line 2). Then, the table is traversed by processing columns in a decreasing order based on the number of free variables in the state adornments (lines 3–17). In contrast, the inner loop (lines 4–16) proceeds in an increasing state cost order starting from the best state $T[i][1]$ in each column $T[i]$. For each present extension operation o in each stored state S (lines 6–15), the next state S' is prepared in a two-phase process, which (1) calculates the search plan $SP_{S'}$, the adornment $a_{S'}$ and the cost $c_{S'}$ of the next state S' immediately in `calculateNextState` (lines 8–9), and (2) updates the search plan, and the sequences of present extension, future extension, and future check operations in

---

[2] Note that past and present check operations need not be stored as they will be immediately processed by the algorithm.

Algorithm 1. The procedure `calculateSearchPlan`(S_0, k)

1: $n := |a_{S_0}|$ //number of free variables in the initial state adornment a_{S_0} is calculated
2: $T[n][1] := S_0$
3: **for** $(i := n$ to $1)$ **do**
4:     **for** $(j := 1$ to $k)$ **do**
5:         $S := T[i][j]$ //current state S
6:         **for each** $(o \in O_S^{pe})$ **do**
7:             // for each present extension operation
8:             $S' := $ `calculateNextState`(S, o) // next state S' is calculated
9:             $i' := |a_{S'}|$ // next state S' has i' free variables in its adornment $a_{S'}$
10:            $(\mathbf{a}, \mathbf{c}) := $ `determineIndices`$(T[i'], S')$
11:            **if** (`checkInsertCondition`$(T[i'], S', \mathbf{a}, \mathbf{c})$) **then**
12:                `updateOperations`(S', S, o)
13:                `insert`$(T[i'], S', \mathbf{a}, \mathbf{c})$
14:            **end if**
15:         **end for**
16:     **end for**
17: **end for**
18: **return** $SP_{T[0][1]}$

a delayed manner in `updateOperation` (line 12), but only if the next state S' passes the insert condition (line 11), which uses indices $\mathbf{a}$ and $\mathbf{c}$ for decision making, which are calculated by `determineIndices` (line 10). In the latter case, the complete next state S' is inserted into the column $T[i']$ by using indices $\mathbf{a}$ and $\mathbf{c}$ (line 13). Finally, the algorithm returns the search plan $SP_{T[0][1]}$ (line 18).

The procedure `calculateNextState`(S, o) partially calculates the new state S' from state S and operation o. The new search plan $SP_{S'}$ is determined by appending operation o to the search plan SP_S of state S. The new adornment $a_{S'}$ is calculated by applying operation o on the adornment a_S of state S (i.e., $a_S \overset{o}{\Longrightarrow} a_{S'}$). The new costs $c_{S'}$ and $\pi_{S'}$ are computed from the costs c_S and π_S of state S, and the weight w_o of operation o according to the cost function f.

The procedure `determineIndices`$(T[i'], S')$ calculates indices $\mathbf{a}$ and $\mathbf{c}$. Index $\mathbf{a}$ marks the position of that stored state $T[i'][\mathbf{a}]$, which has the same adornment $a_{S'}$ as state S'. Index $\mathbf{a}$ is set to $k+1$, if no such stored state exists. Index $\mathbf{c}$ marks the position at which state S' should be inserted based on its cost. Index $\mathbf{c}$ is set to $k + 1$, if state S' is not among the best k adornment disjoint states. Formally, $\mathbf{c}$ is the smallest index for which $c_{S'} < c_{T[i'][\mathbf{c}]}$ holds (or $c_{T[i'][\mathbf{c}]} = $ **null**).

The procedure `checkInsertCondition`$(T[i'], S', \mathbf{a}, \mathbf{c})$ makes a positive decision, (1) if column $T[i']$ does not contain any states with the adornment $a_{S'}$ of new state S' ($\mathbf{a} = k + 1$), new state S' is among the best k adornment disjoint states ($\mathbf{c} < \mathbf{a}$), and a reachability analysis[3] determines that the search plan $SP_{S'}$ can be completed in a valid manner, or (2) if column $T[i']$ already stores a state $T[i'][\mathbf{a}]$ at location $\mathbf{a}$ with the adornment $a_{S'}$ of new state S' ($\mathbf{a} < k + 1$), and this new state S' is better than the stored state $T[i'][\mathbf{a}]$ ($\mathbf{c} \leq \mathbf{a}$).

---

[3] The reachability analysis is only discussed in [15] due to space limitations.

The procedure $\mathtt{updateOperations}(S', S, o)$ processes all operations o^* of present extension O_S^{pe}, future extension O_S^{fe}, and future check O_S^{fc} sequences of state S in an increasing weight order by also recategorizing these operations with respect to the adornment $a_{S'}$ of new state S'.

- **Discard operations causing multiple checks.** If operation o^* originates from the same constraint as the selected operation o, then operation o^* is discarded to avoid checking a constraint more than once. This can be easily checked as each operation maintains a reference to its constraint.
- **Discard past operations.** If operation o^* is a past operation, then it is discarded as it violates the general operation applicability condition.
- **Append present check operations to the search plan.** If operation o^* is a present check operation, then it is immediately appended to the search plan to perform the corresponding checks as soon as possible.
- **Append present extension, future extension, and future check operations to the corresponding list.** If operation o^* is a present extension, a future extension or a future check operation, then it is appended to the corresponding operation sequence $O_{S'}^{pe}$, $O_{S'}^{fe}$, or $O_{S'}^{fc}$ of state S', respectively.

As operation application can only change variables from free to bound, a past operation can never be recategorized in any states derivable from S', (hence, its immediate disposal is justified) while a future operation might eventually become a present or past operation in a later phase of the algorithm.

The procedure $\mathtt{insert}(T[i'], S', \mathbf{a}, \mathbf{c})$ determines $m = \min\{\mathbf{a}, k\}$, removes state $T[i'][m]$, shifts elements between $T[i'][\mathbf{c}]$ and $T[i'][m-1]$ downward, and inserts state S' at position $\mathbf{c}$.

Example. The dashed box of Fig. 5 presents the contents of table T (with 3 empty fields in the second row) after running our algorithm on Model 2 with initial adornment BFFFF. Each arrow represents the derivation of a new state, which was produced by one execution of the innermost cycle (lines 6–15). States with watermark A were temporarily stored in the table (but later discarded due to the appearance of better states). The state with letter B failed the reachability analysis, while states with watermark C were discarded as the corresponding column had already contained a better state with the same adornment.

For instance, the first execution of the innermost cycle processes operation $\mathtt{hasSensors(RO,IDX,SE)}$ with adornment BFF, whose weight is $\frac{\#\mathtt{hasSensors}}{\#\mathtt{Route}} = \frac{2}{1} = 2$ as Model 2 has 2 $\mathtt{hasSensors}$ links, and 1 $\mathtt{Route}$. The corresponding new state is inserted into $T[2][1]$ as its adornment BBBFF has 2 free variables, and column $T[2]$ is empty at this time. In this new state, both costs are 2, operations with constraint $\mathtt{hasSensors(RO,IDX,SE)}$ are discarded, and all other operations are recategorized w.r.t. adornment BBBFF.

5   Measurement Results

In this section, we quantitatively assess the effects of different cost models and various configurations of our proposed search plan generation algorithm on the

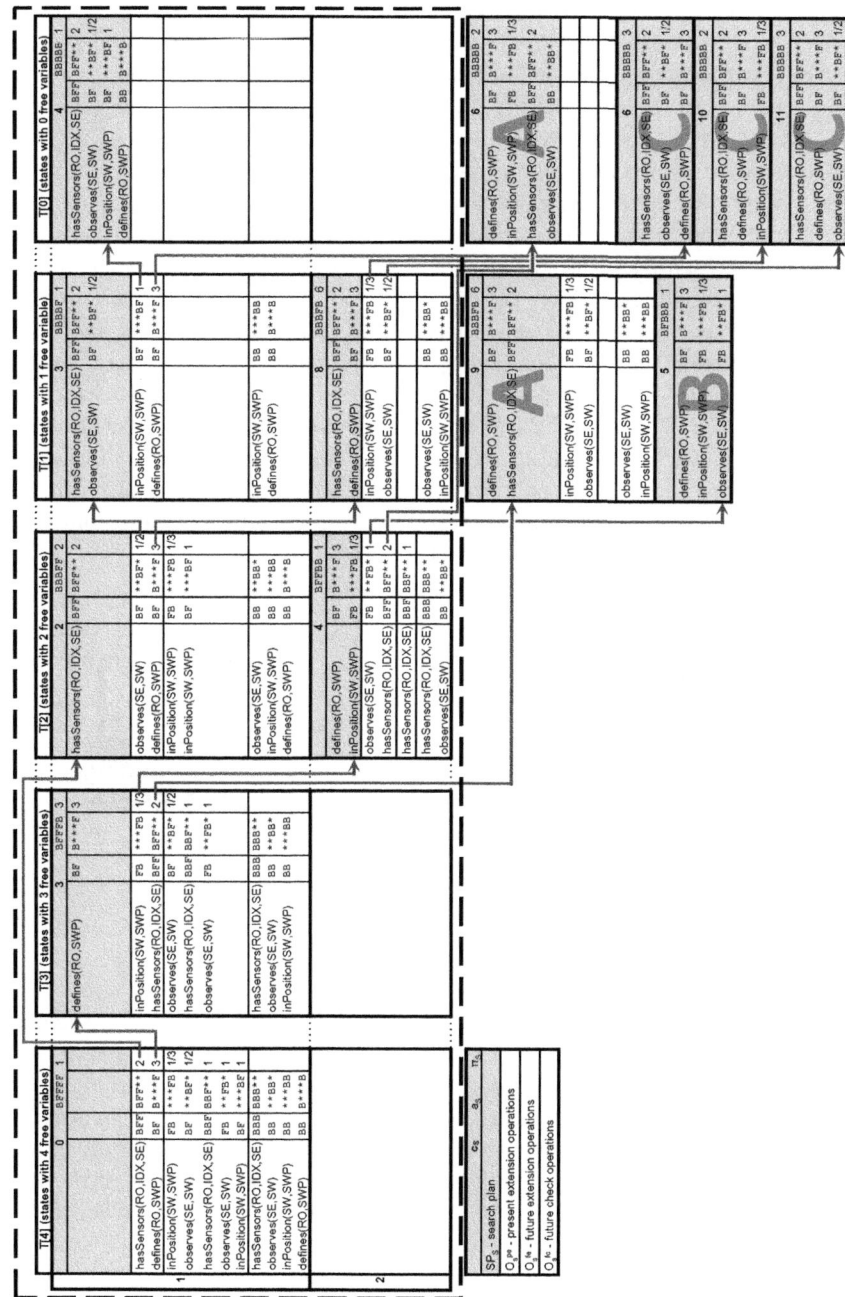

Fig. 5. Execution of the algorithm on Model 2 with $k = 2$

runtime performance of the pattern matching process. More specifically, our model-sensitive (MS) cost model was compared to a domain-specific (DS) approach, which latter used operation weights 1 and 10 for constraints representing structural features with at most one (1) and arbitrary (*) multiplicity, respectively. For configuring our algorithm, its parameter k was set to 1 and 2.

The pattern `routeSensor` of Fig. 2 and 10 models of different size from the case study [12] were used for experimentation purposes. Pattern matching was always restricted to a given `Route` in the model, which was assigned to variable `RO` in the initial match and used as a starting point. The complete process (including search plan generation) was repeated on each distinct `Route`.

Figure 6(a) presents the measured data. The first column indicates the model identifier, the second and third columns the model size and the number of distinct `Routes` in the model, respectively. The remaining columns show the measured values for the different configurations, which independently involve domain-specific (DS) and model-sensitive (MS) cost models, and algorithm parameter values $k = 1$ and 2. The PM columns denote the number of PM states (i.e., elementary pattern matching steps), which was averaged over all distinct `Routes` in the model. The SP columns show the cost of the (model-sensitive) search plan that was considered the best by the search plan generation algorithm and that was actually used to control pattern matching.

	Model size #	Routes #	DS (k=1) PM #	DS (k=2) PM #	MS (k=1) SP $c_{\Pi[0][1]}$	MS (k=1) PM #	MS (k=2) SP $c_{\Pi[0][1]}$	MS (k=2) PM #
1	1450	20	1128.55	579.80	430	579.80	115	118.50
2	2601	40	885.15	456.75	349	456.75	102	104.78
4	5234	80	881.15	454.94	355	454.94	101	105.19
8	10627	160	912.64	470.81	361	470.81	102	106.29
16	21186	320	939.48	483.93	357	483.93	104	107.36
32	42202	640	936.80	482.44	353	482.44	104	107.21
64	85428	1280	960.83	494.65	362	494.65	105	108.51
128	171030	2560	955.14	491.77	362	491.77	105	108.78
256	339490	5120	943.89	486.02	356	486.02	104	107.93
512	685830	10240	953.93	491.18	364	491.18	106	109.18

(a) Comparison of PM state spaces

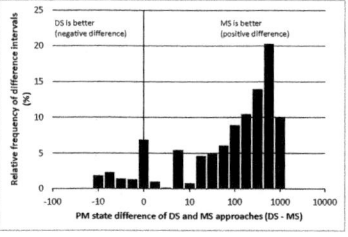

(b) PM state difference profile

Fig. 6. Measurement results

Fig. 6(a) shows that model-sensitive search plans have the capability to clearly outperform domain-specific ones (in this case on all test models by nearly 400 steps in average) when the pattern has many structural feature constraints with arbitrary multiplicity. Our algorithm generated the same search plan for the settings of the fifth and the seventh column, which explains the equal values there. Fig. 6(b) presents the *relative frequency* distribution histogram of the PM state differences of DS and MS approaches (with parameter $k = 2$) when these differences are calculated on a route-by-route basis for each of the 2560 starting points of model 128 (see the thick frames in Fig. 6(a)). Fig. 6(b) shows that the DS approach was better by 6 to 10 steps in 1.875% of the 2560 cases (first column), the MS search plan was faster by 562 to 1000 steps in nearly 10% of the cases (last column), while a draw occured in 6.875% of the cases (fifth column).

In contrast to our preliminary expectations, which assumed that it was sufficient to set parameter k only to 1 in practical cases, it can be seen that a more thorough analysis with $k = 2$ can already pay off for small and simple patterns.

Unfortunately, the models of this case study were structurally similar, since all the MS search plans (irrespectively of the different models) were the same for a given parameter, which should not necessarily be the case. As further general characteristics, the average wall clock time[4] for search plan generation was 50 μs (for all configurations), and a single PM step took 51 ns in average. Neither the search plan generation, nor the pattern matching is affected by the model-sensitive nature of the approach, as object and link counters are initialized and incrementally updated, when the model is loaded and changed, respectively.

6 Related Work

Numerous useful model transformation tools are now surveyed, which internally perform search plan driven pattern matching. A more detailed comparison of pattern matcher engines is provided in [14].

Search plan driven pattern matchers. Fujaba [1] uses a search plan generation strategy that solely exploits type and multiplicity restrictions, which are derived from the metamodel. According to the used strategy, a navigation along an edge with an at most one multiplicity precedes navigations along edges with arbitrary multiplicity. Fujaba originally operated on top of a non-standard model representation, but recent versions can handle EMF models as well.

Pattern matchers driven by model-sensitive search plans. Although Fujaba [16] is a model-sensitive approach and runs on EMF models, it has only a simple greedy strategy to control pattern matching. GrGen [9] and Viatra [10], which employ model-sensitive search plans, operate on a non-standard modeling layer, which has several consequences. On one hand, these tools can use an arbitrary and optimized model representation, which can already have an integrated support for statistical data collection. On the other hand, if these tools aim to manipulate EMF-compliant models, then they have to be converted by import and export mechanisms, which (i) is not always possible for legacy EMF-based systems, and (ii) results in the inherent duplication of the complete model, which has a significant negative impact on the memory consumption. Since all other similarities and distinctions of GrGen, Viatra, and our approach are related to the employed search plan generation algorithms, these are evaluated in the following separate paragraphs.

Analysis of model-sensitive search plan generation algorithms. In contrast to our dynamic programming search plan generation algorithm, GrGen and Viatra use graph based techniques, which are obviously sufficient for sorting

---

[4] A 2.93 GHz Intel Pentium Dual-Core CPU with 3.7 GB RAM was used for all measurements. A 64-bit Ubuntu 11.04 with kernel 2.6.32–33 and Java 1.6.0_20 served as the underlying operating system and virtual machine, respectively. Measurements that result in time values were repeated 50 times for each starting point.

and filtering unary and binary constraints, which are the most widespread restriction types, but these solutions lack the integrated handling of general n-ary constraints, which are required for ordered references and pattern composition [5]. Both GrGen and Viatra support the construction of complex patterns from simpler ones, but the calculation of matches along pattern composition is scheduled by a separate piece of code and not the core search plan algorithm.

Search plan costs are calculated from the operation weights as a sum $\sum_i w_{o_i}$ in Viatra, and as a product $\prod_i w_{o_i}$ in GrGen, which can also be restructured to a sum by using the logarithm operator (i.e., $\sum_i \ln w_{o_i}$). As a graph based algorithm provides a provably optimal solution with these cost functions, they are perfect for filtering operations, but completely useless for sorting due to the insensitivity of these cost functions to the operation order.

A dynamic programming algorithm can cope with more complex cost functions, and it can provably find the optimum, if the whole solution space is explored when $k = \binom{n}{\lfloor \frac{n}{2} \rfloor}$. For a smaller k, the optimality is no longer guaranteed as the optimal search plan might have a prefix that is not among the best k adornment disjoint solutions at some point, and thus, this solution is discarded. In this sense, the selection of k can be considered as a trade-off between the polynomial runtime of the algorithm and the proven optimality of the solution.

Finally, it must be emphasized that the overall success of model-sensitive search plan generation algorithms highly rely on a strong correlation between the search plan cost and the size of the actually traversed state space, which is only a hypothesis that was thoroughly analyzed in [11], but not a provable fact.[5]

7   Conclusion

In this paper, we proposed a novel search plan generation algorithm based on dynamic programming together with a model-sensitive cost function for EMF models to speed up pattern matching in practice. Additionally, performance measurements have been carried out in a hardware and JVM independent manner to assess the effects of search plan generation on the pattern matching process.

Our future tasks are to repeat measurements in additional scenarios, to give a quantitative performance comparison of our approach to other pattern matchers, and to embed the pattern matching framework into different modeling tools.

Acknowledgements. The authors acknowledge the help of Benedek Izsó, István Ráth and Dániel Varró in providing us the railway scenario for the measurements.

References

1. Geiger, L., Schneider, C., Reckord, C.: Template- and modelbased code generation for MDA-tools. In: Giese, H., Zündorf, A. (eds.) Proc. of the 3rd International Fujaba Days, pp. 57–62 (2005),
ftp://ftp.upb.de/doc/techreports/Informatik/tr-ri-05-259.pdf

---

[5] This means that the execution of the optimal search plan does not necessarily result in the traversal of the smallest state space.

2. Jouault, F., Kurtev, I.: Transforming Models with ATL. In: Bruel, J.-M. (ed.) MoDELS 2005. LNCS, vol. 3844, pp. 128–138. Springer, Heidelberg (2006)
3. Anjorin, A., Varró, G., Schürr, A.: Complex attribute manipulation in TGGs with constraint-based programming techniques. In: Hermann, F., Voigtländer, J. (eds.) Proc. of the 1st International Workshop on Bidirectional Transformations, Electronic Communications of the EASST (accepted paper, 2012)
4. Rozenberg, G. (ed.): Handbook of Graph Grammars and Computing by Graph Transformation. Foundations, vol. 1. World Scientific (1997)
5. Horváth, Á., Varró, G., Varró, D.: Generic search plans for matching advanced graph patterns. In: Ehrig, K., Giese, H. (eds.) Proc. of the 6th Int. Workshop on Graph Transformation and Visual Modeling Techniques. ECEASST, vol. 6 (2007)
6. Zündorf, A.: Graph Pattern Matching in PROGRES. In: Cuny, J., Engels, G., Ehrig, H., Rozenberg, G. (eds.) Graph Grammars 1994. LNCS, vol. 1073, pp. 454–468. Springer, Heidelberg (1996)
7. Fischer, T., Niere, J., Torunski, L., Zündorf, A.: Story Diagrams: A New Graph Rewrite Language Based on the Unified Modeling Language and Java. In: Ehrig, H., Engels, G., Kreowski, H.-J., Rozenberg, G. (eds.) TAGT 1998. LNCS, vol. 1764, pp. 296–309. Springer, Heidelberg (2000)
8. Rensink, A.: The GROOVE Simulator: A Tool for State Space Generation. In: Pfaltz, J.L., Nagl, M., Böhlen, B. (eds.) AGTIVE 2003. LNCS, vol. 3062, pp. 479–485. Springer, Heidelberg (2004)
9. Geiß, R., Batz, G.V., Grund, D., Hack, S., Szalkowski, A.M.: GrGen: A Fast SPO-Based Graph Rewriting Tool. In: Corradini, A., Ehrig, H., Montanari, U., Ribeiro, L., Rozenberg, G. (eds.) ICGT 2006. LNCS, vol. 4178, pp. 383–397. Springer, Heidelberg (2006)
10. Varró, G., Varró, D., Friedl, K.: Adaptive graph pattern matching for model transformations using model-sensitive search plans. In: Karsai, G., Taentzer, G. (eds.) Proc. of International Workshop on Graph and Model Transformation. ENTCS, vol. 152, pp. 191–205. Elsevier (2005)
11. Batz, G.V., Kroll, M., Geiß, R.: A First Experimental Evaluation of Search Plan Driven Graph Pattern Matching. In: Schürr, A., Nagl, M., Zündorf, A. (eds.) AGTIVE 2007. LNCS, vol. 5088, pp. 471–486. Springer, Heidelberg (2008)
12. Izsó, B.: Ontology based verification of system models. Master's thesis, Budapest University of Technology and Economics (2011) (in Hungarian)
13. The MOGENTES project, http://www.mogentes.eu/
14. Varró, G., Anjorin, A., Schürr, A.: Unification of compiled and interpreter-based pattern matching techniques. Technical Report 2922, Technische Universität Darmstadt (March 2012), http://tuprints.ulb.tu-darmstadt.de/2922/
15. Deckwerth, F.: Model-sensitive search plan algorithm for EMF models. Master's thesis, Technische Universität Darmstadt (2012)
16. Giese, H., Hildebrandt, S., Seibel, A.: Improved flexibility and scalability by interpreting story diagrams. In: Margaria, T., Padberg, J., Taentzer, G. (eds.) Proc. of the 8th Int. Workshop on Graph Transformation and Visual Modeling Techniques. ECEASST, vol. 18 (2009)

Paisley: Pattern Matching à la Carte

Baltasar Trancón y Widemann[1,2] and Markus Lepper[2]

[1] University of Bayreuth, Germany
[2] <semantics/> GmbH
baltasar@trancon.de, post@markuslepper.eu

Abstract. Professional development of software dealing with structured models requires more systematic approach and semantic foundations than standard practice in general-purpose programming languages affords. One remedy is to move to domain-specific environments. Here, instead, we present a tool for the implementation of pattern matching as fundamental means of automated data extraction from complex models in a general-purpose programming language. The interface is simple but, thanks to elaborate and rigorous design, is also light-weight, portable, non-invasive, type-safe, modular and extensible. It is compatible with object-oriented data abstraction and has full support for nondeterminism by backtracking. The tool comes as a library consisting of two levels: elementary pattern constructs (generic, highly reusable) and pattern bindings for particular data models (specific, fairly reusable, user-definable). Applications use the library code in a small number of idiomatic ways, making pattern-matching code declarative in style (yet retaining richer host-language semantics), easily writable, readable and maintainable. Library and idiom together form a tightly embedded domain-specific language; no extension of the host language is required. The current implementation is in Java, but assumes only standard object-oriented features, and can hence be ported to other mainstream languages.

1 Introduction

Whether models and model transformations are expressed in problem-specific tools that need to be implemented, or in general-purpose programming environments, at the end of the day a model is a data structure describing an actual or potential system, and a model transformation is a program that either creates such data or extracts relevant information from them. Whereas declarative (functional or logical) languages are more or less equally powerful on the creation and extraction sides, object-oriented languages are notorious for the relative clumsiness of their default extraction idiom, in terms of type cases, casts and getter methods, compared to the creation idiom of compositional constructor calls.

In earlier work, we have pleaded the cause of a more advanced, specifically object-oriented technique of data extraction, namely the *visitor* style pattern: in [5], we have demonstrated how visitor-based extraction can be optimized using a combination of static and dynamic analyses. Here, we turn to a more paradigm-neutral technique that enjoys great popularity, e.g. in functional programming and in string processing, namely *pattern matching*.

Z. Hu and J. de Lara (Eds.): ICMT 2012, LNCS 7307, pp. 240–247, 2012.

We present a generic programming aid for data extraction by pattern matching that unifies desirable features of declarative paradigms with an object-oriented approach to data abstraction, making no assumptions about the data metamodel other than the host-language semantics. It comes in two parts: a basic library and a programming idiom that uses the library operations as its core vocabulary. Problem-specific composite operations can be provided by the user by extending the library cleanly through subclassing. Our implementation is hosted in Java, but nothing prevents the same technique to be used in other strongly typed object-oriented environments such as C++ or .NET.

2   Standards of Pattern Matching

Pattern matching, in the wide sense, plays an important role in many different kinds of programming environments. The techniques applied differ substantially regarding theoretical foundation and expressiveness, the treatment of nondeterminism, type discipline, etc. The following are the relevant role models, positive or negative, for our approach.

String processing with regular expressions. Typing is not an issue, since patterns refer to character strings only. Theoretical foundation is sound, but only as long as certain pragmatic extensions are excluded. Nondeterminism is not supported except for top-level substring search. Conceptually nondeterministic forms are disambiguated explicitly by different flavours (greedy etc.) of constructors.

Functional programming with algebraic datatypes. Inverse constructors are a central means of data extraction and equational function definition in functional programming languages (Hope, Haskell, ML, Opal), and shares the full type discipline of the language. Nondeterminism arises not within one pattern, but rather between overlapping patterns of equations, and is usually resolved implicitly by a first-fit rule.

XML processing with XPath and XSLT. XPath pattern definition has become practically relevant as a central part of the XSL transformation system. There is no commonly agreed theory for the full scope of the pattern language. Nondeterminism is supported by the "for-each" statement of the XSLT language, although control is limited to complete enumeration.

Logic programming with goals & unification. Logic programming languages (most famous: Prolog) offer a distinct quality by making nondeterminism, unification of variables and values, and automated exploration of solution spaces first-class constructs of the language. They are usually weakly typed.

Model query and transformation. In dedicated model query languages pattern matching is a central functionality as well: the evaluation of a query delivers a subset of model nodes. Selection criteria range from simple checks on attribute values to complex relational constraints among nodes. In graph transformation systems, graph patterns feature prominently as the left hand sides of rewrite rules. The pervasive nondeterminism in graphs is often handled by explicit control flow.

3   Requirements

Our approach is distinguished by following a carefully selected canon of rigorous design requirements:

1. *Statically type-safe variables.* No need to down-cast variable bindings.
2. *Statically type-safe patterns.* Detect ill-typed pattern matching attempts as often as possible.
3. *No language extension: independent of host compiler/interpreter.* Solution can be used with off-the-shelf programming platforms.
4. *No assumptions on host language beyond standard OOP.* Solution can be re-implemented in any comparable host environment.
5. *No adaptation of model datatypes required.* Data models can be developed without pattern matching in mind; no source code access required.
6. *Support for multiple views per type.* Different collections of pattern elements can expose different structural aspects of data models.
7. *Declarative, readable, writable, customizable.* Patterns express the programmer's intention of data extraction with as little formal noise as possible.
8. *Full reification: no parsing/compilation overhead at runtime.* Patterns are typed host-language objects; ill-defined usage is detected at compile time.
9. *Support for continuation-style nondeterminism.* Patterns allow multiple successive matches postponable indefinitely, even across persistent serialization.
10. *Nondeterminism incurs no significant cost unless actually used.* No central storage or control mechanism; lazy exploration of alternatives.

Space does not permit us to give either a full *verification* that our implementation satisfies these requirements, nor a full *validation* that the requirements actually yield a good solution to a particular class of problems. With respect to the latter, the reader is invited to judge on the grounds of own practical experiments with the downloadable demo distribution; see Section 6 below.

4   Design

The requirements for static type safety and reification rule out a domain-specific language for pattern matching that is used at run-time *syntactically*, in the style of textually encoded regular expressions. On the other hand, the requirements for host language independence rule out implicit compile-time embedding of pattern matching code. The remaining middle ground is covered by generative approaches, which embed domain-specific notation in an explicit translation, and by library approaches, where patterns are constructed at run-time, but in terms of *semantical* host language objects. Here, we pursue the library approach because it is more lightweight and flexible. Of course, complex fragments of code using the library can be generated from a more concise domain-specific notation, as for instance done by our umod tool [5].

The classical semantics of patterns as the inverse of constructor terms of algebraic datatypes, de-facto standard in declarative languages, does not carry over

```
abstract class Pattern<A> {
  // instance methods
  public abstract boolean match(A target);
  public boolean matchAgain();
  // factory methods
  public static <A> Pattern<A>
    both(Pattern<? super A> first, Pattern<? super A> second);
  public static <A> Pattern<A>
    either(Pattern<? super A> first, Pattern<? super A> second);
  public Pattern<Object> forInstancesOf(Class<? extends A> cls);
  public static <A> Pattern<A> eq(A constant);
  public static <A> Pattern<A> equal(A constant);
}

class Variable<A> extends Pattern<A> {
  public A getValue();
  public <B> List<A>     eagerBindings(Pattern<? super B> root, B target);
  public <B> Iterable<A> lazyBindings(Pattern<? super B> root, B target);
}

abstract class Transform<A, B> extends Pattern<A> {
  protected final Pattern<? super B> body;
  protected abstract B apply(A target);
  protected abstract boolean isApplicable(A target);
}
```

Fig. 1. Interface synopsis (Core)

smoothly to the object-oriented paradigm, because object constructors generally lack the mathematical benevolent properties of their algebraic counterparts,
namely extensional equality, injectivity, disjointness and completeness. A looser
notion of pattern matching, more appropriate to the abstraction style of object
orientation, is to consider it the reification and composition of certain classes of
data-extraction operations, namely *testing* (classifying objects as either acceptable or not), *projection* (extracting values and subobjects) and *binding* (assigning
data to variables). The design of our library is such that these three concerns
are separated as much as possible, but can be composed as freely as possible.

The main interface of the library is the abstract base class Pattern<A> of
patterns that can process objects of type A. The success of all implied *testing*
is indicated by the Boolean return value of the match method. *Projection* is
accomplished by invoking getter methods on the argument; for a compositional
approach see below. *Binding* occurs as a side effect of successful matches; unsuc-
cessful matches leave the corresponding variables in unspecified state. A variable
is simply a pattern of subclass Variable<A> that matches always, and binds
to the matched object for later retrieval via the getValue method.

Note that the (recursive) invocation of match is the primary concept, and the
return values govern the matching control flow. Variables are plain von Neumann
variables (assignable slots) rather than the richer ones of logic programming: No
unification or implicit equality test is performed; successive matches for the
same variable simply overwrite its value. This makes both data extraction and

backtracking extremely light-weight. Thus variables should be used linearly, but can of course be related by constraints embedded in a composite pattern.

The variable interface is unique in the sense that its type parameter occurs in a return type. All other generic pattern instance methods have type parameters only in argument types, making them *contravariant*: patterns that accept a subtype form a supertype and vice versa. Patterns are used only via their contravariant interface in many places. This can be expressed in Java by wildcards with lower bounds, of the form `Pattern<?` **super** `A>`.

As mentioned before, the binding aspect of pattern matching is conditional on the testing aspect: Bound values are only meaningful if the matched object is deemed acceptable. For reasons of simplicity and efficiency, our library does not provide automatic means to detect whether a variable has been bound; the responsibility to bind occurring variables on success is part of the contract of composite pattern classes. The basic idiom of pattern matching is thus:

```
Variable<C> vc = new Variable<C>();
Variable<D> vd = new Variable<D>();
Pattern<A> p = myPattern(vc, vd);     // known to bind vc, vd
if (p.match(x))
  doSomething(vc.getValue(), vd.getValue());
```

It is not an accident that the pattern variables vc and vd in this example have local declarations with precise static type (first two lines): This style enables the full use of static type information for bound values, even if the matching pattern has been constructed from generic building blocks that are defined independently of the type of occurring variables.

Generic building blocks are the key to reuse and concise notation. For instance, operators to lift patterns from elements to data structures of the Java collection framework, thereby encapsulating ubiquitous but repetitive search procedures, are part of the Paisley library. Fine control over searching is supported by explicit nondeterminism.

Nondeterminism is realized by backtracking. After a successful call to match, additional matches *for the same target* may be attempted by subsequent calls to matchAgain. Patterns may also be reused, although not concurrently, by calling match with a new target. The semantics of return value and binding side effects is the same for both methods. Note however that the argument is not repeated with matchAgain: it is the responsibility of each pattern subclass to store the relevant information. Storage should be strictly local, such that memory leaks can be avoided by allowing patterns to become garbage after use.

Iteration over all matches of a nondeterministic pattern is effected by extending the above idiom slightly by a **do** ... **while** loop, with minimal redundancy.

```
if (p.match(x)) do
  doSomething(vc.getValue(), vd.getValue()) ;
while (p.matchAgain());
```

The default implementation of matchAgain always returns **false**, specifying a deterministic pattern. For patterns with a single variable, bindings for all matches can be collected either eagerly or lazily with eagerBindings and `lazyBindings`, respectively, thus effecting fully reified encapsulated search as strongly typed objects of the Java collection framework (collecting bindings of multiple variables is awkward in Java due to the lack of Cartesian products). The iteration pattern for all matches simplifies accordingly.

```
for (C c : vc.lazyBindings(p, x))
    doSomething(c) ;
```

Patterns can be combined in a logically generic way conjunctively (`both`) and disjunctively (`either`), giving rise to Cartesian product and disjoint union of matches, respectively: The pattern `both(p, q)` succeeds where both `p` and `q` succeed, and binds all variables bound by either `p` or `q`; no guarantees of non-interference are given for variables occuring in both. The pattern `either(p, q)`, dually, succeeds where either `p` or `q` succeeds, and binds all variables bound by both `p` and `q`; variables bound only by one of `p` or `q` are contingent on information not available from logical disjunction, but other pattern control flow constructs implemented in our library.

For dynamic type case distinction, every pattern can be guarded by an instance test by calling `forInstancesOf`, which delegates to the primitive `Class.isInstance`, resulting in a pattern widened to type `Object`.

Since there is no semantically natural and unique notion of equality in object-oriented systems, there is no direct equivalent of pattern constants, with the exception of primitive types. However, comparing targets to a fixed object by either `==` or `equals` is a ubiquitous testing pattern and supported by the `eq` and `equal` pattern factory methods, respectively.

As a generic construct for both testing and projection, patterns can be lifted contravariantly over partial functions: Given a transformation that produces an object of type B from a known subset of objects of type A, implemented as a subclass of `Transform<A, B>`, each pattern with target type B induces a pattern with target type A, which transforms its argument and delegates the matching of the result to its constructor argument. Testing can be accomplished in the `isApplicable` method, projection in the `apply` method. Hence both case distinctions and getters can be reified as atomic pattern building blocks, which can then be combined by logical operators and nesting. Constructor patterns à la Scala **case class** can be assembled in this way from components.

Where more appropriate, complex composite testing/projections operations can also be hand-coded as monolithic subclasses of `Transform`. Note however that correct implementation of nondeterministic backtracking is not a trivial issue, and the generic realization in `both` and `either` is quite helpful.

Generic constructions for more complex pattern applications, such as searching in aggregate data structures with various degrees of nondeterminism, are either already available as extensions to the basic library or can be user-defined tailored to problem-specific needs. Specific constructions that encapsulate a view, or systematic mode of extraction, of a certain data type should be bundled for use as specific pattern matching idioms. Such views can be developed without changes or even access to the internal definitions of data classes, and they can coexist and be used on a completely case-by-case basis.

5   Related Work

Limited space allows only the discussion of the most relevant related work.

In specialized model transformation systems the application of pattern matching is often supervised by a dedicated control language. See for instance the "Rule Application Control Language" of GrGen.NET [1]. This role is taken in our approach by the hosting imperative programming language in the natural way.

Pattern notations take a vast number of theoretically and pragmatically different forms in the multitude of existing model transformation systems. For instance, the query language of the GReTL system [3], GReQL, offers regular path expressions to express complex patterns. Our approach lacks a "Kleene star" primitive for the iteration of a fixed access operation, but this can easily be accomplished using the recursion mechanism of the host language.

A theoretically elegant design of pattern matching capabilities for Java, JMatch, is presented in [6]. While it has had much impact, and is cited heavily by later work, there are severe drawbacks: The approach assumes a perspective on pattern matching that is very much like logical programming. As a result, their nondeterminism is rather heavy-weight, requiring CPS transformation of certain program parts. Furthermore, the solution is a host language extension and requires a special academic compiler. All such experiments are eventually doomed to oblivion unless some big vendor adopts the technology.

The multi-paradigm language Scala [7] incorporates a powerful pattern matching idiom with clean semantics and user-defined extensibility, via singleton objects and the `unapply` method. Being part of the core Scala design, it is better integrated with the host language than our approach can ever hope to be. On the other hand, we find the lack of nondeterminism a significant weakness.

The approach most similar to ours, Kiama [8], is hosted on Scala. It makes pervasive use of the pattern matching functionality of its host language. This project is a carry-over of Stratego [2], which is a transformation system as a dedicated DSL. Their relationship corroborates our views on the tradeoff between the lightweight nature of embedding vs. analysis and optimisation opportunities from separation of languages.

Hosaya and Pierce [4] show the significance of pattern matching for XML processing, a standpoint we strongly support, in particular with regard to readability and maintainability of real-world code. Their work focuses on type inference and DSLs for XML processing, two tasks we deliberately delegate to the hosting programming language.

6   Conclusion

We have outlined how our requirements for effective pattern matching are fulfilled by a simply and concisely designed library. The separation of the concerns of testing, projection and binding, together with powerful generic pattern constructs ensures static type safety, compositionality and a flexible notation allowing for arbitrary user-defined extensions. From the engineering viewpoint, the three most valuable features are

1. the type safety of patterns and their variables,
2. the cleanly compositional behaviour of nondeterminism, which may save the user a substantial amount of cluttered and hardly reusable code fragments consisting of loops and recursions, and
3. the full reification of patterns, whereby data extraction procedures can be organized, aggregated, serialized and shared like any other objects of the host language.

The light-weight nature of the language embedding has proven a blessing rather than a curse, in particular with respect to nondeterminism, which is semantically straightforward, easy to use, and nondisruptive with respect to the rest of the host language, especially compared with heavy-weight solutions such as JMatch.

The Paisley toolkit is being developed and extended continually. An up-to-date version of the library and a number of example applications can be downloaded from http://bandm.eu/metatools/paisley.

References

1. Blomer, J., Geiß, R., Jakumeit, E.: The GrGen.NET User Manual (2011), http://www.grgen.net
2. Bravenboer, M., Kalleberg, K.T., Vermaas, R., Visser, E.: Stratego/XT Tutorial, Examples, and Reference Manual (latest). Department of Information and Computing Sciences, Universiteit Utrecht, Utrecht, The Netherlands (2006), http://www.strategoxt.org
3. Horn, T., Ebert, J.: The GReTL Transformation Language. In: Cabot, J., Visser, E. (eds.) ICMT 2011. LNCS, vol. 6707, pp. 183–197. Springer, Heidelberg (2011)
4. Hosoya, H., Pierce, B.C.: Regular expression pattern matching. In: ACM SIGPLAN–SIGACT Symposium on Principles of Programming Languages (POPL), London, England (2001); full version in Journal of Functional Programming, 13(6), 961–1004 (November 2003)
5. Lepper, M., Trancón y Widemann, B.: Optimization of Visitor Performance by Reflection-Based Analysis. In: Cabot, J., Visser, E. (eds.) ICMT 2011. LNCS, vol. 6707, pp. 15–30. Springer, Heidelberg (2011)
6. Liu, J., Myers, A.C.: JMatch: Iterable Abstract Pattern Matching for Java. In: Dahl, V. (ed.) PADL 2003. LNCS, vol. 2562, pp. 110–127. Springer, Heidelberg (2002)
7. Odersky, M., Spoon, L., Venners, B.: Programming in Scala. artima, 2nd edn. (2010)
8. Sloane, A.M., Kats, L.C.L., Visser, E.: A pure object-oriented embedding of attribute grammars. Electronic Notes in Theoretical Computer Science 253 (2010), http://wiki.kiama.googlecode.com/hg/papers/LDTA09.pdf

Constraint-Driven Modeling through Transformation

Andreas Demuth, Roberto E. Lopez-Herrejon, and Alexander Egyed

Institute for Systems Engineering and Automation
Johannes Kepler University (JKU)
Linz, Austria
{andreas.demuth,roberto.lopez,alexander.egyed}@jku.at

Abstract. In model-driven software engineering, model transformations play a key role since they are used to automatically generate and update models from existing information. However, defining concrete transformation rules is a complex task because the designer has to cope with incompleteness, ambiguity, bidirectionality, and rule dependencies. In this paper, we propose a vision of Constraint-driven Modeling in which transformation is used to automate the generation of model constraints instead of generating entire models. Three illustrative scenarios show how this approach addresses common transformation issues and how designers can benefit from using model constraints and guidance. We developed a proof-of-concept implementation that covers an important part of this vision and thus demonstrates its feasibility. The implementation also suggests that a constraint-driven transformation is efficient and scales even with increasing numbers of involved models.

1   Introduction

With the increasing use of *Model-Driven Engineering (MDE)* [1] for complex software systems, the generation of models from existing artifacts through *model transformation* [2] is a vital necessity. Various classifications and taxonomies have been published to compare the state-of-the-art (e.g., [3,4]) and rich transformation languages are available, such as ATL [5] or QVT [6], which define *transformation rules* that are executed by a *transformation engine* to generate models. Since the source models of transformations are likely to be manually edited during development, re-transformations are necessary to update the corresponding generated models. However, such a re-transformation of non-trivial models can be time expensive and may affect the modeler's normal workflow [7,8]. Hence, incrementality is required to allow partial model updates without complete re-transformations in order to achieve acceptable performance when working with large, non-static models [8]. To date, various sophisticated transformation techniques exist that produce excellent results as long as the generated models are static and there are no uncertainties.

However, problems arise when these requirements are not fulfilled. For example, a common issue with re-transformations – even when performed incrementally – is the inevitable loss of manual changes to the generated models.

Z. Hu and J. de Lara (Eds.): ICMT 2012, LNCS 7307, pp. 248–263, 2012.

The issue is similar with *bidirectional transformations* [9,10], which are often used to synchronize models or to keep them consistent, when both involved models are edited concurrently. Furthermore, there are situations where it cannot be ensured that traditional approaches will generate the desired models because of ambiguity, uncertainties, and the fact that certain information neither is available at the time the transformation rules are written nor can be derived from the involved models when those rules are executed.

In this paper, we propose *Constraint-driven Modeling (CDM)*, a generic approach that guides the construction of new models while conserving consistency with the related models and eliminates issues arising with re-transformations, uncertainties, and bidirectionality. CDM relies on incremental model transformations to generate *constraints* from existing models that represent the invariants that the generated models should meet. Such constraints, written in a *constraint language* (e.g., the *Object Constraint Language (OCL)* [11]), are validated by a *consistency checker* on a given model. The provided *guidance*, which is derived from the generated constraints and existing inconsistencies, helps designers to stepwise transform the initially generated model to a version that matches the desired characteristics by pointing out inconsistencies (i.e., aspects of the model that do not satisfy invariants). Such guidance can be either the information which elements are causing inconsistencies, or suggestions of model changes (*options*) that can be performed to restore consistency. To obtain an initial, yet incomplete version of the desired model to start working with, a traditional batch model transformation with unambiguous rules can be used to generate a skeleton. Thus, CDM can be seen as a complement to traditional model transformation.

We evaluated our approach and showed its feasibility by implementing a prototype that generates constraints, enforces them incrementally, and informs the user about existing inconsistencies. Performance tests with large industrial models of up to 162,237 model elements previously showed the scalability of constraint validation [12]; our tests with these models show that the median times for incremental transformation and constraint generation are under .07 milliseconds. Thus, the approach scales and provides instant user feedback when involved models are edited.

2 Running Example

To illustrate our work, we first present two incremental changes that are challenging for common model transformation approaches.

Let us consider the sequence and class diagrams shown in Fig. 1(a) and Fig. 1(b) respectively. In Fig. 1(a), the unnamed instance of class LightSwitch receives a message named activate. According to the semantics of UML sequence diagrams, this message requires that the instance of LightSwitch provides a method also named activate. At first glance, it looks like a simple transformation can be used to automatically add the method activate to class LightSwitch in Fig. 1(b) whenever a message is added to a sequence diagram

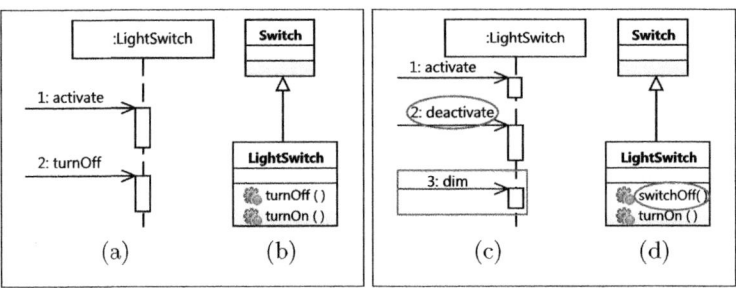

Fig. 1. Two UML models (a) and (b), and evolved versions (c) and (d)

whose name does not match any method in the class. An example for such a transformation rule written in ATL-like syntax is shown in Listing 1.

However, there is an issue with this approach: Should the method `activate` be added to `LightSwitch` or would it make more sense for the system to add it to the superclass `Switch`?

Obviously, this question cannot be answered automatically. The only possibility would be to make an assumption (e.g., always add the method to the specified class to be on the safe side), which leads to the generation of potentially unintended models where methods are not declared in the desired place or where methods are unnecessarily overridden.

3   Constraint-Driven Modeling

Common transformation languages usually describe the steps that have to be performed to generate new models from existing ones. The previous section illustrated that it can be difficult or even impossible to writing transformation rules that automate complex decisions or always lead to desired results.

Intuitively, and in contrast with standard model transformations, we propose to generate constraints on a model (to guide designers) rather than generating the model itself whenever precise transformation results cannot be derived. For example, the added message `activate` on the source model should impose a constraint that a same named method should be available to class `LightSwitch` rather than saying it should be owned by it. If the method is already there then the constraint is instantly satisfied. If the method does not exist then further

```
from
    s  :  SequenceDiagram ! Message
to
    t  :  ClassDiagram ! Method  (
          name <-- s.name ,
          owner <-- getClass ( s.receiver.className )
    )
```

Listing 1. Sample transformation to generate methods in class diagrams

(a) Model transformation.   (b) Constraint transformation.

Fig. 2. From ambiguous model transformation (a) to constraint transformation (b)

actions are required to deal with this problem – actions that must either come from a human or be derivable from other transformations.

When traditional model transformation approaches are used, the transformation process can be regarded as:

$$A \xrightarrow{T_m} B_g \tag{1}$$

where A is called the source model, consisting of an arbitrary number of model elements. T_m is the transformation model, consisting of transformation rules, that is used to transform A to the generated model B_g.

We expanded this notation and define our approach as:

$$A \xrightarrow{T_c} C \rightsquigarrow B_r \tag{2}$$

where the variable A denotes the source model and T_c is a set of model transformation rules. However, as the solid arrow from A to C and the changed subscript of T suggest, this set of rules no longer generates a model (i.e., B_g), but instead it contains transformation rules that are applied to A in order to generate constraints (i.e., the constraint model C). This constraint model consists of a set of constraints that are enforced by an incremental consistency checker on the model B_r, as indicated by the curvy arrow from C to B_r. The model B_r is no longer the generated model but is now called the restricted model, as indicated by the subscript r, that is either consistent or inconsistent with the constraint model C, and therefore a valid or invalid solution of the modeling problem.

Note that an initial version of B_r may be generated through a traditional transformation (analogous to B_g) or even built manually by a designer. However, once generated, this proposed approach can detect inconsistencies if both A and B_r are evolved concurrently. Thus, our approach should not be seen as replacing traditional transformation approaches but instead complementing them in case of co-evolution, uncertainties, complex rule-scheduling issues or even model merging as will be demonstrated below. Next, we present how it is applied.

3.1   Application: Uncertainties

Let us come back to our running example from Section 2 where we illustrated that choosing the right class for a required method cannot be fully automated. The traditional approach shown in Fig. 2(a) automatically generates one of several possible models and we could at most use heuristics for deciding on which

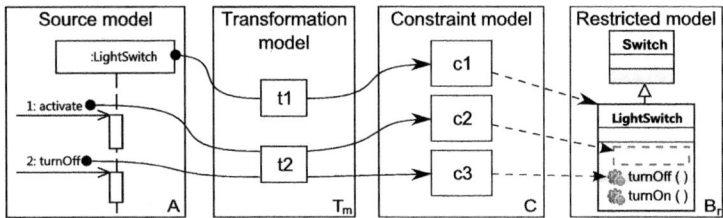

Fig. 3. Application of approach to models from Fig. 1(a) and Fig. 1(b)

transformation to use (which never guarantees correctness). However, while the knowledge contained in Fig. 1(a) is insufficient to generated a correct update to the class diagram, it is sufficient to generate a correct constraint on said diagram. Such constraints can be generated by transformation rules that are triggered by the addition/removal of class instances or messages in sequence diagrams that can be efficiently validated by state-of-the-art consistency checkers.

To automate constraint generation, we provide two transformation rules that are triggered by class instances or messages in sequence diagrams and that use information from the sequence diagram to generate very specific and expressive constraints. These rules are shown in Listing 2.

```
rule t1
  from
    s : SequenceDiagram!Instance
  to
    t : ConstraintModel!Constraint (
      context <- "Package",
      inv <- "self.classes->exists(c|c.name='" + s.className + " ')"
    )
rule t2
  from
    s : SequenceDiagram!Message
  to
    t : ConstraintModel!Constraint (
      context <- "Class",
      inv <- "self.name='" + s.receiver.className + " ' implies self.
        providedMethods->exists(m|m.name='" + s.name + " ')"
    )
```

Listing 2. Transformation rules to generate class (t1) and method (t2) constraints

Note that, even though we use ATL-like syntax for this example, our approach can be used with any transformation language. After applying these rules to the motivating example from Section 2 as illustrated in Fig. 3 and according to (2), C consists of the following OCL constraints:

c1 context Package inv: self.classes->exists(c|c.name='LightSwitch')
c2 context Class inv:
    self.name='LightSwitch' implies
    self.providedMethods->exists(m|m.name='activate')
c3 context Class inv:
    self.name='LightSwitch' implies
    self.providedMethods->exists(m|m.name='turnOff')

In Fig. 3 we can see that the method required by the constraint $c2$ is not present in B_r, as indicated by the empty, dashed rectangle in the `LightSwitch` class, meaning that this particular model will be marked inconsistent. Note that we use OCL as the constraint language in our example because it is a well known and accepted language for writing constraints and we have existing tool support for incrementally validating OCL constraints. Nonetheless, in principle any constraint language and consistency checker may be used.

Fig. 2(b) illustrates the basic concept of the constraint-driven modeling approach. It is noteworthy that the approach does not modify the restricted model. It simply restricts it. The generated restriction – depicted as partial frame with rounded corners around the restricted model B_r – may be light in that there are various options on how to change the restricted model. In such as case, the designer has the freedom to decide which of the options is the desired one (e.g., add `activate` to `LightSwitch` or `Switch`) with the knowledge that the approach notifies/prevents options that are invalid. In the most extreme case, the restrictions may be severe enough to allow for one option only. In such a case, the approach could automatically select this option with the knowledge that it is the one and only right option (e.g., if `LightSwitch` had no parent class then there is a single option only).

3.2   Incremental Constraint Model Management

Let us take a closer look at the transformation that generates the constraint model C. As shown in (2) and Fig. 3, applying the transformation rules of the transformation model to the source model generates the constraint model.

Source model update. The transformation approach we use supports incrementality to allow updates of the constraint model without performing a complete re-transformation of the source model. When A is updated to A', we can write this as

$$A \xrightarrow{\Delta A} A' \tag{3}$$

where ΔA is a sequence of modifications done to elements in A (e.g., add a new model element). ΔA is used as input for the transformation model to generate the set ΔC, as shown in (4).

$$\Delta A \xrightarrow{T_c} \Delta C \tag{4}$$

ΔC includes pairs of constraints and actions (i.e., $\{add, remove\}$) that define whether the constraint should be added or removed from the existing constraint model C. By applying ΔC on C, the updated constraint model C' is generated:

$$C \xrightarrow{\Delta C} C' \tag{5}$$

Let us consider the evolution of the models shown in Fig. 1(a) and Fig.1(b) to the versions shown in Fig. 1(c) and Fig. 1(d) where the name of the message #2 was updated to `deactivate`, the message #3 was introduced, and the name of the method `turnOff` was changed to `switchOff`. For the changes in the source model, the corresponding ΔA is $\langle\langle Message2, update\rangle, \langle Message3, new\rangle\rangle$.

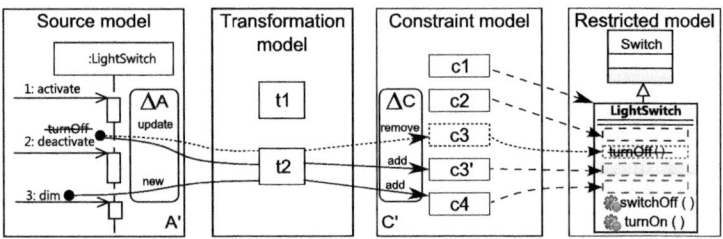

Fig. 4. Update of constraint model after changes in source model

To build ΔC, the transformation engine executes the applicable transformation rules for the elements in ΔA (i.e., message #2 and message #3) to generate the corresponding constraints, as defined in (4) and shown in Fig. 4. For $\langle Message2, update\rangle$, the constraint $c3'$ is generated and the information $\langle c3', add\rangle$ is added to ΔC.

```
c3' context Class inv:
    self.name='LightSwitch' implies
    self.providedMethods->exists(m|m.name='deactivate')
```

Since the constraint $c3$ was already generated from the same element as $c3'$, message #2, $\langle c3, remove\rangle$ is also added to ΔC in order to remove the now outdated constraint $c3$. For $\langle Message3, new\rangle$, the transformation rule $t2$ is executed to generate a new constraint $c4$ and $\langle c4, add\rangle$ is added to ΔC.

```
c4 context Class inv:
    self.name='LightSwitch' implies
    self.providedMethods->exists(m|m.name='dim')
```

At this point, ΔC is $\{\langle c3, remove\rangle, \langle c3', add\rangle, \langle c4, add\rangle\}$. When these changes are applied to $C = \{c1, c2, c3\}$ as defined in (5) and shown in Fig. 4, the resulting updated constraint model is $C' = \{c1, c2, c3', c4\}$. We used dotted lines for removed elements, that is $c3$ and the corresponding inconsistency in LightSwitch. As Fig. 4 indicates, the constraints $c3'$ and $c4$ are violated by the restricted model since the class LightSwitch does not provide the required methods deactivate and dim.

Ultimately, changes of the source model A affect the constraints that are enforced by the consistency checker:

$$C' \rightsquigarrow B_r \tag{6}$$

Next, we describe how such constraint model changes can affect the consistency status of the restricted model B_r.

3.3   Constraint Validation and Solution Space

We define the solution space of a modeling problem to initially include all possible instances of a metamodel (there are likely infinite). When a constraint is validated, it determines whether a model meets those characteristics. Therefore, applying a constraint decreases the size of the solution space and the validation result shows whether a specific model is part of the solution space.

We define the validation of a constraint c for a specific model m as $val :$ $(m, c) \rightarrow \{false, true\}$ where $false$ is returned if m violates c, $true$ otherwise. For a model B_r and a constraint model C, the result of a total validation (i.e., a validation of all available constraints, written as val_T) would then be equal to:

$$val_T(B_r, C) = \bigwedge_{1 \leq i \leq |C|} val(B_r, c_i) \qquad (7)$$

If at least one constraint validation $val(B_r, c_i)$ returns $false$, the overall status of B_r is also $false$ and therefore outside the solution space. It is easy to see that **the order of constraint validation does not affect the final result**. However, the execution order determines when the overall inconsistency of a model B_r is discovered during the validation and the order in which inconsistencies are corrected can of course be important when deriving stepwise adaptations.

Since constraints are composed of expressions that are evaluated on only the restricted model, direct dependencies among constraints typically do not exist and are not considered here. The addition of a new constraint thus does not affect the validity of existing constraints. This leads us to the conclusion that **constraints are independent of each other**. Furthermore, the used transformation rules do only access the source model to construct constraints and add the constraint to the constraint model without accessing other constraint model elements, thus the **transformation rules for generating constraints are independent** and dependencies between them that require a certain order of execution cannot occur. These observations have interesting benefits to model transformation discussed next.

3.4   Providing Guidance

When an inconsistency is detected, the minimum amount of guidance provided to the designer is a notification about the inconsistency's occurrence and its location (i.e., which model element is violating which constraint). Based on data captured during constraint validation, the consistency checker can determine which model elements are actually causing the inconsistency. Hence, it can inform the designer about the locations of error-causing elements.

Constraint-driven modeling may appear inferior to traditional transformation in that it never generated model elements in the restricted model. However, there is currently considerable progress in automatically suggesting repairs to inconsistencies in design models. Based on a specific constraint and the inconsistent parts, it is thus possible to derive modifications – like specialized transformations – that lead to a consistent model. If such modifications can be derived, they are proposed to the user as a list of options. If the restrictions are unambiguous, only a single option remains and it could be applied automatically (much like transformation). For example, the action <add method "dim" to

class "LightSwitch"> is an option for removing the inconsistency caused by the absence of the method dim in the LightSwitch class and the constraint $c4$. Thus, using constraints does not only expose inconsistencies but it also enables user guidance to help understanding and solving them. Note that incorporating source model data makes a constraint much more specific and expressive when presented to the user than a manually written, generic constraint that relies on metamodel data and functions.

Nevertheless, dependencies between constraints in terms of required model characteristics and corresponding model elements can occur (e.g., $c1$ requires a class LightSwitch and $c2 - c3$ require specific methods in this class). Creating additional inconsistencies can therefore be necessary to achieve overall model consistency. More research in automated fixing of design models based on constraint violations is needed to automate this. However, we believe that this problem is solvable and the focus of our future work.

Guidance is however not limited to inconsistencies. For each constraint, its source as well as the locations where it is validated are available and can be presented to the user. When the source model is edited during development, the constraints that are affected by those changes can also be highlighted. When a designer, for example, adds a new message to a sequence diagram with a name that already has a matching method in a class diagram, the highlighted constraint shows him or her the existing method immediately. The designer can then easily decide whether this existing method should be used (i.e., the message means the existing method) or if a naming conflict was introduced (i.e., a new method was planned).

4   Additional Benefits of Constraint-Driven Modeling

Now, we want to show several additional scenarios that benefit from constraint-driven modeling in context of rule-scheduling, model merging, and bidirectionality.

Rule-scheduling and race conditions. Now let us consider an example where two transformation rules t_{m1} and t_{m2} are working with the same generated model and the order of rule execution is important. For example, the sequence diagram in Fig. 1(a) contains an instance of the class LightSwitch. Therefore, let us assume that transformation rule t_{m1} generates a corresponding class if no such class exists in the diagram in Fig. 1(b). As we have discussed in Section 2, the sequence diagram requires the class LightSwitch to provide a method activate. Let transformation rule t_{m2} generate this method in LightSwitch[1]. When the transformations are performed, it is crucial that t_{m1} is executed before t_{m2} to ensure that the class LightSwitch exists before the method activate is added. This issue is illustrated in Fig. 5(a) where the bottom transformation encounters an error after the execution of t_{m2}. If the rule t_{m1} is still executed,

---

[1] We ignore the fact that such a transformation will not always lead to satisfying results – as discussed above – for this example.

(a) Model transformation.   (b) Constraint transformation.

Fig. 5. From dependent transformation rules (a) to independent ones (b)

the resulting model B will contain an empty `LightSwitch` class because only t_{m1} was executed successfully. If the execution of rules is stopped after the error, no model is generated at all. Defining the order of rule execution manually is tedious and a constant source of error. Moreover, support for defining an execution order is not a standard feature of all transformation languages or systems [3].

The constraining approach, shown in Fig. 5(b), is free of scheduling issues because constraints cannot directly depend on other constraints and the order of transformation is not relevant for the transformation results, as discussed in Section 3.3. Hence, the rules t_1 and t_2 we have previously defined can be applied in any order. If a model does not provide the required information for constraint validation (e.g., the class that should be checked is not present), the validation fails and an inconsistency is detected.

Bidirectionality and model merging. When models should be synchronized automatically, transformations are often used to propagate changes from one model to the other and perform the corresponding changes. Let us assume that we have established transformation rules that keeps message names and method names synchronized and that a link between messages and corresponding methods exists. In Fig. 1(c), the name of the highlighted message has been changed from `turnOff` (see Fig. 1(a)) to `deactivate`. Concurrently, the corresponding method in the class diagram was changed from `turnOff()` (see Fig. 1(b)) to `switchOff()`, as highlighted in Fig. 1(d). Since both synchronized model elements were changed (indicated by the bold arrows), there is no way to determine in which direction the required synchronization should be performed. Performing a synchronization in this situation will always lead to the loss of the changes in the generated model (i.e., either B'' overrides changes in B' or A'' overrides changes in A' that cannot be used for a transformation in the opposite direction afterwards). A possible solution would be the concurrent execution of the transformations followed by a merge of the updated models (A' and B') and the resulting generated models (A'' and B''), as illustrated in Fig. 6(a), that generated A''' and B'''. However, this requires a complex merging strategy and is likely to produce models that still require manual adaptation.

The solution of the constraint transformation approach is shown in Fig. 6(b). We can see that our approach still has to decide which change to process first. However, because only constraint models are updated, the restricted models A'_r and B'_r are not changed and can therefore still be processed to perform constraint updates in the opposite direction, leading both constraint models ca and cb

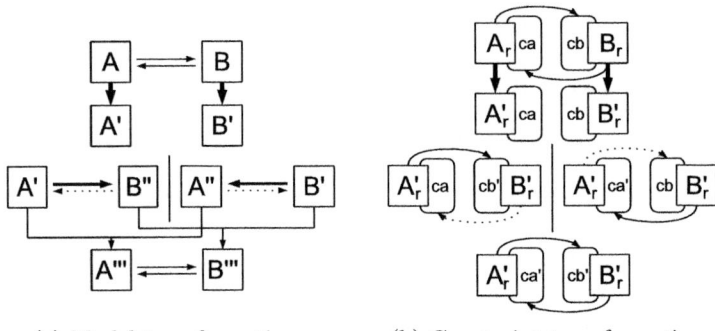

(a) Model transformation.   (b) Constraint transformation.

Fig. 6. From bidirectionality (a) to unidirectional constraint transformation (b)

being updated. With our approach, no immediate merging (either automated or manual) is required when restricted models are edited and following source model changes lead to constraint updates.

After the constraint model updating took place in the example, there are two new constraints: i) message number 2 in Fig. 1(c) should be named switchOff (from Fig. 1(d)) and ii) the name of the method switchOff in Fig. 1(d) should be changed to deactivate (from Fig. 1(c)). The designer can then decide which of the elements should be renamed.

5   Validation

In this section we first discuss various aspects regarding the correctness of our approach and its results. Then, we present the results of a performance evaluation and finally discuss possible threats to validity.

5.1   Correctness

Based on the presented scenarios and the properties of constraints we showed that common transformation issues like rule-scheduling, race conditions, and model merging do not arise when constraint models are generated through transformation and that those models can be updated easily.

Nevertheless, the correctness of the applied constraints and the provided user guidance is determined by the correctness of the manually written transformation rules, the used source models, and the transformation language implementation – as with traditional approaches.

Errors in both the source models and the applied rules lead to errors in the generated model. As with traditional approaches, such errors also affect the generated model (i.e., the constraints) in our approach. However, designers can inspect arbitrary constraints and decide whether the constraints are correct. By using a transformation mechanism that creates traceability links between source model elements, transformation rules and the generated constraints automatically – as we did in our prototype implementation – designers can use faulty constraints to detect errors in the source model or the transformation rules and fix them. Moreover, faulty constraints can simply be ignored or deactivated, which means that contradicting constraints do not prevent designers from constructing the desired model. Generally, existing errors are never incorporated in the restricted model automatically when our approach is used.

5.2 Implementation

To support the vision proposed in this work, we implemented a proof-of-concept prototype tool and applied it to several domains. The prototype, based on the *Cross-Layer Modeler (XLM)* [13], investigates the constraint generation/validation and the presentation of consistency information and constraints, but not the inconsistency repair. The latter is future work. The tool employs the *Model/-Analyzer* [14] consistency checker to validate constraints that are automatically generated through incremental transformation from templates and are managed and updated incrementally. We added components to the XLM to support multiple different source, constraint and restricted models simultaneously, which requires the management of multiple, parallel running consistency checkers.

5.3 Performance Evaluation

Basically, our approach has two phases: i) generating constraints, and ii) validating them. For the latter, the performance of the employed consistency checker was thoroughly evaluated with 34 large-scale industrial models of up to 162,237 model elements and complex constraints in [12,15]; it was shown that most changes in restricted models are processed in less than one millisecond. To show that also the former is fast and scalable, two different test setups were used:

Test I. Replacing ambiguous transformations – as discussed in Section 3
Test II. Replacing merges – different sources restrict a single model

Test I simulated simple unidirectional transformations in the scenario we described in Section 3 and was performed with 20 of the industrial models we previously used in [12,15]. Test II determined the performance of our approach in scenarios where multiple source models are used to generate various constraints that are restricting the same model (i.e., merges of generated models would be required with traditional approaches). This test shows the behavior of the approach when complexity is gradually increased and more models become involved. For Test II, we generated random models and constraints with similar characteristics as those we used for constraint validation testing because the scenario required multiple, related source models and our industrial models were designed as independent models.

For all tests, single model elements were added to or removed from a source model, which forced an incremental constraint model update (i.e., the addition or

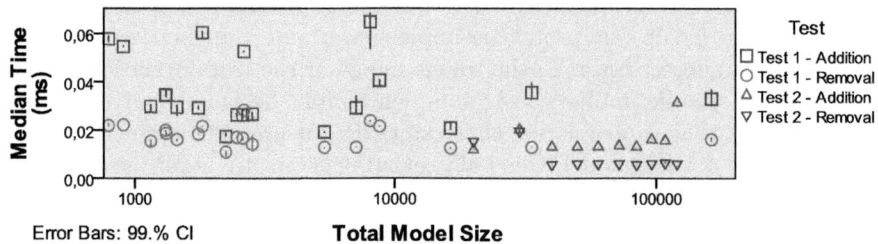

Fig. 7. Median processing times for constraint updates

removal of exactly one constraint). Between 1 and 10 source models were used for Test II, the required time for processing the source model change and performing constraint model updates was measured. For the evaluation we used an Intel Core i5-650 machine with 8GB of memory running Windows 7 Professional. In Fig. 7 the median times for 1,000 runs per test with a 99% confidence interval are shown. Note that the increasing number of source models in Test II does not affect the processing time significantly and that the median times for the addition of elements are between .01 and .07 milliseconds in both tests. Element removal takes between .006 and .03 milliseconds and is indeed faster than element addition because no transformations are required. These numbers show that our approach can update constraint models instantly and does scale for increasing numbers of source models. The similar results for Test I and Test II suggest that our random models for Test II were a valid choice for testing scalability. Note that bidirectional transformations are split into unidirectional ones in our approach, thus there was no need for testing bidirectional transformations explicitly.

5.4   Threats to Validity

Although it seems intuitive that decisions made by domain experts in situation with very specific problems and with guidance are more trustworthy than automated decisions based on generalized knowledge or heuristics, we have yet to show that the quality of the resulting models is higher or that our approach leads to quicker results. Additionally, we have not investigated to which degree guidance and suggested options reduce the time needed for design decisions or finding inconsistencies. Another threat to the validity of our vision is the automated derivation and execution of options to remove existing inconsistencies. Even though basic traceability information – which is always available – provides a certain amount of guidance, a key aspect of constraint-driven modeling is the automated suggestion of valid options to remove inconsistencies. However, this is still an open research question that we want to address in future work. Finally, we have yet to develop an efficient strategy for finding contradictions between constraints and fixing them automatically.

6   Related Work

Model Transformation is a very active field of research and several topics related to our work have been discussed. Regarding bidirectionality, Sasano et al. [16] developed a system to perform bidirectional transformations with ATL, and Stevens [17] focused on bidirectionality for QVT. Cicchetti et al. [18] developed the bidirectional transformation language *JTL* that supports the specification of non-bijective transformations so that one model can be mapped to a set of other models. We tackle the complexity of bidirectional transformations by using unidirectional transformations and constraints.

In terms of incrementality and execution speed, Jouault and Tisi [19] proposed an approach to make ATL transformations incremental. They achieve incrementality by using scopes built during OCL expression execution to determine which rules have to be re-executed after source model changes. We make use of automatically created scopes in the same way to determine which constraints have to be re-created in our prototype and also for finding constraints that have to be re-validated by the consistency checker [14]. In [20], Tisi et al. propose the lazy execution of transformations, which eliminates the need for an initial transformation of the entire source model to speed up the process for large source models, which is also the performance bottleneck of our prototype.

Regarding automated design error fixes, the generation of fixing actions was discussed by Xiong et al. [21]. They developed a language called *Beanbag* that allows the definition of constraints and fixing behavior at the same time. With our approach, such Beanbag programs can be generated automatically. Saxena and Karsai [22] published a MDE-based approach for design space exploration in which constraints are used to describe invariants of valid models. Our approach is ideal to generate constraints for design space exploration algorithms.

7   Conclusions and Future Work

In this paper we presented an incremental and generic approach that uses model transformation to automatically generate constraint models. We showed that constraints are independent, and constraint validation does not require a fixed order of execution. We then discussed how model transformation issues like ambiguity, rule-scheduling, model merging, and bidirectionality are addressed and how the approach enables user guidance and encourages the use of domain-knowledge to solve specific modeling problems. We believe this work contributes a novel complement to existing state-of-the-art on model transformation.

We validated the approach by developing a prototype implementation. Performance tests showed that our approach is scalable and provides instant guidance for designers. For future work we plan to further investigate the usability of the approach and to implement the automated derivation of options for removing inconsistencies and the checking for contradicting constraints.

Acknowledgments. The research was funded by the Austrian Science Fund (FWF): P21321-N15 and the EU Marie Curie Actions – Intra European Fellowship (IEF) through project number 254965.

References

1. Schmidt, D.C.: Guest editor's introduction: Model-driven engineering. IEEE Computer 39(2), 25–31 (2006)
2. Sendall, S., Kozaczynski, W.: Model transformation: The heart and soul of model-driven software development. IEEE Software 20(5), 42–45 (2003)
3. Czarnecki, K., Helsen, S.: Feature-based survey of model transformation approaches. IBM Systems Journal 45(3), 621–646 (2006)
4. Mens, T., Gorp, P.V.: A taxonomy of model transformation. Electr. Notes Theor. Comput. Sci. 152, 125–142 (2006)
5. Jouault, F., Allilaire, F., Bézivin, J., Kurtev, I.: ATL: A model transformation tool. Sci. Comput. Program. 72(1-2), 31–39 (2008)
6. Object Management Group, Query/View/Transformation (QVT), http://www.omg.org/spec/QVT/
7. Vierhauser, M., Grünbacher, P., Egyed, A., Rabiser, R., Heider, W.: Flexible and scalable consistency checking on product line variability models. In: ASE, pp. 63–72. ACM (2010)
8. van Amstel, M., Bosems, S., Kurtev, I., Ferreira Pires, L.: Performance in Model Transformations: Experiments with ATL and QVT. In: Cabot, J., Visser, E. (eds.) ICMT 2011. LNCS, vol. 6707, pp. 198–212. Springer, Heidelberg (2011)
9. Stevens, P.: A Landscape of Bidirectional Model Transformations. In: Lämmel, R., Visser, J., Saraiva, J. (eds.) GTTSE 2007. LNCS, vol. 5235, pp. 408–424. Springer, Heidelberg (2008)
10. Czarnecki, K., Foster, J.N., Hu, Z., Lämmel, R., Schürr, A., Terwilliger, J.F.: Bidirectional Transformations: A Cross-Discipline Perspective. In: Paige, R.F. (ed.) ICMT 2009. LNCS, vol. 5563, pp. 260–283. Springer, Heidelberg (2009)
11. Object Management Group, Object Constraint Language (OCL), http://www.omg.org/spec/OCL/
12. Egyed, A.: Automatically detecting and tracking inconsistencies in software design models. IEEE Trans. Software Eng. 37(2), 188–204 (2011)
13. Demuth, A., Lopez-Herrejon, R.E., Egyed, A.: Cross-layer modeler: A tool for flexible multilevel modeling with consistency checking. In: ESEC/SIGSOFT FSE, pp. 452–455 (2011)
14. Reder, A., Egyed, A.: Model/analyzer: a tool for detecting, visualizing and fixing design errors in UML. In: ASE, pp. 347–348. ACM (2010)
15. Groher, I., Reder, A., Egyed, A.: Incremental Consistency Checking of Dynamic Constraints. In: Rosenblum, D.S., Taentzer, G. (eds.) FASE 2010. LNCS, vol. 6013, pp. 203–217. Springer, Heidelberg (2010)
16. Sasano, I., Hu, Z., Hidaka, S., Inaba, K., Kato, H., Nakano, K.: Toward Bidirectionalization of ATL with GRoundTram. In: Cabot, J., Visser, E. (eds.) ICMT 2011. LNCS, vol. 6707, pp. 138–151. Springer, Heidelberg (2011)
17. Stevens, P.: Bidirectional Model Transformations in QVT: Semantic Issues and Open Questions. In: Engels, G., Opdyke, B., Schmidt, D.C., Weil, F. (eds.) MODELS 2007. LNCS, vol. 4735, pp. 1–15. Springer, Heidelberg (2007)

18. Cicchetti, A., Di Ruscio, D., Eramo, R., Pierantonio, A.: JTL: A Bidirectional and Change Propagating Transformation Language. In: Malloy, B., Staab, S., van den Brand, M. (eds.) SLE 2010. LNCS, vol. 6563, pp. 183–202. Springer, Heidelberg (2011)
19. Jouault, F., Tisi, M.: Towards Incremental Execution of ATL Transformations. In: Tratt, L., Gogolla, M. (eds.) ICMT 2010. LNCS, vol. 6142, pp. 123–137. Springer, Heidelberg (2010)
20. Tisi, M., Martínez, S., Jouault, F., Cabot, J.: Lazy Execution of Model-to-Model Transformations. In: Whittle, J., Clark, T., Kühne, T. (eds.) MODELS 2011. LNCS, vol. 6981, pp. 32–46. Springer, Heidelberg (2011)
21. Xiong, Y., Hu, Z., Zhao, H., Song, H., Takeichi, M., Mei, H.: Supporting automatic model inconsistency fixing. In: ESEC/SIGSOFT FSE, pp. 315–324 (2009)
22. Saxena, T., Karsai, G.: MDE-Based Approach for Generalizing Design Space Exploration. In: Petriu, D.C., Rouquette, N., Haugen, Ø. (eds.) MODELS 2010, Part I. LNCS, vol. 6394, pp. 46–60. Springer, Heidelberg (2010)

The Impact of Class Model Redesign on State Machines

Piotr Kosiuczenko

Institute of Information Systems, WAT, Warsaw

Abstract. This paper examines the effect of class diagram transformation on state machines, a subject which has not been properly investigated. It is demonstrated that structural relations between states can be interpreted as a logical relation between the corresponding formulas and the preservation of the latter corresponds to the preservation of state machine structure. A sufficient condition, based on the form of the underlying transformation and proofs, is provided which guarantees that class structure transformation preserves the structure of state machines. The goal is to automatically transform state-invariants and to identify those state machines which need to be manually transformed after class model redesign.

Keywords: UML, OCL, State Machines, redesign, refactoring, design patterns.

1 Introduction

Unified Modeling Language [19] provides textual and diagrammatic means for system specification. During the development process, a system specification undergoes several changes due to a number of factors including changed or new client requirements, new technology enablers and so on. For example if an interface is changed, then the constraints describing it may no longer make sense. In such cases extensive manual reengineering of system specification and design is needed, which is very time consuming and error prone. One of the major problems is the fact that after changing a UML diagram several other diagrams, of the same or other kinds, need to be reconsidered and possibly modified. This is due to the multiplicity of UML diagrams and their partial redundancy. The same aspect can be covered by different diagrams. Thus one needs methods to automatize synchronization of updates of different diagrams. Model transformations is a new area of research (cf. e.g. [4,17]). They are at the core of Model Driven Engineering. However, not much attention has been paid to the impact of refactoring patterns on software models. In particular, it is not clear which properties are preserved and what is the impact on other UML diagrams such as state machines. It should be pointed out that such a refactoring does not have to preserve an underlying metamodel, as is usually assumed in the case of model transformations.

In this paper we investigate the impact of class transformations on state machines with the help of the notion of interpretation functions. Those functions

Z. Hu and J. de Lara (Eds.): ICMT 2012, LNCS 7307, pp. 264–279, 2012.
© Springer-Verlag Berlin Heidelberg 2012

were defined in the paper [13] to formally study the redesign of UML class diagrams with OCL constraints and to transform and trace those constraints. We establish a sufficient condition for a class diagram modification to preserve the structure of associated state machines. UML interprets states as conditions satisfied during objects' life-cycles [19]; such conditions are called in OCL state-invariants [18]. Consequently states in a state machine can be interpreted as OCL formulas. They describe values of object attributes and inter-relationships between different objects. We show that structural relationships between states of a state machine can be specified by logical relations between the corresponding OCL-invariants. These relations are defined in terms of first order formulas and entailment relations. We define a generic method of deriving such relations. UML metamodel and OCL are used to define schemata generating the corresponding logical formulas. We show that preservation of state machines structure corresponds to preservation of those logical relations. We provide a sufficient condition guaranteeing that interpretation functions preserve structure of state machines.

Interestingly, several class transformation patterns can be formalized using interpretation functions (cf. [13]) and consequently these functions can be used to automatically transform the corresponding state-invariants (see subsection 3.2). Our approach is consistent with the use of inductive theorem provers, such as PVS, and logic programming since interpretation functions preserve proofs using resolution, induction and proposition calculi. Consequently, it is not necessary to redo proofs after transformation. Similarly, this approach allows us to avoid the rewriting of propositional temporal logic formulas after the modification of class diagrams. Automatic model-checking is not a problem, but it is usually hard to write temporal formulas. Class diagrams and state machines are different and seemingly incomparable. Therefore it is interesting that the class refactoring has a well defined impact on state machines. For the transformation of class diagrams and constraints very powerful term rewriting tools, such as Elan/Thom [9] and Maude [2], can be used. Our method allows us to automatically rewrite state invariants when an interpretation function is applicable, to identify state-invariants which need to be rewritten and state machines which need to be reconsidered. This saves the clerical work of reconsidering all state machines and state invariants and performing changes by hand. To our best knowledge, this is the first approach investigating the impact of class model modification on state machines.

This paper is organized as follows. In section 2, we study the relation between state machines structure and formulas, in particular OCL formulas. In section 3, we present briefly the idea of interpretation functions, the accompanying results and apply them to SM transformation. In section 4, we illustrate our approach with an application of the so called State Pattern. In section 5, we discuss the related work. Section 6 concludes this paper.

2   Logical Structure of State Machines

In this section, we consider the so called behaviour state machines describing behaviour of model elements, in particular objects [19]. In the first subsection, we consider state-invariants and relations between them. We argue that those relations correspond to logical relations. In the second subsection, we define the logical relations in a formal way. In the third one, we present a generic way of defining logical relations with the help of the UML metamodel and OCL constraints. In the fourth one, we explain those ideas using a simple example.

2.1   States and Regions

A state machine (SM) is composed of a number of states connected by edges corresponding to transitions. States can be structured; one state can contain several other states called substates. In UML, "a state models a situation during which some (usually implicit) invariant condition holds" ([19], subsection 15.3.11). Those formulas are called state-invariants and can be expressed for example in OCL. They describe values of object attributes and inter-object relationships.

The expressive power of state machines is due to the concept of structured states which allow one to describe system behaviour in a hierarchical way. UML distinguishes between atomic/indecomposable states and composite states which contain other states. Basically a composite state (i.e., state with attribute *isComposite* being true) can be either an or-state with exclusive substates or a concurrent state (so called and-state). A state is an or-state if it contains one region. A state is an and-state if it contains more than one region. In UML it is called orthogonal composite state and *isOrthogonal* has value true. UML standard imposes several constraints on structured states [19]:

1. "If a composite state is active and not orthogonal, at most one of its substates is active."
2. "If the composite state is active and orthogonal, all of its regions are active, with at most one substate in each region."
3. "Each region has a set of mutually exclusive disjoint subvertices and a set of transitions."
4. "If the state machine is in a simple state that is contained in a composite state, then all the composite states that either directly or transitively contain the simple state are also active."

For every system configuration, the set of active states contains a particular state if, and only if, the corresponding invariant is satisfied. If the invariants describe precisely the reachable configurations, then the conditions listed above can be explained in terms of logical relations such as implication.

2.2   Structural Relationships versus Invariants

State-invariants concern objects or in general UML model elements [19]. They are formulas of the form $F(self)$ where *self* is the only free variable of F; this variable is bound to the object (or model element) in question. Below we assume that every state of a state machine has an associated invariant. Moreover, for a state s (or s_i) we denote by F (F_i, resp.) the corresponding invariant.

It is natural to assume that the invariant corresponding to a substate logically implies the invariant corresponding to its superstate, since the invariant corresponding to the substate should be more restrictive. We call this condition invariant monotonicity. Formally, let s_1 and s_2 be two different states and let F_1 and F_2 be the corresponding invariants. If s_1 is a substate of s_2, then $F_1 \implies F_2$ (i.e., the invariant corresponding to s_1 implies the invariant corresponding to s_2). Due to the transitivity of implication, this property holds also for indirect substates, i.e., substates of a substate and so on (see the constraint (4) above).

As a running example, we consider a simple, but illustrative SM containing composite states (see Fig. 1).

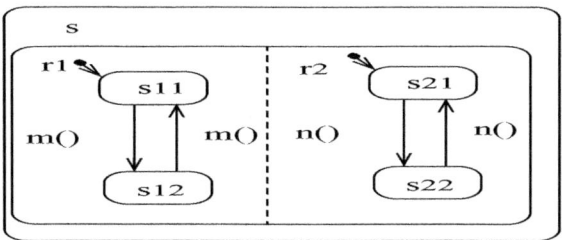

Fig. 1. SM with nested states

Regions r_1 and r_2 are parts of state s. s_{11} and s_{12} are direct substates of r_1 (i.e., s_{11} and s_{12} are elements of $r_1.subvertex$); similarly s_{21} and s_{22} are substates of r_2. In the above example, the invariant monotonicity condition for states s_{11} and s has the form ($F_{11} \implies F$).

One can ask, when states of a SM cover all possibilities. We say that direct substates $s_1, ..., s_n$ of state s are exhaustive if, and only if, the disjunction $F_1 \vee ... \vee F_n$ of invariants corresponding to the substates logically implies the invariant F corresponding to s. UML requires that only one direct substate of an or-state can be active at a given time. We say that states in a state machine are non-overlapping if for every or-state s and for every two different direct substates s_1 and s_2 of s, the corresponding conjunction $F_1 \wedge F_2$ is logically false. Note that this property concerns only or-states and does not exclude the use of and-states, since according to UML [19] substates of an and-substate cannot be direct substates of an or-state. In the above example, this means that $F_{11} \wedge F_{12}$ and $F_{21} \wedge F_{22}$ are logically false.

It is also natural to require that for every and-state s, for the corresponding invariant F and for every set $A = \{F_1, ..., F_n\}$ including invariants corresponding to all states in a certain sub-region of s, the disjunctions $\bigvee A$ is logically implied by F. We use $\bigvee\{F_1, ..., F_n\}$ as an abbreviation for $F_1 \vee ... \vee F_n$ (similarly for $\bigwedge$). We call this requirement region-exhaustivity condition. It says that disjunction of state-invariants corresponding to a region of a concurrent state is logically implied by the invariant of the concurrent state. In the above example, this corresponds to the requirement that F implies $F_{11} \lor F_{12}$ and $F_{21} \lor F_{22}$.

The notion of orthogonal states is used in the UML standard without a precise definition. Orthogonality usually means independence. For example if a point has two Cartesian coordinates, then those coordinates are considered orthogonal. One can generalize this idea to object attributes and say that two sets of object attributes are orthogonal if they are disjoint. Logically, satisfaction of an invariant including only attributes from the first set is independent from satisfaction of an invariant including only attributes from the second set. This has an analogy in boolean algebra. We say that regions $r_1, ..., r_n$ are truly orthogonal if for every state s_i from r_i, for $i = 1, ..., n$, the conjunction of the corresponding invariants $F_1 \land ... \land F_n$ is not logically false. Thus an arbitrary combination of substate-invariants is logically possible.

UML defines so called initial pseudo states for marking SM states. A state s is initial if it is marked by an initial pseudo state. We define invariants specifying initial system states with the help of an auxiliary function f. We associate *false* with states which are not initial. Then we move from substates to composite states; we take disjunctions of or-states and conjunctions of compound and-substates.

- If s is not initial, then $f(s) = \textit{false}$
- Else if s is an atomic state, then $f(s) = F$ where F corresponds to s
- Else s is a composite state; let $s_1, ..., s_n$ be all its direct substates
  - If s is an or-state, then $f(s) = F \land (f(s_1) \lor ... \lor f(s_n))$
  - Else s is an and-state and then $f(s) = F \land (f(s_1) \land ... \land f(s_n))$

For a topmost initial state s we can identify the condition $f(s)$ which is satisfied when the corresponding system starts to operate. It is stronger than the invariant corresponding to s. One can express this as the conjunction of invariants corresponding to initial state configurations, i.e., trees of states active when the system starts to operate [19]. In the running example, the initial state is characterized by the formula $F \land F_{11} \land F_{21}$. The invariant monotonicity condition implies that it is enough to consider invariants corresponding to atomic states, i.e., $F_{11} \land F_{21}$.

There are other structural properties which can be expressed by logical relations between formulas. It is hard to describe more involved structural properties of state machines in general terms. Therefore in subsection 2.4, we use UML metamodel and OCL to define such properties in a fully formal way.

2.3   Logical Relations

In the previous subsection we have shown that structural relations between states in a SM correspond to logical relations between formulas. In this section, we define the notion of the logical relation using the entailment relation. The novel thing is the exposure of the relation of state machine structures to the entailment.

Constraints considered in the previous section are first order formulas. The relations between them are considered at the level of propositions and the only thing that matters is whether some formulas are implied by other formulas (e.g. in the monotonicity condition). This kind of relationship can be studied in general terms using entailment relations, i.e., a binary relation of the form $A \vdash F$ where A is a set of formulas, which we sometimes call axioms, and F is a single formula. Usually, it corresponds to the relation of being provable (logical consequence) or being satisfied (semantic consequence). In this paper we use the notion of entailment relation in the first sense. We write $\vdash F$ for $\emptyset \vdash F$; in this case we say that F is a tautology with respect to the entailment relation $\vdash$, or equivalently that it logically holds.

We define logical relation constraining state-invariants to be a propositional formula of the form $C(Y_1, ..., Y_n)$ where $Y_1, ..., Y_n$ are propositional variables ranging over first order formulas, e.g, state-invariants. As mentioned above, state-invariants are formulas with one free variable *self* which is bound to the underlying object. All the logical relations defined in the previous section have to be considered in respect to an entailment relation. In case of the running example, we say that states s_{11} and s satisfy the invariant monotonicity condition in respect to $\vdash$ if, and only if, $\vdash F_{11} \implies F$ holds.

In general, let A be a set of formulas. We say that states $s_1, ..., s_n$ of a SM satisfy propositional formula $C(Y_1, ..., Y_n)$ in respect to specification A and entailment relation $\vdash$ if, and only if, $A \vdash C(F_1, ..., F_n)$ holds. A can be seen as a set of domain specific axioms specifying domain's semantics. It can be also an axiomatization of OCL. Of course A can be empty.

2.4   Derivation of Logical Formulas from SM

State machines may have different shapes and can be very complex. Consequently deriving logical dependencies from the topology of a SM and writing the corresponding logical formulas 'by hand' would be very time consuming and error prone. Therefore, we define a generic method for deriving such formulas from abstract syntax of SM diagrams. We use OCL formulas referring to elements of the UML metamodel [19]. They can be seen as schemata allowing us to generate a concrete logical relation for a concrete SM.

State-invariants are associated with states via the association *stateInvariant* specified by the UML metamodel. As mentioned above, we assume that every state s has one associated state-invariant, i.e., $s.stateInvariant->size() = 1$. In the following we use *smst* as an abbreviation for an OCL formula specifying all states of a SM. It can be defined with the help of so called iterators. The invariant monotonicity condition can be expressed in OCL in respect to UML metamodel as follows ([19], Fig. 15.2):

$smst->forAll(s \mid s.region.subvertex->forAll(v \mid$
$v.stateInvariant \text{ implies } s.stateInvariant))$

This formula says that for an arbitrary state s of the underlying SM (i.e., for every $s \in smst$), all the direct substate-invariants imply the invariant of s. The expression $s.region.subvertex$ means all direct substates of s. The corresponding first order formula (more precisely - schema) has the following form (cf. subsection 2.2):

$\bigwedge\{v.stateInvariant \implies s.stateInvariant \mid s \in smst \land v \in s.region.subvertex\}$

In case of the running example, instantiation of this generic form results in the following formula:

$$(F_{11} \implies F) \land (F_{12} \implies F) \land (F_{21} \implies F) \land (F_{22} \implies F)$$

The generic forms defined below can be translated to first order logic in a similar way. The non-overlappingness property can be expressed in OCL as follows:

$smst.region{\rightarrow}forAll(r \mid r.subvertex{\rightarrow}forAll(v_1, v_2 \mid$
$\quad not(v_1 = v_2) \; implies \; not(v_1.stateInvariant \; and \; v_2.stateInvariant)))$

This formula says that for every two different states in the same region the corresponding invariants contradict one another. In particular if s is an or-state, i.e., has one subregion and if it has two different direct substates, then they are not overlapping.

The substate exhaustiveness condition says that for every region its substate-invariants cover all possibilities. OCL does not provide the disjunction operator over sets $\bigvee$ (cf. subsection 2.2), but it can be defined using iterators.

$smst{\rightarrow}forAll(s \mid s.region{\rightarrow}forAll(r \mid$
$\quad s.stateInvariant \; implies \; \bigvee(r.subvertex.stateInvariant)).$

This constraint says that for every composite state and every region of that state, the state-invariant implies the disjunction of region's substate-invariants. Thus the substates of a region cover all possibilities described by its state-invariant. In the case of the running example, this results in the formula:

$$(F \implies (F_{11} \lor F_{12})) \land (F \implies (F_{21} \lor F_{22}))$$

State orthogonality can be defined by assuming that the following schema does not generate counter-tautologies (see subsection 2.2):

$smst{\rightarrow}forAll(s \mid s.isOrthogonal \; implies \; s.region{\rightarrow}forAll(r_1, r_2 \mid not(r_1 = r_2)$
$\quad implies \; r_1.subvertex{\rightarrow}forAll(s_1 \mid r_2.subvertex{\rightarrow}forAll(s_2 \mid$
$\quad\quad s_1.stateInvariant \; and \; s_2.stateInvariant))$

The instantiation of this frame for the SM shown on Fig. 1 results in formulas $F_{1i} \land F_{2j}$, for i, j = 1, 2, which, by assumption, are not counter-tautologies.

3   Transformation of State Machines

In this section we investigate the impact of class model redesign on the structure of state machines. Such redesign can be described with the help of interpretation functions (IF) [13]. They differ from the other term rewriting based approaches [3,22,2] in that they are oriented at transforming formulas, in particular OCL constraints, and that they preserve proofs [12]. We reduce the problem of SM structure preservation to the problem of logical relations preservation. In the first subsection, we present the notion of interpretation function and explain how it can be used. In the second subsection, we introduce the notion of logical invariants.

3.1   Interpretation Functions

Interpretation functions proved to be a very useful vehicle for an automatic transformation of OCL constraints when changes to class diagrams are performed [13] (see also [11]). For example, if an attribute a of type Integer is replaced by a path $b.v$ where b is an association pointing to another class and v is an attribute of that class of integer type, then every OCL constraint or state-invariant containing a has to be modified. This kind of modifications can be performed using interpretation functions, i.e., partial functions generated by mappings satisfying conditions analogous to orthogonal term rewriting systems (cf. e.g. [21]). In the above mentioned case, roughly speaking, they replace the attribute a by the term $b.v$ in all OCL-constraints in a correctness preserving way.

Basically, IFs are generated by mappings on sets of independent terms, like linear functions are generated by mapping defined on a set of base vectors, they preserve types and variables. The composition of terms corresponds to the linear composition. Model transformations generate functions on terms, but it has to be checked if they are interpretation functions. By definition, IFs preserve quantifiers and propositional operators such as negation, conjunction and implication. To some extent, they can be seen as a formal counterpart of dependency relationships in UML (cf. [13]). Those relationships are used to compare different specifications [19]; selected model elements in one specification are mapped to the related model elements in another one. In case of IFs, model elements are formalized by terms, thus we obtain a mapping on terms. We use term rewriting (cf. [21]) for the model transformation, more precisely orthogonal term rewriting systems. Orthogonal term rewriting systems are the most regular ones. Roughly speaking, a set of term rewriting rules is orthogonal if its rewrite rules do not overlap and consequently an application of one rule does not exclude an application of another one. Sets of orthogonal terms play a role similar to base sets in case of linear algebras.

Two algebraic terms u and v overlap if, and only if, one of the following conditions is satisfied:

- u and v are different and u can be unified with v
- u can be unified with a non-variable, proper subterm of v, or vice versa

By unification we mean the existence of a substitution making both terms equal. We say that a set of terms is overlapping free if it does not contain terms which overlap. We say that a term t is linear if for every variable x, t contains x at most once. We say that a set of terms is orthogonal if it does not contain variables, it is overlapping free and contains only linear terms.

In this paper we use a variant of the so called order sorted algebras [8]. Term algebra of an order sorted signature has the form $T(S, F, \leqslant, X, \tau)$ where S is the set of sorts, F is a set of function symbols, X is a set of variables, $\leqslant$ is a sort comparison relation saying whether a sort is a subsort of another one and τ is a typing function for terms. We assume that F contains functions for the property names in the model (e.g., it contains functions corresponding to above mentioned attribute a), and also boolean and arithmetic operations such as conjunction, addition and multiplication. Let $\psi : T(S, F, \leqslant, X, \tau) \to T(S', F', \leqslant', X', \tau')$ be a partial function. ψ is compositional iff for every term t the following conditions hold:

(i) $\psi(x)$ is defined for every variable $x \in X \cap X'$
(ii) If $\psi(t)$ is defined, mapping $\sigma : X \to X$ preserves types, i.e., $\tau(\sigma(x)) = \tau(x)$, then $\psi(t^\sigma) = \psi(t)^\sigma$
(iii) $var(\psi(t)) \subseteq var(t)$ if $\psi(t)$ is defined
(iv) If ψ maps term t_i to the term t'_i, for i $= 0,...,$ n, $x_1, ..., x_n \in X \cap X'$ and term t has the form $t_0[t_1/x_1, ..., t_n/x_n]$, then the substitution $t'_0[t'_1/x_1, ..., t'_n/x_n]$ is well defined and $\psi(t)$ has the form $t'_0[t'_1/x_1, ..., t'_n/x_n]$

Conditions (i) and (ii) imply that compositional functions are defined on common variables and that they do not depend on names of variables; t^σ denotes the renaming of variables of t according to σ. Condition (iii) is the variable inclusion requirement made in case of term rewriting systems (cf. e.g. [21]). (iv) is a compositionality condition. It allows us to scale up a mapping to complex terms.

Let $A \subseteq T(S, F, \leqslant, X, \tau)$ be a set of terms. Mapping $\psi : A \to T(S', F', \leqslant', X', \tau')$ is orthogonal if there exists a partial function on sorts $\rho : S \to S'$ such that the following conditions are satisfied:

(a) For every variable x, $\rho(\tau(x))$ is defined iff $x \in X \cap X'$
(b) If $x \in X \cap X'$, then $\rho(\tau(x)) = \tau'(x)$
(c) $var(\psi(v)) \subseteq var(v)$, for $v \in A$
(d) If $v \in A$, then $\rho(\tau(v))$ is defined and $\rho(\tau(v)) = \tau'(\rho(v))$
(e) If $\rho(s_1)$, $\rho(s_2)$ are defined and $s_1 \leqslant s_2$, then $\rho(s_1) \leqslant' \rho(s_2)$
(f) A (i.e., $Dom(\psi)$) is orthogonal

Conditions (a) and (b) say that the sort mapping ρ is determined by types of common variables; they are analogous to conditions (i) and (ii). Condition (c) is analogous to (iii). Condition (d) says that ρ commutes with the typing function τ. Condition (e) says that ρ is monotone. Note that in case of a single-sorted algebra, every orthogonal term rewriting system determines the corresponding orthogonal identity mapping and vice versa any orthogonal mapping determines an orthogonal term rewriting system.

A compositional function can be extended to boolean valued terms in the following way: If $\psi(\Phi) = \Phi'$ and $\psi(\Psi) = \Psi'$, then $\psi(\Phi \wedge \Psi) = \Phi' \wedge \Psi'$. If $\psi(\Phi) = \Phi'$ and $x \in X \cap X'$, then $\psi(\forall_x \Phi) = \forall_x \Phi'$. Similarly we can define the extension for other propositional operators.

As proved in [13], an orthogonal mapping can be uniquely extended to a compositional function; a function generated by an orthogonal mapping is called interpretation function. The idea of IFs is that a mapping has to be defined on its 'generators' and that such an implicitly defined mapping can then be extended to an interpretation function on the level of OCL and on the level of formal specifications if certain assumptions are satisfied. Those functions allow us to transform OCL specifications. In [14], we have demonstrated how to implement interpretation functions using term rewriting tool Elan [9]. The above mentioned conditions can be checked automatically. In particular, the orthogonality of a set of terms can be checked using standard unification algorithms. In the paper [12], we have shown that IFs preserve proofs using propositional tautologies, the resolution rule and induction, as well as simple equational proofs. This paper builds on that results. IFs allow one to save the clerical work of redoing such proofs after transformation of class diagrams. We refer the interested reader to [12,13] for more details. Thus, we can automatically rewrite state invariants after class model redesign if a corresponding interpretation function is applicable to the invariant. If not, then the invariant has to be rewritten manually. Moreover, if the proofs do not have the above mentioned forms, then the state machines need to be reconsidered. Such cases can be automatically identified.

3.2 Logical Invariants

As explained above, a transformation of a class diagram requires the transformation of the corresponding class constrains as well as state-invariants of a SM. Subsection 2.2 explains that structural relationships between states impose logical relation between the corresponding state-invariants. In general nothing hinders us from arbitrary rearrangement of states into a new SM apart of the logical relations between the corresponding state-invariants.

Let there be a transformation which maps state s_j on state s'_j, for $j = 1, ..., n$. We say that this transformation preserves a logical relation $C(Y_1, ..., Y_n)$ between those states in respect to entailment relation $\vdash$ if, and only, if $\vdash C(F_1, ..., F_n)$ implies $\vdash C(\psi(F'_1), ..., \psi(F'_n))$. Similarly, we say that an interpretation function ψ preserves a logical relation $C(Y_1, ..., Y_n)$ between those states in respect to axioms A and entailment relation $\vdash$ if, and only if, $A \vdash C(F_1, ..., F_n)$ implies $\psi(A) \vdash C(\psi(F_1), ..., \psi(F_n))$.

For example, we say that IF ψ preserves the invariant monotonicity condition in respect to $\vdash$ if, and only if,

$$\vdash \bigwedge \{v.stateInvariant \implies s.stateInvariant \mid s \in smst \wedge v \in s.region.subvertex\}$$

implies that

$$\vdash \bigwedge \{\psi(v.stateInvariant) \implies \psi(s.stateInvariant) \mid s \in smst \wedge$$
$$v \in s.region.subvertex\}$$

Preservation of other conditions is defined in a similar way.

The statement below follows from the fact that by definition interpretation functions preserve quantifiers and propositional operators such as negation, conjunction and implication (see subsection 3.1).

Statement

Let $C(Y_1, ..., Y_n)$ be a propositional formula and let ψ be an interpretation function. Then $\psi(C(Y_1 ..., Y_n)) = C(\psi(Y_1), ..., \psi(Y_n))$.

As mentioned above, interpretation functions preserve proofs using resolution rule, propositional tautologies and induction. In the following, $\vdash_{pri}$ denotes entailment relation defined in the following way: Let A be a set of first order formulas and let F be a formula. $A \vdash_{pri} F$ if, and only if, there is a proof of F containing only formulas from A and using propositional reasoning, resolution rule and induction.

IFs are partial functions. The following corollary says that if a logical relation is like the one defined in the previous statement and if an interpretation function is defined on state-invariants and underlying domain specific axioms A, then it preserves the logical relation between them.

Corollary

Let C be an OCL formula and let A be a set of OCL formulas. Let ψ be an interpretation function which is defined on A and on state-invariants F_j, for $j = 1, ..., n$. If $A \vdash_{pri} C$, then $\psi(A) \vdash_{pri} \psi(C)$.

This corollary implies that if an interpretation function is defined on state-invariants, then it preserves the structural relationships between the corresponding states in respect to $\vdash_{pri}$. Consequently we can say that if the above mentioned properties are proved using above mentioned kinds of reasoning, then they are preserved by interpretation functions. If for example, there is a proof of the invariant monotonicity condition using those kinds of reasoning, then after transformation via an interpretation function this property remains valid. Therefore the topology of a state machine is preserved. Despite the fact that SM structure is preserved, the corresponding state-invariants have to be modified. But to do this it is enough to apply the corresponding IF.

4   Example: Application of the State Pattern

In this section we apply ideas developed so far to a concrete example. We consider a refactoring of a class diagram according to the so called State Pattern (cf. e.g. [7]). The following example is rather simple, but the primary goal of this approach is not to transform very sophisticated mathematical proofs. This method is meant to eliminate doing proofs and refactoring by hand. In case of many, even very simple proofs, manual transformation may prove to be very laborious and lead to several errors. Note that in general, the method preserves proof structure in respect to propositional operators such as conjunction and negation, but it changes those parts of a proof which corresponds to induction and resolution. It should be also noted that proofs like the one below can be done manually or with the help of interactive as well as automatic theorem provers. Our method is applicable to all these cases.

First we show an implementation of a state machine by enumeration types; then we redesign this implementation using the State Pattern. (We refer the interested reader to [13] for the details of this pattern application.)

Fig. 2 shows class *FlipFlop* with attribute *state* of an enumeration type and operation *next()*. This attribute is formalized by an equally named function

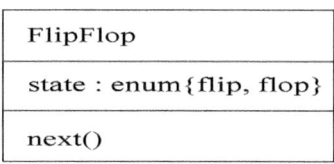

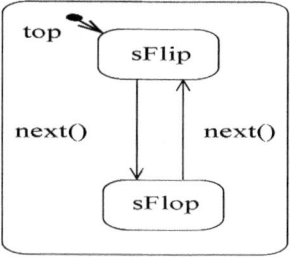

Fig. 2. Implementation with enumeration types

symbol in the set F (see subsection 3.1). The state machine on the right hand side corresponds to this class. An object of this class can be either in the state *sFlip* or *sFlop*. The elements of the enumeration type define these states. The state-invariant associated with state *top* has the form:

$self.state = \#flip \lor self.state = \#flop$

Invariants corresponding to states *sFlip* and *SFlop* have the form:

$state = \#flip$ and $state = \#flop$, respectively.

Note that the invariant monotonicity and the exhaustiveness condition are satisfied. The operation *next()* transposes these states.

The non-overlappingness condition for this state machine has the form:

$\neg(self.state = \#flop \land self.state = \#flip)$

We prove it formally by a contradiction using propositional tautologies, resolution as well as symmetry and transitivity axioms for equations. In general, such a reasoning may include applications of the resolution rule, induction and the propositional reasoning.

We apply the and-congruence tautology and the symmetry axiom

$(a \implies b) \implies (c \land a \implies c \land b), x = y \implies y = x$

to the negation of the non-overlappingness condition

$self.state = \#flop \land self.state = \#flip$

and obtain the formula

$\#flop = self.state \land self.state = \#flip$

Then we apply the transitivity axiom $x = y \land y = z \implies x = z$ to the formula

$\#flop = self.state \land self.state = \#flip$

and obtain the formula $\#flop = \#flip$.

The two elements of the enumeration type are different, thus $\#flip \neq \#flop$ holds. Finally, we apply propositional tautologies concerning the transitivity and negation of implication

$$(p \implies q) \wedge (q \implies r) \implies (p \implies r), (p \implies q) \wedge \neg q \implies \neg p$$

and obtain the formula

$$\neg(self.state = \#flop \wedge self.state = \#flip)$$

Consequently, the states $sFlip$ and $sFlop$ are non-overlapping.

The State Pattern application yields the class diagram shown on Fig. 3. In the redesigned version, the states are implemented by objects instantiating classes $Flip$ and $Flop$ which in turn subclass the class $State$ (see Fig. 3). Thus we do not use here objects instead of elements of an enumeration type. We map the elements of the enumeration type to the corresponding classes; i.e., $flip$ and $flop$ are mapped to classes $Flip$ and $Flop$, respectively. Moreover, the attribute $state$ is mapped to the term $lnkState.oclType$. This mapping is defined on the level of model elements; it induces an interpretation function (cf. [13]).

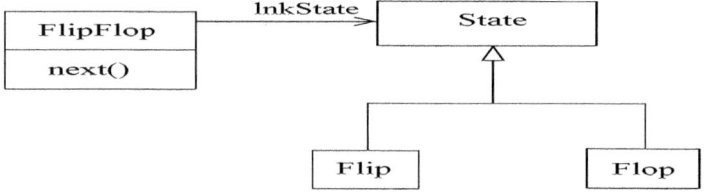

Fig. 3. States as objects

Note that the invariant monotonicity and the exhaustiveness condition are satisfied for the transformed SM. The results proved in [12] guarantee that the interpretation function preserves proofs like the one above. We show that the property of non-overlappingness is indeed preserved by the interpretation function. The transformed proof has the same form as the previous one, in respect to propositional formulas.

As in the previous case, we apply the and-congruence tautology and the symmetry axiom for equations to derive the formula

$$Flop = self.lnkState.oclType \wedge self.lnkState.oclType = Flip$$

from the formula

$$self.lnkState.oclType = Flop \wedge self.lnkState.oclType = Flip$$

Similarly the formula

$$Flop = self.lnkState.oclType \wedge self.lnkState.oclType = Flip$$

implies *Flop* = *Flip*, thanks to the transitivity axiom for equations. The classes *Flip* and *Flop* are different. Thus, as in the previous case, we derive

$$\neg(self.lnkState.oclType = Flop \wedge self.lnkState.oclType = Flip)$$

Consequently, in the transformed SM state-invariants corresponding to states *sFlip* and *sFlop* exclude each other and thus the application of State Pattern preserves the property of non-overlappingness.

5   Related Work

Today's software engineering processes embrace change as a constant factor. There exist a number of approaches to redesign of UML class diagrams. The best known is the refactoring [5]. It provides simple patterns for code and class structure modification to extend and modify a system without altering its behaviour. There are a number of model transformation languages, one of the most popular is ATL [1]. This language transforms models specified in XMI, at the metalevel.

The term rewriting theory proved to be a very natural means for defining the semantics of metamodels [3]. There exists a formal semantics of ATL based on this theory [22]. There exists also a graph and term rewriting based approach to the formalization and verification QVT-like transformations [2]. It formalizes model transformations as theories in so called rewriting logic. However, these approaches do not deal with class constraint transformations. It should be mentioned that there are a number of very powerful term rewriting tools such as for example Elan/Thom [9] and Maude [2].

The first formal approach to the class diagram and constraint transformation was proposed in [10]. It is based on term rewriting and is related to the method proposed by Egyed for abstracting away class diagram details if they are irrelevant from a particular point of view [6]. In a previous paper [13], we have refined the approach proposed in [10] and studied the redesign of UML class diagrams with OCL constraints, as well as the transformation and tracing of constraints. Those constraints may concern dependencies between classes, associations, attributes or generalization relationships. We have introduced the notion of interpretation function for redesign of class diagrams and constraint transformation.

An analogous concept to [13] and to [6] was proposed in [16]. It allows one to formalize refactoring rules for class diagrams and to classify them with respect to their impact on annotated OCL constraints. It describes how the OCL constraints have to be refactored to preserve their syntactical correctness. The focus is on refactoring patterns rather than entailment preservation. We have shown [12], that different kinds of entailment relations are preserved by interpretation functions, in particular proofs using propositional tautologies, the resolution rule and the induction, and also a simple form of equational proofs. The possibility of applying interpretation functions to formalize the State Pattern has bee pointed out in [11].

We should mention here an approach applying type theory and the proofs-as-programs paradigm to transform platform independent models into platform specific ones in a contract-preserving way [20]. The difference between our approach and that one is that interpretation functions do not preserve formulas, but the entailment relation between them.

6 Concluding Remarks

In this paper we demonstrated that SM structure can be expressed by logical relations between state-invariants. Those relations are preserved by a refactoring if it corresponds to an interpretation function and the underlying proofs are performed using induction, resolution and propositional tautologies. Moreover, interpretation functions corresponding to refactorings can be used to rewrite state invariants. For the automatic invariant transformation very powerful term rewriting tools can be used. Our results can lessen the burden of UML diagrams modifications after a class diagram modification. In the future we are going to build a tool based on term rewriting for the automatic transformation of state invariants and for the identification of state machines which have to be reconsidered.

References

1. The AtlanMod Team: ATL (2011), http://www.eclipse.org/m2m/atl/doc/
2. Boronat, A., Heckel, R., Meseguer, J.: Rewriting Logic Semantics and Verification of Model Transformations. In: Chechik, M., Wirsing, M. (eds.) FASE 2009. LNCS, vol. 5503, pp. 18–33. Springer, Heidelberg (2009)
3. Boronat, A., Meseguer, J.: An Algebraic Semantics for MOF. In: Fiadeiro, J.L., Inverardi, P. (eds.) FASE 2008. LNCS, vol. 4961, pp. 377–391. Springer, Heidelberg (2008)
4. Czarnecki, K., Van Gorp, P.: Classification of Model Transformation Approaches. In: OOPSLA 2003 Workshop on Generative Techniques in the Context of Model-Driven Architecture (2003)
5. Fowler, M.: Refactoring: Improving the Design of Existing Code. Addison-Wesley, Reading (2000)
6. Egyed, A.: Compositional and Relational Reasoning During Class Abstraction. In: Stevens, P., Whittle, J., Booch, G. (eds.) UML 2003. LNCS, vol. 2863, pp. 121–137. Springer, Heidelberg (2003)
7. Gamma, E., Helm, R., Johnson, R., Vlissides, J.: Design Patterns. Addison-Wesley, Reading (1995)
8. Goguen, J., Meseguer, J.: Order Sorted Algebra. Theoretical Computer Science 105(2), 167–215 (1992)
9. Kirchner, C., Moreau, P.-E., Tavares, C.: A Type System for Tom. In: Proceedings of The 10th International Workshop on Rule-Based Programming, Brazil (2009)
10. Kosiuczenko, P.: Formal Redesign of UML Class Diagrams. In: Evans, A., France, R., Moreira, A., Rumpe, B. (eds.) Proc. of pUML Workshop on Practical UML Based Rigorous Development Methods, Toronto. Lecture Notes in Informatics (2001)

11. Kosiuczenko, P.: Proof Transformation via Interpretation Functions: Results, Problems and Applications. In: Proceedings of the SETra 2004 Workshop. ENTCS, vol. 127(3), pp. 139–145. Elsevier, Amsterdam (2005)
12. Kosiuczenko, P.: Proof Transformation via Interpretation Functions, Technical Report Nr. CS-05-002, Dep. of Comp. Sc., University of Leicester (2005)
13. Kosiuczenko, P.: Redesign of UML Class Diagrams: A Formal Approach. Journal of Software & System Modeling 8(2), 165–183 (2009)
14. Kosiuczenko, P.: Term Rewriting as a Unifying Basis for Graphical Modelling of Object-Oriented Systems, 144 pages. WAT Publishing (2009)
15. Lano, K.: Formal Object-Oriented Development. Springer, Berlin (1995)
16. Marković, S., Baar, T.: Refactoring OCL annotated UML class diagrams. Software and System Modeling 7(1), 25–47 (2008)
17. Mens, T., Van Gorpa, T.: A Taxonomy of Model Transformation. ENTCS, vol. 152, pp. 125–142. Elsevier, Amsterdam (2006)
18. OMG: Object Constraint Language, Version 2.3.1, Formal/2011-09-02 (2011)
19. OMG: Unified Modeling Language Specification, Version 2.4, Formal/2011-09-22 (2011)
20. Poernomo, I.: Proofs-as-Model-Transformations. In: Vallecillo, A., Gray, J., Pierantonio, A. (eds.) ICMT 2008. LNCS, vol. 5063, pp. 214–228. Springer, Heidelberg (2008)
21. Terese, et al.: Term rewriting systems. Cambridge University Press, Cambridge (2003)
22. Troya, J., Vallecillo, A.: Towards a Rewriting Logic Semantics for ATL. In: Tratt, L., Gogolla, M. (eds.) ICMT 2010. LNCS, vol. 6142, pp. 230–244. Springer, Heidelberg (2010)

Fact or Fiction – Reuse in Rule-Based Model-to-Model Transformation Languages*

Manuel Wimmer[1], Gerti Kappel[1], Angelika Kusel[2],
Werner Retschitzegger[2], Johannes Schönböck[1], and Wieland Schwinger[2]

[1] Vienna University of Technology, Austria
{lastname}@big.tuwien.ac.at
[2] Johannes Kepler University Linz, Austria
{firstname.lastname}@jku.at

Abstract. Model transformations are mostly developed from scratch. For increasing development productivity as well as quality of model transformations, reuse mechanisms are indispensable. Although numerous mechanisms have been proposed, no systematic comparison exists making it unclear, which reuse mechanisms may be best employed in a certain situation. Therefore, this paper provides an in-depth comparison of reuse mechanisms in rule-based model-to-model transformation languages and categorizes them along their intended scope of application. For this, a systematic comparison framework for reuse mechanisms is proposed to highlight commonalities as well as differences. Finally, current barriers to model transformation reuse are outlined.

Keywords: Reuse Mechanisms, Model Transformations, Comparison.

1 Introduction

Model transformations are crucial for the success of Model-Driven Engineering, being comparable in role and importance to compilers for high-level programming languages. Nevertheless, most of today's transformation designers still follow an ad-hoc manner to specify model transformations [10]. For increasing development productivity as well as quality of model transformations, the application of appropriate reuse mechanisms is indispensable. This need has been recognized by the research community as a plethora of proposed reuse mechanisms reveals [6–9, 13, 16, 18, 22–24, 26–30]. Nevertheless, there exists no survey providing an overview of the proposed mechanisms to deeper understand their commonalities and differences. Thus, it is unclear which reuse mechanisms may be employed in a certain situation and which barriers exist in applying them.

Therefore, this paper provides an in-depth comparison of proposed reuse mechanisms in rule-based model-to-model transformation languages to highlight when to apply a certain reuse mechanism and how reuse mechanisms complement each other. In this respect, reuse mechanisms are categorized along their intended scope of application, ranging from *reuse in the small*, e.g., functions,

---

* This work has been funded by the FWF under grant P21374-N13.

Z. Hu and J. de Lara (Eds.): ICMT 2012, LNCS 7307, pp. 280–295, 2012.

to *reuse in the large*, e.g., orchestration of model transformations. For this, a systematic comparison framework for reuse mechanisms is proposed comprising comparison criteria along four different dimensions analogous to the main phases in software reuse [15], being *abstraction, selection, specialization* and *integration.* On the basis of this framework, the categorized reuse mechanisms are compared and for each reuse mechanism corresponding supporting representatives are given. To illustrate the different mechanisms, example reuse scenarios on the basis of a running example are given.

Outline. Section 2 introduces the running example and presents the comparison framework with its four dimensions. In Section 3, the comparison framework is used to compare the reuse mechanisms along their different scopes of reuse. Section 4 presents barriers to reuse in model transformations and finally, Section 5 concludes the paper.

2 Comparison Framework

This section introduces scopes of reuse based on an example which is used throughout the paper to illustrate the different reuse mechanisms. Furthermore, a comparison framework is presented to characterize the reuse mechanisms.

2.1 Scopes of Reuse

Different scopes of reuse exist which possess different reuse potentials, e.g., within/ across transformations or between the same/different metamodels (MMs). In this respect, we identified five different scopes ranging from *reuse in the small* to *reuse in the large* which are depicted in Fig. 1 on basis of the Class2ER example [2]:

- *Scope 1:* To avoid code duplication, reuse of logic within *a single transformation* is needed, i.e., the scope is to reuse the same transformation logic between the *same MMs* in the *same transformation* (cf. (1) in Fig. 1).
- *Scope 2:* To realize similar transformation logic, e.g., to pursue different OR-mapping approaches – a "one table per hierarchy" approach instead of a "one table per class" approach (cf. (2) in Fig. 1) – reuse of transformation logic between the *same MMs* in *different transformations* is needed.
- *Scope 3:* The transformation logic of the Class2ER example might be needed in an Ontology2XML transformation as well, requiring that the same transformation logic could be reused in the context of *different MMs* and thus *different transformations* (cf. (3) in Fig. 1).
- *Scope 4:* Since cross-cutting concerns, e.g., debugging or tracing (cf. (4) in Fig. 1), should be reusable throughout transformations, mechanisms are needed that allow to reuse *logic* irrespective of *MMs* and *transformations.*
- *Scope 5:* Reuse in the large is achieved when existing transformations can be applied *without changing transformation logic or MMs*, as is the case for chaining a ER2Relational transformation after the Class2ER transformation in our running example (cf. (5) in Fig. 1).

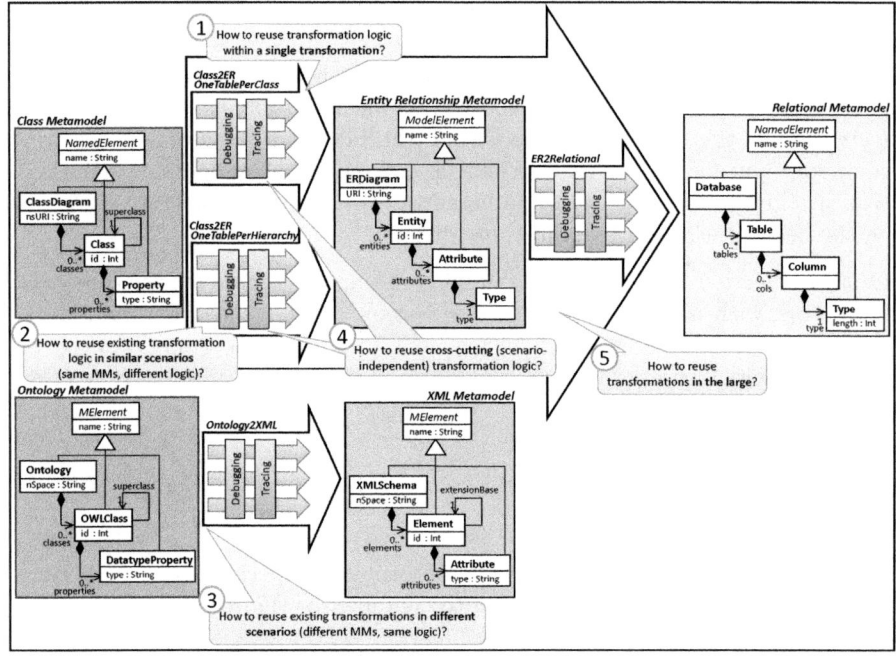

Fig. 1. Running Example - Different Scopes of Reuse

2.2   Comparison Criteria

In order to highlight commonalities as well as differences between reuse mechanisms across their scopes, a comparison framework (cf. Fig. 2) based on the common phases (i) *abstraction*, (ii) *selection*, (iii) *specialization*, and (iv) *integration* of reuse mechanisms according to [15] is proposed in the following.

Abstraction. To enable reuse, abstraction is the key of any reuse mechanism. According to [14], one might distinguish between *abstraction by generalization* and *abstraction by simplification*. Abstraction by generalization allows to make an artifact reusable in different situations. To achieve this in the context of model transformations, it should be possible to decouple transformation logic from *type information*, i.e., the source and the target *MMs*. Furthermore, reuse of transformation logic across platforms should be possible by generalizing from a certain *transformation language*. Abstraction by simplification allows to emphasize the information necessary for reuse, i.e., the *visible part* (e.g., interface of a function to reuse), but to hide the actual realization of the artifact, i.e., the *hidden part* (e.g., the implementation of the function) [15].

Selection. Provided that repositories of reusable artifacts exist, mechanisms are needed to efficiently find the artifacts. Such mechanisms range from *metainformation*, e.g., documentation or pre-/post-conditions, to *automatism* in the form of wizards or more advanced techniques from information retrieval [19].

Specialization. To adapt an abstracted artifact to a specific transformation, specialization is needed. Ideally, only *knowledge* of the signature of the abstracted artifact, but not of the realization is needed (i.e., *reuse in the black-box view*). In contrast, *reuse in the white-box view* demands additional knowledge of the realization. For specialization, typically certain *mechanisms* are needed, e.g., passing of parameter values in functions or overriding/extending parts in the context of inheritance. Finally, *language-inherence* states if a transformation designer stays in the same formalism for specialization or not.

Integration. Whereas specialization solely configures an artifact, integration focuses on how reusable artifacts interact with the remaining parts of the specified transformation. Reuse mechanisms in software engineering are typically categorized into *composition* and *generation* mechanisms [3, 20]. Thereby, composition implies that integration must take place whereas generation implies that an executable transformation without further need for integration is produced. Therefore, the first criterion *ability* distinguishes between composition and generation, whereas the second

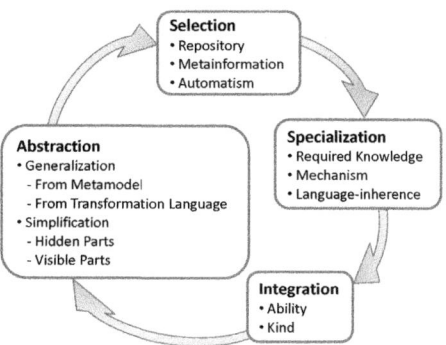

Fig. 2. Comparison Framework

criterion *kind* differentiates potential ways of composition. In this respect, according to [17], composition can be realized by (i) *containment*, i.e., the specified transformation nests the reusable artifact, (ii) *connection*, i.e., the specified transformation reuses the artifact by delegation, (iii) *extension*, i.e., the reusable artifact is extended and refined, and (iv) *coordination*, i.e., a synchronization language is used to coordinate the reusable artifacts.

3   Comparison of Reuse Mechanisms

Based on the identified scopes and the introduced comparison framework, proposed reuse mechanisms for model transformations are compared in the following (cf. Table 1 for an overview and Table 2 for details). For each reuse mechanism, representative transformation languages are listed, irrespective of their paradigm (declarative, imperative or hybrid) or scenario (e.g., inplace or model-to-model, or uni- or bidirectional transformations). To illustrate the different mechanisms, sample transformations for different facets of the running example are provided. To enhance understandability, ATL as a *single* transformation language has been used to exemplify the reuse mechanisms since it supports most of them.

Table 1. Categorization of Reuse Mechanisms

Scope of Reuse	Reuse Mechanisms	Supporting Representatives
Reuse of Transformation Logic within a Single Transformation	Functions Inheritance	All languages ATL, ETL, TGGs, Tefkat
Reuse of Transformation Logic in Similar Scenarios (same MMs, different logic)	Superimposition Transformation Product Lines	ATL, QVT, RubyTL [12], [24]
Reuse of Transformation Logic in Different Scenarios (different MMs, same logic)	Genericity DSL	SDM, VIATRA2, Tefkat, [6] External: [9], [30], Internal: [7], Epsilon, RubyTL
Reuse of Transformation Logic Independent of the Scenario	AOP HOT Reflection	Xtend All languages providing an explicit metamodel SDM, MISTRAL
Reuse of Transformation Logic in the Large	Orchestration	[13], [22], [27], ATLFlow, QVT

3.1 Reuse of Transformation Logic within a Single Transformation

Mechanisms to avoid code duplication and thus to enhance readability and maintainability within a single transformation include *functions* and *inheritance*, since both depend on concrete MM types.

Functions. All known transformation languages provide means to extract and then reuse recurring transformation logic in functions. As can be seen in Fig. 3(a) which realizes the running example in a "one table per class approach", the concatenation of the `name` with `_translated` is realized by an ATL helper which is invoked several times in the transformation specification. Nevertheless, the gained *abstraction* of this reuse mechanism is low, since functions typically depend on concrete MM types, e.g., `NamedElement` in the example. Abstraction by simplification is gained since the implementation is hidden after being developed once. For *selection*, no repository exists since functions are specific to a single transformation. *Specialization* is done black-box based, i.e., functions are specialized in a language-inherent manner by parameter values. Concerning *integration*, functions are a connection-based composition mechanism.

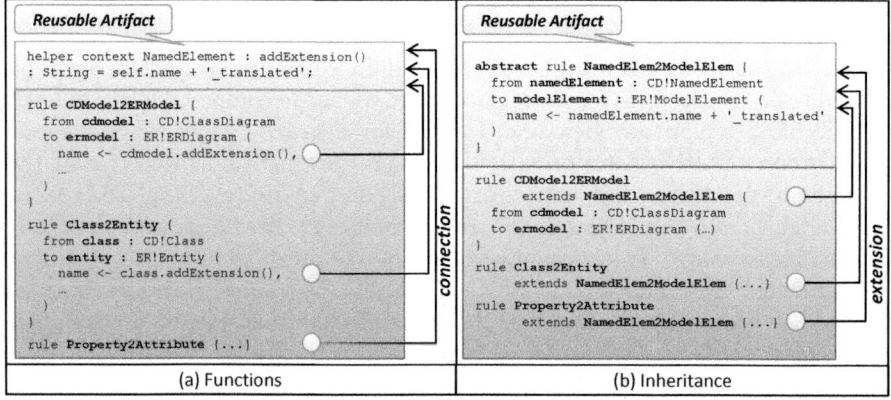

Fig. 3. Reuse Mechanisms Within a Single Transformation

Inheritance. In MMs it is common to employ inheritance between classes to reuse feature definitions from previously defined classes. Since classes and their respective objects are typically input for transformation rules, inheritance between transformation rules dealing with classes that are in an inheritance hierarchy in the MM may be applied in order to avoid code duplication, i.e., attribute assignments. As can be seen in Fig. 3(b), rule inheritance in our running example is used in order to avoid the re-specification of the `name` assignment. Therefore, all rules inherit from the common base rule `NamedElem2ModelElem`. Nevertheless, inheritance does neither achieve *abstraction* from the actual MM types nor from the underlying transformation language, since the superrules are bound to concrete MM types. Furthermore, no abstraction by simplification takes place, since the whole implementation of the superrules is exposed to transformation designers. For *selection*, no repository exists since inheritance is currently specific to a single transformation. Superrules might be *specialized* by overriding them in a white-box, language-inherent manner. With respect to *integration*, inheritance represents an extension-based composition mechanism.

Synopsis. Functions as well as inheritance are both mechanisms to avoid code duplication within a single transformation. Nevertheless, they complement each other, since functions reuse arbitrary transformation logic whereas inheritance reuses assignments provided that the MM incorporates inheritance and thus allows for rule inheritance. Although inheritance is an important reuse mechanism in OOP, not all transformation languages support inheritance, or if they do, they offer different semantics as we already investigated in [32]. For example there are differences in how overridden assignments are incorporated in the overriding assignment or the way type substitutability is supported.

3.2   Reuse of Transformation Logic in Similar Scenarios

Provided that a similar transformation scenario has to be realized on the basis of an existing transformation, i.e., a transformation between the *same source and target MMs*, but with *different transformation logic*, mechanisms are needed that allow to either alter the existing transformation, e.g., superimposition, or to configure an existing transformation such that it meets the changed requirements, e.g., transformation product lines.

Superimposition. Superimposition allows to build the union of transformation rules from different transformations. Thereby rules can be redefined, i.e., a rule is replaced by a new one if their signatures are identical, and added, whereby it is impossible to reuse the original rule. Superimposition has been proposed for ATL and QVT Relations [29] and is applicable in our running example to provide a new transformation that implements a "one table per hierarchy" approach on basis of the existing "one table per class" transformation. Thereby, the superimposed transformation redefines the rule `Class2Entity` and adds an additional helper `Closure` to calculate the transitive closure (cf. Fig. 4(a)). The so-called phasing mechanism and refinement rules in RubyTL [5] extend the idea of superimposition in the way that superimposed rules may refine the

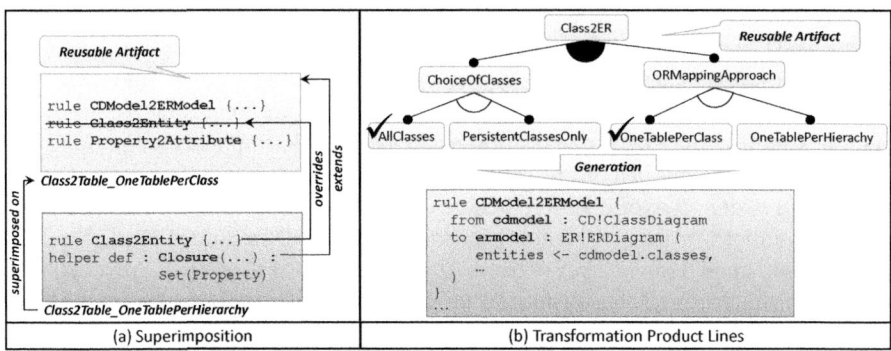

Fig. 4. Mechanisms to Reuse Transformation Logic in Similar Scenarios

results of the original rules. Nevertheless, superimposition *abstracts* neither from the MMs (since old and redefined rules are bound to concrete MM types) nor from the transformation language. Superimposition also does not abstract by simplification, since the whole original transformation is visible to the transformation designer. Concerning *selection*, existing transformations in the "ATL Model Transformation Zoo"[1] could be reused by superimposition. Nevertheless, the selection process is supported by documentation only. *Specialization* is done in a language-inherent, black-box manner since redefining existing transformation rules has to be done in the same language and requires to know the exact signatures of the to be redefined transformation rules only. Regarding *integration*, superimposition represents again an extension-based composition mechanism.

Transformation Product Lines. To deal with variabilities in model transformations, approaches [12, 24] arose that allow transformation designers to explicitly specify potential variabilities in model transformations, which we call *Transformation Product Lines (TPLs)* (inspired by Software Product Lines). These approaches typically use some variability model, e.g., feature models, to guide the generation of a specific transformation. Fig. 4(b) shows a simplistic feature model for our running example, allowing to choose the classes to be translated as well as the applied OR-mapping approach. In this respect, the reusable artifact is not only the already existing transformation but additionally the feature model, which models interdependencies and constraints of a model transformation. Since TPLs realize a set of related transformations, they are bound to concrete MM types and thus, *abstract* neither from MMs nor from the transformation language. Currently, no repository is available for *selecting* a certain TPL. *Specialization* is done by configuring the feature model, thus, no internals of the transformation are needed, being a black-box, non-language-inherent mechanism. Concerning *integration*, TPLs represent a generation-based reuse mechanism on basis of the configured feature model.

---

[1] http://www.eclipse.org/m2m/atl/atlTransformations

Synopsis. Both superimposition and TPLs allow to realize related transformation scenarios. Nevertheless, superimposition follows an ad-hoc development approach, i.e., a transformation may be incrementally modified on demand whereas TPLs represent a planned development approach, i.e., all potential variabilities of a transformation have to be modeled in advance. Although changes in TPLs themselves are challenging since the feature model, the transformation code as well as the code generator have to be adapted accordingly, TPLs have the advantage, that even domain experts without profound knowledge of a transformation language might develop transformations by just selecting values from the feature model. In contrast to TPLs, superimposition requires profound knowledge of the transformation language but allows flexible changes of transformations.

3.3   Reuse of Transformation Logic in Different Scenarios

Assuming that the same transformation logic should be reused in a different scenario, i.e., different source/target MMs, mechanisms are needed that allow to decouple transformation logic from concrete MM types. In this respect, generic transformations and domain-specific languages (DSLs) have been proposed as detailed in the following.

Genericity. Genericity allows to parameterize transformation logic with types to *abstract* from concrete MMs. Thereby, approaches have been proposed for *fine-grained* genericity [18, 28], i.e., on the level of rules or functions, and *coarse-grained* genericity [6], i.e., on the level of transformations. Fig. 5 shows an example for coarse-grained genericity whereby the whole `Class2ER` transformation should be reused for an `Ontology2XML` transformation. This is possible, since the

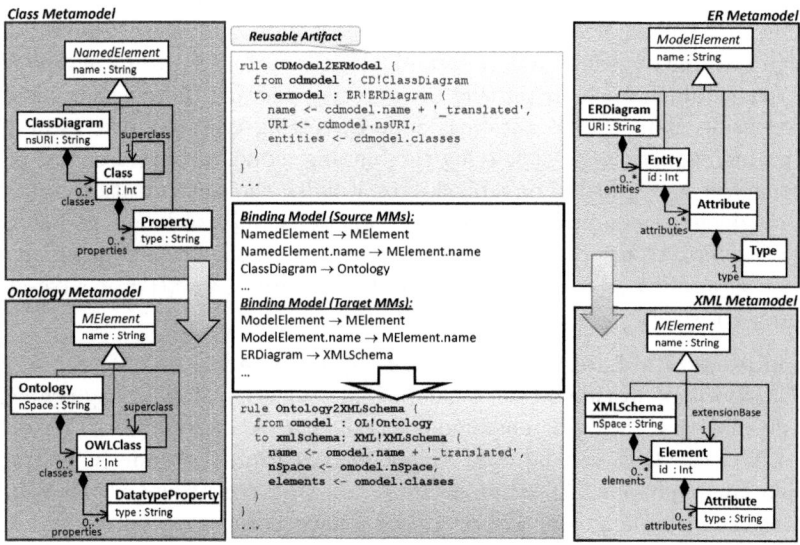

Fig. 5. Genericity On the Level of Transformations

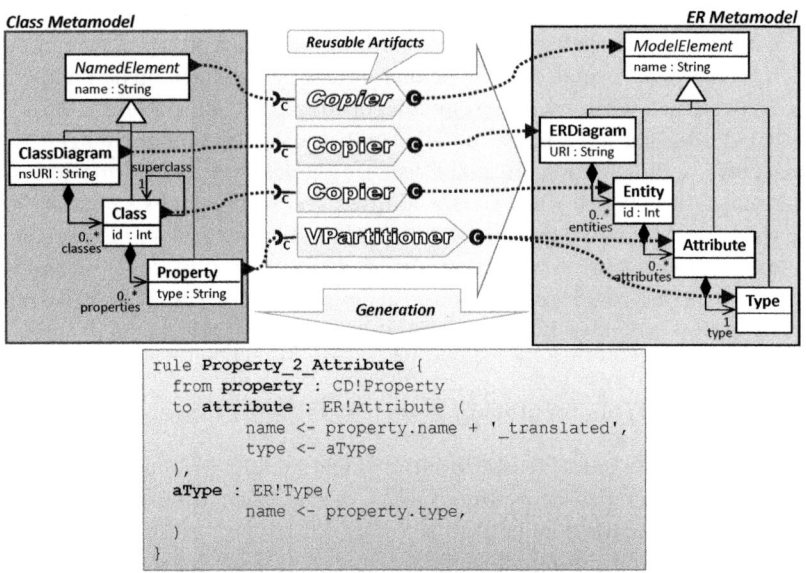

Fig. 6. External DSL Example [30]

new MMs (Ontology and XML) are structurally similar to the old MMs (Class and Relational). In this case, the transformation designer only has to specify a binding model, denoting which types of the old MMs correspond to which types of the new MMs. This binding model is then used to modify the original transformation by means of a higher-order transformation (HOT) (cf. below). Thus, the transformation designer only has to care about the source/target MMs representing the visible part, whereas the implementation is hidden. Nevertheless, although the idea of generic functions and transformations is promising, no library has been established so far putting the question whether there is support for *selection* aside. Since *specialization* is done by setting type parameters in case of fine-grained genericity or specifying the binding model in case of coarse-grained genericity, it is considered as a black-box. Finally, the specialization process occurs language-inherent in case of generic functions and non-language-inherent in case of generic transformations. Whereas in case of fine-grained genericity connection is applied as a composition mechanism for *integration*, coarse-grained genericity resembles a generation-based mechanism.

Domain-specific Languages. Another way to reuse logic in different scenarios are DSLs, which provide means to simplify specification of recurring problems in transformations. Two different kinds of DSLs can be distinguished: (i) *external* DSLs, i.e., the DSL can be used independently from the underlying transformation language, and (ii) *internal* DSLs, i.e., DSL constructs are embedded in a transformation language. External DSLs have been proposed by [9] and [30] which focus on the resolution of structural heterogeneities, e.g., the splitting of the class Property into the classes Attribute and Type represents a *vertical* *partitioning heterogeneity.* Fig. 6 shows the solution of the running example using the DSL presented in [30]. In order to execute a DSL-based specification, it has to be translated into a certain executable transformation language. Internal DSLs follow the same principles but differ in the fact that DSL constructs are tightly integrated in a certain transformation language. A representative for internal DSLs is the High Level Navigation Language (HNL) [7] which hides complex OCL navigation expressions using ATL as the host language. Internal DSLs are also supported by RubyTL [8] and Epsilon[2], whereby specialized DSLs, e.g., Epsilon Flock for model migration, build upon the host language Epsilon Object Language. Although both kinds of DSLs *abstract* from concrete MMs, only external DSLs also abstract from the underlying transformation language. Concerning *simplification*, the provided DSL syntax, i.e., the visible part, abstracts from the operational semantics, i.e., the hidden part. *Selection* of a certain reusable artifact, i.e., a DSL construct, is typically semi-automatically supported by editors, e.g., by means of code completion based on the DSL's grammar. DSLs are *specialized* in a black-box, language-inherent manner, since specialization is done by binding a certain grammar element to MM types, e.g., so-called ports need to be bound to a certain MM element in [30] (cf. Fig. 6). Since DSL constructs are compiled to ordinary transformation code, generation based *integration* takes place.

Synopsis. Genericity as well as DSLs allow both to decouple transformation logic from concrete MM types. Genericity is a promising approach to reuse transformation logic for structurally similar MMs, either on the fine-grained level of rules or the coarse-grained level of transformations. Although especially in case of coarse-grained genericity, large parts of transformation logic are reusable, it has the drawback that it requires structural similarity resulting in a low probability for application. In contrast, DSL constructs abstract from structural similarity to a certain extent, e.g., in [30] structural flexibility is supported by providing fixed parts as well as configurable parts. Thus, although the DSL constructs do not allow to reuse whole transformations, DSLs have a higher probability for application.

3.4 Reuse of Transformation Logic Independent of the Scenario

Parts of transformation logic might be independent of any concrete scenario and might thus occur in series of transformations, e.g., cross-cutting concerns like tracing or debugging (cf. Fig. 7). To reuse such cross-cutting concerns, several mechanisms have been proposed, including higher-order transformations (HOTs), aspect-orientation (AO), and reflection.

Higher-order Transformations, Aspect Orientation and Reflection. As detailed in [26], HOTs can be applied in several ways to achieve reuse in model transformations, being (i) *transformation composition*, (ii) *transformation synthesis*, and (iii) *transformation modification*. *Transformation composition,*

---

[2] http://www.eclipse.org/gmt/epsilon

```
                                  rule addDebugMessage{
 ┌─Reusable Artifact─┐              from oldAssignment : ATL!Binding
                                    to assignmentWithDebug : ATL!Binding (
                                        propertyName <- oldAssignment.propertyName,
                                        value <- debugger
                                    ),
                                    debugger : ATL!OperationCallExp (
                                        source <- oldAssignment.value,
                                        operationName <- 'debug',
 rule CDModel2ERModel {             arguments <- Sequence {arg}
   from cdmodel : CD!ClassDiagram   ),
   to ermodel : ER!ERDiagram (     arg : ATL!StringExp (
     name <- cdmodel.name + '…',       stringSymbol <-
     URI <- cdmodel.nsURI,              (oldAssignment.outPatternElement.outPattern."rule".name
     entities <- cdmodel.classes        + '.' + oldAssignment.outPatternElement.varName + '.'
   )                                     + oldAssignment.propertyName)
 }                                  )
 …                    Class2ER }                              Higher-Order Transformation
```

```
 URI := cdmodel.nsURI.debug(CDModel2ERModel.ermodel.URI),
 entities := cdmodel.classes.debug(CDModel2ERModel.ermodel.entities)
```

Fig. 7. Higher-order Transformations

meaning that a HOT takes at least one transformation and potentially other configuration models as input and produces a transformation as output, can be used, e.g., to achieve genericity as described above. *Transformation synthesis*, meaning that a transformation is generated from other artifacts, is often applied in the context of DSLs to generate transformations from DSL constructs as mentioned above. Therefore, in this subsection, HOTs in the sense of *transformation modification* are covered. The HOT takes a transformation as input to, e.g., introduce cross-cutting concerns like debugging or tracing into an existing transformation. Similar goals might be achieved by AO, e.g., supported in Xtend[3] or discussed in [21] and reflection provided that the target of the reflection is the transformation itself as, e.g., in MISTRAL [16].

Considering these three mechanisms, the reusable artifact might either be the transformation, the introduced cross-cutting concerns or even both, depending on what is newly developed. All mechanisms *abstract* from concrete MM types, but none of them abstracts by simplification, since no parts are explicitly hidden. With respect to *selection*, several ATL-based HOTs are available in the Model Transformation Zoo. There are, however, no repositories for AO or reflection. *Specialization* happens typically as a black-box, provided that only transformation-independent modifications take place, e.g., for each assignment, add a debug message. The specialization mechanism is either the HOT itself, the so-called join point model in AO or the meta-rules in case of reflection [16], describing where to introduce cross-cutting code. If the transformation to be specialized and the HOT are written in the same transformation language, the HOT is considered to be language-inherent. If AO and thus the specification of the join point model is supported by the transformation language, specialization occurs language-inherent. The same is true for reflection. Concerning *integration*, all mechanisms are composition-based reuse mechanisms in terms of extension.

---

[3] http://www.eclipse.org/workinggroups/oaw

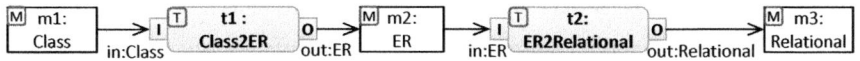

Fig. 8. Orchestration Example

Synopsis. Although the three mechanisms pursue similar goals, i.e., introducing cross-cutting concerns into transformation languages without changing the underlying transformation, the main difference lies in the kind of specialization. Since HOTs are defined on the abstract syntax of a transformation language, a transformation designer must have profound knowledge thereof (cf. Fig. 7). In contrast, AO allow specialization on basis of the concrete syntax and reflection on basis of the provided reflective API.

3.5   Reuse of Transformation Logic in the Large

To achieve reuse in the large, whole transformations might be reused *without adaptations*. Thus, mechanisms are needed to orchestrate model transformations, e.g., describing sequential or conditional executions of model transformations.

Orchestration. Orchestration languages have been proposed to replace low-level descriptions, e.g., in terms of Ant[4] tasks. Basically, they can be divided into approaches allowing to orchestrate model transformations written in different languages [13, 22, 27] or in a specific language only (Wires* [23], ATLFlow[5]). Fig. 8 shows a simple example in Wires* sequentially executing two ATL model transformations, first the Class2ER transformation and then a ER2Relational) transformation. In this respect, no *abstraction* from the MMs is achieved since the transformations to be reused still operate on concrete MMs. No abstraction from the underlying transformation language is achieved, except the orchestration allows for transformations written in different languages. Concerning simplification, the hidden parts comprise the implementation, since for orchestration only the source and target MMs of the transformations are of interest. ATL transformations might be *selected* again from the Model Transformation Zoo. Since transformations must be reused without adaptation, no *specialization* might occur. *Integration* happens by means of the orchestration language, thus it is classified as coordination.

Synopsis. Orchestration is a promising approach for reusing large portions of transformation logic. Nevertheless, the frequency of occurrence is constrained by the specificity of the reused transformations since each one is bound to concrete source and target MMs. Thus, it might be beneficial to combine orchestration with generic model transformations as proposed by [6].

---

[4] http://ant.apache.org/
[5] http://opensource.urszeidler.de/ATLflow/

Table 2. Comparison of Reuse Mechanisms

		Scope 1: Single Transformation		Scope 2: Similar Scenario		Scope 3: Different Scenario		Scope 4: Scenario-independent	Scope 5: In the Large
		Functions	Inheritance	Superimposition	Transformation Product Lines	Genericity	DSL	HOT, AO, Reflection	Orchestration
Reusable Artifact		Function	Base Rules	Base Transformation	Feature Model, Transformation	Generic Rules, Transformations	DSL constructs, Generator	Base Transformation, Concern or Both	Transformation
Abstraction									
Generalization	From MM	x	x	x	x	✓	✓	✓	x
	From TL	x	x	x	x	x	yes (external), no (internal)	x	x
Simplification	Hidden Parts	Implementation	none	none	Transformation	Implementation	Operational Semantics	none	implementation
	Visible Parts	Signature	all	all	Features	Signature with Type Parameters	DSL syntax	all	Signature (Source/Target MMs)
Selection									
Repository		-	-	ATL Model Transformation Zoo	x	x	DSL constructs	ATL Model Transformation Zoo (HOTs)	ATL Model Transformation Zoo
Metainfo		-	-	Documentation	-	-	Documentation, Grammar	Documentation	Documentation
Automatism		-	-	manual	-	-	semi-automatic (code completion)	manual	manual
Specialization									
Required Knowledge		black-box	white-box	black-box	black-box	black-box	black-box	black-box	-
Mechanism		parameter binding	overriding rule	redefining rules	configuration on basis of the feature model	type parameter binding	parameter binding	Transformation (HOT), Join Point Model (AO), Metarules (Reflection)	-
Language-inherent		✓	✓	✓	x	yes (fine-grained genericity), no (coarse-grained genericity)	✓	✓	-
Integration									
Ability		Composition	Composition	Composition	Generation	Composition	Generation	Composition	Composition
Kind		Connection	Extension	Extension	-	Connection	-	Extension	Coordination

4   Barriers to Model Transformation Reuse

Although numerous reuse mechanism have been proposed, barriers to model transformation reuse exists which hinder the adaptation of the mechanisms in practice. In the following, the main barriers derived from our comparison are presented, identifying further research potentials.

Insufficient Abstraction from Metamodels. Although some mechanisms, e.g., genericity, allow to decouple transformation logic from concrete MM types, the transformation logic is still dependent on the structure of the MMs. Thus, reuse of transformation logic between different MMs is hampered. To improve this situation, domain-specific standardized MMs would be beneficial, where standardized transformations might be defined on. Specific MMs might then extend the standardized MMs. Thus, the standardized transformations might also be reused to realize specific transformations, resembling the idea of frameworks in software engineering. This way, reuse of larger portions of transformation logic would be enabled. Nevertheless, additional costs for connecting them to the actual MMs may occur.

Insufficient Abstraction from Platform. Except external DSLs, all reuse mechanisms target at a single transformation language, but there is little work on how to reuse transformation logic in general. A first step in this direction is presented in [31] where a classification of structural heterogeneities in model-to-model transformations is given, which may serve as a pattern library for model transformations. Furthermore, reusable transformation patterns have been presented in [1] for graph transformation languages and idioms for QVT in [11]. Thinking this one step further, reuse should be enabled during the whole development cycle including requirements analysis, design, implementation, and testing as also stated in [10].

Missing Repositories for Selection. As can be derived from our comparison, hardly any repository of reusable artifacts has been established so far, except the Model Transformation Zoo comprising a collection of ATL-based model transformations. This is in contrast to software engineering, where different kinds of repositories of reusable artifacts exists, ranging from fine-grained class-libraries (being delivered with any programming language) over components to coarse-grained frameworks.

Lack of Meta-information in Selection. As Table 2 reveals, there is little meta-information available for selecting a reusable artifact without having to know its internals. Therefore, it would be important to provide transformations with according meta-information, comprising source/target MMs, test models, pre- and postconditions, and documentation. Preconditions may be used, e.g., to check if input models conform to the implemented transformation logic [4]. More abstract models for model transformations, e.g., requirements, would provide an additional source of meta-information.

Challenging Specialization Mechanisms. Although most reuse mechanisms allow for specialization, they are sometimes challenging to be applied in practice. This includes especially HOTs as also stated in [25] where the user must be familiar with the abstract syntax of the transformation language. In case of inheritance, specialization has potential for improvement, since none of the approaches allows to define reuse policies, e.g., to disallow rule inheritance (cf. `final` keyword in Java) or to define some access rights (cf. keywords `private`, `protected` or `public`). However, one important step in this direction has been the introduction of the "module" concept in transformation languages [5].

Insufficient Support for Integration in the Large. Although orchestration languages have been proposed to chain transformations to build larger ones, a main issue is the compatibility of source/target MMs between the orchestrated transformations. Thus, mechanisms are needed that ensure type compatibility in transformation chains similar to type checks in ordinary programs. This would incorporate compilation errors, if compatibility between MMs in the context of a specific transformation is violated.

5   Conclusion

In this paper, we provided an overview on proposed reuse mechanisms in rule-based model-to-model transformations. The comparison has been conducted on the basis of a framework covering the main phases in reuse, comprising abstraction, selection, specialization and integration. Although the comparison showed that a variety of mechanisms for reuse have been proposed, several barriers hindering their successful application have been identified. Furthermore, currently there is a strong focus on reuse in the implementation phase but reuse across all development phases would be urgently needed, e.g., general guidelines on how to design transformations. Thus, in our opinion further research is needed to make model transformation reuse more a fact than a fiction.

References

1. Agrawal, A., Vizhanyo, A., Kalmar, Z., Shi, F., Narayanan, A., Karsai, G.: Reusable Idioms and Patterns in Graph Transformation Languages. Electronic Notes in Theoretical Computer Science 127(1), 181–192 (2005)
2. Bézivin, J., Rumpe, B., Schürr, A., Tratt, L.: Model Transformations in Practice Workshop. In: Bruel, J.-M. (ed.) MODELS 2005. LNCS, vol. 3844, pp. 120–127. Springer, Heidelberg (2006)
3. Biggerstaff, T.J., Richter, C.: Reusability Framework, Assessment, and Directions. In: Software Reusability. Concepts and Models, vol. 1, pp. 1–17 (1989)
4. Cariou, E., Belloir, N., Barbier, F., Djemam, N.: OCL contracts for the verification of model transformations. ECEASST 24 (2009)
5. Cuadrado, J.S., Molina, J.G.: Approaches for Model Transformation Reuse: Factorization and Composition. In: Vallecillo, A., Gray, J., Pierantonio, A. (eds.) ICMT 2008. LNCS, vol. 5063, pp. 168–182. Springer, Heidelberg (2008)
6. Cuadrado, J.S., Guerra, E., de Lara, J.: Generic Model Transformations: *Write Once, Reuse Everywhere*. In: Cabot, J., Visser, E. (eds.) ICMT 2011. LNCS, vol. 6707, pp. 62–77. Springer, Heidelberg (2011)
7. Cuadrado, J.S., Jouault, F., Molina, J.G., Bézivin, J.: Experiments with a High-Level Navigation Language. In: Paige, R.F. (ed.) ICMT 2009. LNCS, vol. 5563, pp. 229–238. Springer, Heidelberg (2009)
8. Cuadrado, J.S., Molina, J.G.: A Model-Based Approach to Families of Embedded Domain-Specific Languages. IEEE Trans. Softw. Eng. 35, 825–840 (2009)
9. Del Fabro, M., Valduriez, P.: Towards the Efficient Development of Model Transformations using Model Weaving and Matching Transformations. Journal on SoSyM 8(3), 305–324 (2009)
10. Guerra, E., de Lara, J., Kolovos, D.S., Paige, R.F., dos Santos, O.M.: transML: A Family of Languages to Model Model Transformations. In: Petriu, D.C., Rouquette, N., Haugen, Ø. (eds.) MODELS 2010. LNCS, vol. 6394, pp. 106–120. Springer, Heidelberg (2010)
11. Iacob, M.-E., Steen, M.W.A., Heerink, L.: Reusable Model Transformation Patterns. In: Proc. of EDOCW 2008, pp. 1–10 (2008)
12. Kavimandan, A., Gokhale, A., Karsai, G., Gray, J.: Templatized Model Transformations: Enabling Reuse in Model Transformations. Technical report, Vanderbilt University (2009)
13. Kleppe, A.: MCC: A Model Transformation Environment. In: Rensink, A., Warmer, J. (eds.) ECMDA-FA 2006. LNCS, vol. 4066, pp. 173–187. Springer, Heidelberg (2006)
14. Kramer, J.: Is Abstraction the Key to Computing? Commun. ACM 50, 36–42 (2007)

15. Krueger, C.W.: Software Reuse. ACM Comput. Surv. 24(2), 131–183 (1992)
16. Kurtev, I.: Application of Reflection in a Model Transformation Language. SoSyM 9(3), 311–333 (2010)
17. Lau, K., Rana, T.: A Taxonomy of Software Composition Mechanisms. In: Proc. of SEAA 2010, pp. 102–110. IEEE (2010)
18. Legros, E., Amelunxen, C., Klar, F., Schürr, A.: Generic and Reflective Graph Transformations for Checking and Enforcement of Modeling Guidelines. Visual Language Computing 20(4), 252–268 (2009)
19. Mili, A., Mili, R., Mittermeir, R.: A Survey of Software Reuse Libraries. Annals of Software Engineering 5, 349–414 (1998)
20. Mili, H., Mili, F., Mili, A.: Reusing software: Issues and Research Directions. IEEE Transactions on Software Engineering 21(6), 528–562 (1995)
21. Moha, N., Mahé, V., Barais, O., Jézéquel, J.-M.: Generic Model Refactorings. In: Schürr, A., Selic, B. (eds.) MODELS 2009. LNCS, vol. 5795, pp. 628–643. Springer, Heidelberg (2009)
22. Oldevik, J.: Transformation Composition Modelling Framework. In: Kutvonen, L., Alonistioti, N. (eds.) DAIS 2005. LNCS, vol. 3543, pp. 108–114. Springer, Heidelberg (2005)
23. Rivera, J.E., Ruiz-Gonzalez, D., Lopez-Romero, F., Bautista, J., Vallecillo, A.: Orchestrating ATL Model Transformations. In: Proc. of MtATL 2009, pp. 34–46 (2009)
24. Sijtema, M.: Introducing Variability Rules in ATL for Managing Variability in MDE-based Product Lines. In: Proc of MtATL 2010, pp. 39–49 (2010)
25. Tisi, M., Cabot, J., Jouault, F.: Improving Higher-Order Transformations Support in ATL. In: Tratt, L., Gogolla, M. (eds.) ICMT 2010. LNCS, vol. 6142, pp. 215–229. Springer, Heidelberg (2010)
26. Tisi, M., Jouault, F., Fraternali, P., Ceri, S., Bézivin, J.: On the Use of Higher-Order Model Transformations. In: Paige, R.F., Hartman, A., Rensink, A. (eds.) ECMDA-FA 2009. LNCS, vol. 5562, pp. 18–33. Springer, Heidelberg (2009)
27. Vanhooff, B., Ayed, D., Van Baelen, S., Joosen, W., Berbers, Y.: UniTI: A Unified Transformation Infrastructure. In: Engels, G., Opdyke, B., Schmidt, D.C., Weil, F. (eds.) MODELS 2007. LNCS, vol. 4735, pp. 31–45. Springer, Heidelberg (2007)
28. Varró, D., Pataricza, A.: Generic and Meta-transformations for Model Transformation Engineering. In: Baar, T., Strohmeier, A., Moreira, A., Mellor, S.J. (eds.) UML 2004. LNCS, vol. 3273, pp. 290–304. Springer, Heidelberg (2004)
29. Wagelaar, D., Van Der Straeten, R., Deridder, D.: Module Superimposition: A Composition Technique for Rule-based Model Transformation Languages. SoSyM Journal 9, 285–309 (2010)
30. Wimmer, M., Kappel, G., Kusel, A., Retschitzegger, W., Schönböck, J., Schwinger, W.: Surviving the Heterogeneity Jungle with Composite Mapping Operators. In: Tratt, L., Gogolla, M. (eds.) ICMT 2010. LNCS, vol. 6142, pp. 260–275. Springer, Heidelberg (2010)
31. Wimmer, M., Kappel, G., Kusel, A., Retschitzegger, W., Schönböck, J., Schwinger, W.: Towards an Expressivity Benchmark for Mappings based on a Systematic Classification of Heterogeneities. In: Proc. of MDI 2010 @ MoDELS 2010, pp. 32–41 (2010)
32. Wimmer, M., Kappel, G., Kusel, A., Retschitzegger, W., Schönböck, J., Schwinger, W., Kolovos, D., Paige, R., Lauder, M., Schürr, A., Wagelaar, D.: A Comparison of Rule Inheritance in Model-to-Model Transformation Languages. In: Cabot, J., Visser, E. (eds.) ICMT 2011. LNCS, vol. 6707, pp. 31–46. Springer, Heidelberg (2011)

Author Index

GPSR Compliance

*The European Union's (EU) General Product Safety Regulation (GPSR)
is a set of rules that requires consumer products to be safe and our
obligations to ensure this.*

*If you have any concerns about our products, you can contact us on
ProductSafety@springernature.com*

In case Publisher is established outside the EU, the EU authorized
representative is:

Springer Nature Customer Service Center GmbH
Europaplatz 3
69115 Heidelberg, Germany

Batch number: 09490872

Printed by Printforce, the Netherlands